The Need for Sustainable Tourism in an Era of Global Climate Change

The Need for Sustainable Tourism in an Era of Global Climate Change: Pathway to a Greener Future

EDITED BY

PARDEEP SINGH
PGDAV College, India

SHIKHA DAGA
PGDAV College, India

KIRAN YADAV
PGDAV College, India

AND

ABHAY JAIN
Shri Ram College of Commerce, India

emerald PUBLISHING

United Kingdom – North America – Japan – India – Malaysia – China

Emerald Publishing Limited
Emerald Publishing, Floor 5, Northspring, 21-23 Wellington Street, Leeds LS1 4DL.

First edition 2025

British Library Cataloguing in Publication Data
A catalogue record for this book is available from the British Library

ISBN: 978-1-83608-669-7 (Print)
ISBN: 978-1-83608-668-0 (Online)
ISBN: 978-1-83608-670-3 (Epub)

Printed and bound by CPI Group (UK) Ltd, Croydon, CR0 4YY

INVESTOR IN PEOPLE

Contents

List of Contributors

Elizabeth Andrews	Sunway University, Malaysia
Sonam Angmo	Independent Researcher, India
Mahua Basu	St. Xavier's College (Autonomous), Kolkata, India
Raffaele Boffard	University of Naples L'Orientale, Italy
Carlotta D'Alessandro	University of Messina, Italy
Shikha Daga	University of Delhi, India
Gabriela Dione Florêncio de Lima	Federal University of Campina Grande (UFCG), Brazil
Luisa De Simone	University of Naples L'Orientale, Italy
Kaniz Fatema	Sunway University, Malaysia
Renu Ghosh	Netaji Subhas University of Technology, India
Lucky Gupta	Raj Kumar Goel Institute of Technology, India
Syed Monirul Hossain	Sunway University, Malaysia
Anju Jain	University of Delhi, India
Pooja Jain	Google, USA
Pranjal Jain	Lovely Professional University, India
Ishika Jaiswal	Indian Institute of Management (IIM), India
Gyanendra B. S. Johri	Indira Gandhi National Tribal University, India
Noreen Kanwal	Sunway University, Malaysia
Jyoti Kasana	University of Delhi, India
Rajinder Kaur	Chandigarh University, India
Adjnu Damar Ladkoo	University of Mauritius, Mauritius
Kettrin Farias Bem Maracajá	Federal University of Campina Grande (UFCG), Brazil
Chong Soon Meng	Sunway University, Malaysia
Tiago Savi Mondo	Federal Institute of Santa Catarina (IFSC), Brazil
Kim Yoke Ong	Sunway University, Malaysia
Bhartrihari Pandiya	National Forensics Sciences University, India
Nishi Pathak	Raj Kumar Goel Institute of Technology, India
Amrita Saini	Chandigarh Business School of Administration, India
Jolyne Khor Kuan Siew	Sunway University, Malaysia
Mohan Singh	Indira Gandhi National Tribal University, India
Pardeep Singh	University of Delhi, India

Sandeep Singh	Punjabi University, Patiala, Punjab, India
Punitha Sinnappan	Sunway University, Malaysia
Sharad Srivastava	Indira Gandhi National Tribal University, India
Shehnaz Tehseen	Sunway University, Malaysia
Hitesh Tripathi	Indira Gandhi National Tribal University, India
Motoki Watabe	Sunway University, Malaysia
Bhavna Yadav	University of Delhi, India
Kiran Yadav	University of Delhi, India
Nasaruddin Zainal	Sunway University, Malaysia

Chapter 1

Navigating Towards a Greener Tomorrow: The Imperative of Sustainable Tourism in the Era of Global Climate Change

Shikha Daga[a], *Kiran Yadav*[a], *Pardeep Singh*[b] *and Bhavna Yadav*[a]

[a]*Department of Commerce, PGDAV College, University of Delhi, India*
[b]*Department of Environmental Science, PGDAV College, University of Delhi, India*

Abstract

As the world advances towards a sustainable and more environmentally friendly future, the need for protecting the environment becomes more and more apparent in every area. The tourist business has seen remarkable expansion, making a considerable contribution to the worldwide Gross Domestic Product (GDP). Nevertheless, this expansion has coincided with an increase in the release of greenhouse gases and the deterioration of natural ecosystems. In order to tackle these difficulties, there is an increasing acknowledgment of the need for a fundamental change in approach towards implementing sustainable tourism practices. This chapter analyses the present state of tourism and its contribution to the increase of carbon footprints. It also investigates possible remedies, including the implementation of environmentally friendly hotels, the use of cleaner energy sources, and the adoption of ideas from the circular economy. In addition, by using bibliometric analysis, the chapter offers valuable information on current research patterns in sustainable tourism, such as prominent authors, nations, keywords, and topics. This chapter provides a comprehensive analysis of the current state and future potential of sustainable tourism. It is a vital reference for policymakers, academics, and industry professionals, offering guidance on how to develop a more sustainable and resilient tourist business.

Keywords: Sustainable tourism; climate change; green development; overtourism; bibliometric analysis

---

The Need for Sustainable Tourism in an Era of Global Climate Change: Pathway to a Greener Future, 1–12

doi:10.1108/978-1-83608-668-020241002

1. Introduction

Rapid economic growth, industrialisation, trade openness, and unscrupulous use of energy have led to a disastrous change in the climate that has manifested into the destruction of natural flora and fauna along with an exponential rise in health crises leading to environmental emergency (Balsalobre-Lorente et al., 2023). Economies in an urge to enhance their economic wealth promote their economies as tourist destinations that no doubt add to the wealth kitty but at the same time, it calls for increased energy usage and adds the burden on natural resources (Fauzel et al., n.d.). The rising trend of travel and tourism has in fact led to the issue of over-tourism which has been disastrous to many places in India as well as abroad (Khan et al., 2020). The need to address the issue of climate change has led to the initiation of various reforms at the country as well as global level amongst which the most popular is the Sustainable Development Goals by the United Nations (Ratten, 2023). A brief overview of the contribution of the tourism sector to the global GDP is highlighted in Fig. 1.1 which provides how one-tenth of the total GDP comes from Tourism receipts, it thus being an important contributor to employment, wealth generation, and hence increasing the living standard of the various economies. Tourism sector is an important constituent that helps in determining the economic status of a country as it not only adds to the wealth of an economy but also builds a good image at the macro level (Boháč & Drápela, 2022; Kryczka, 2019). The positive effects of tourism are however not without the associated drawbacks (Weaver, 2011; Merli et al., 2019). The rapid expansion of tourism and its impact on climate change along with the steps undertaken to mitigate its impact is the agenda of the present chapter. This chapter also highlights the academic contribution in the domain of sustainable tourism using research papers from the SCOPUS database and analysing the same using bibliometric analysis based on its performance and science mapping. The scientific literature thus would identify the existing status of sustainable tourism along with its future scope. This chapter is divided into subsections wherein Section 2 discusses the present situation of tourism at the global level along with highlighting its role in the changing climatic condition, its contribution towards carbon emission and the carbon footprints associated with various tourism activities namely accommodation, infrastructure, transportation and energy usage. Section 3 draws upon successful tourism case studies that have helped in reducing carbon emissions and hence have contributed towards a sustainable future. Section 4 discusses about challenges that the tourism industry faces in mitigating its environmental impact. Section 5 highlights the academic trend in the domain of sustainable finance and Section 6 provides the future scope and conclusion of the chapter.

2. Current Scenario of Global Tourism and Carbon Footprints

Tourism has led to the generation of economic wealth by contributing hugely to employment generation, community development and an increase in the GDP (León-Gómez et al., 2021). The pros, however, are not free of the associated cost which is the rising pollution, unscrupulous use of energy resources, destruction of natural flora and fauna and adding substantially to the carbon footprints (sustainabletravel.org). Fig. 1.1 provides the contribution of the tourism sector towards global GDP over the last few years.

Emission from the tourism sector comes from multiple sources, be it travelling, lodging, construction or the food and beverage all contribute significantly to the carbon footprints (Lenzen et al., 2018). Various studies have highlighted the role of tourism-related transportation in contributing significantly to increased greenhouse gas emissions (Safaa et al., 2023) as depicted in Fig. 1.2, wherein the expected emissions are bound to cross the 1990 million metric tons mark by the end of 2030. The data is highly contradictory to the sustainable development goals that are expected to be achieved by 2030. Rising carbon emissions through travel and tourism will make it all the more difficult for the countries to attain the target of SDGs. The problem is all the more tricky to handle as tourism cannot be curtailed it being a major contributor of economic wealth. What is required is varied alternatives like making a shift towards greener alternatives be its green hotels, green energy use, paving the concept of circular economy and most importantly educating and spreading awareness about its necessity.

To ensure that the carbon emission related to tourism is reduced it is also suggested that tourists should choose places that are closer to their homes and make more use of public transportation. Additionally, governments could give incentives to tourist providers in order to encourage them to improve their energy efficiency. However, neither strategy has been very effective up to this point. Rising tourism and associated greenhouse gas emissions thus call for sustainable measures at the behest of Government, policymakers, academicians and all

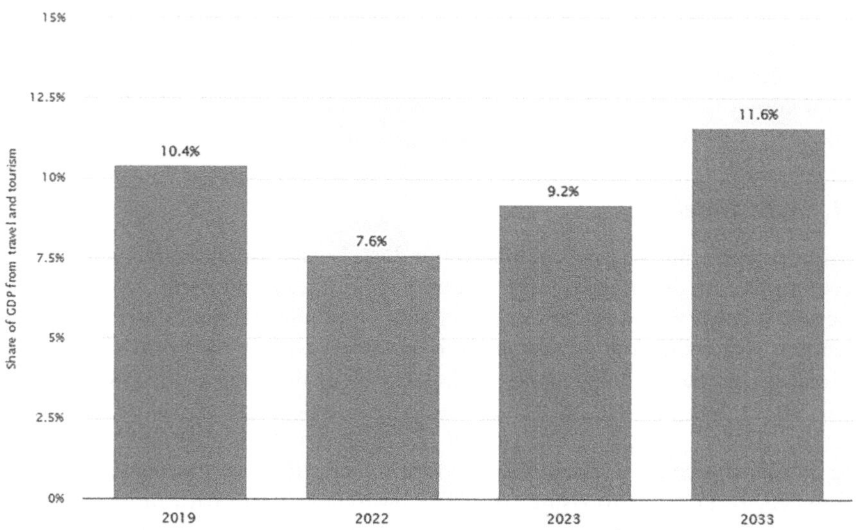

Fig. 1.1. Share of Travel and Tourism's Total Contribution to GDP Worldwide in 2019 and 2022, with a Forecast for 2023 and 2033. *Source*: Statista.com.

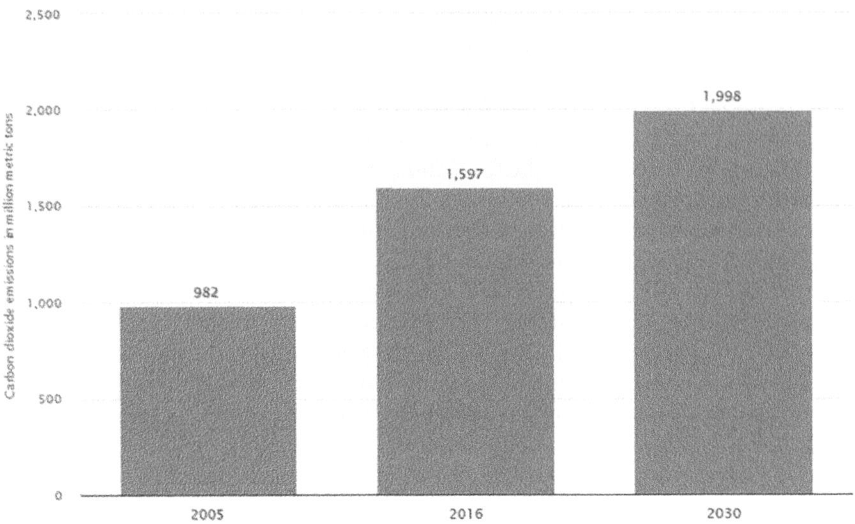

Fig. 1.2. Carbon Dioxide Emissions from Tourism-related Transport Worldwide in 2005 and 2016, with a Forecast for 2030 (in Million Metric Tons of Carbon Dioxide). *Source*: Statista.

others associated to ensure that there is neither compromise of travelling opportunities nor destruction of the environment.

3. Successful Tourism Case Studies and its Contribution to a Better Tomorrow

Increased burden on the land and resources because of an exponential rise in global tourism has brought significant changes to traditional tourism practices. A few contemporary mechanisms that have been introduced to ensure that tourism only promotes positive externality are provided in this section.

3.1. Green Hotels

Rising tourism across geographical boundaries calls for increased demand of accommodation and related infrastructure. It produces a dual impact wherein on one side it helps in community development, employment generation and wealth creation, and at the same time, it leads to increased carbon footprints and unscrupulous use of energy that leads to the destruction of the natural eco-system (Merli et al., 2019). The problem of over-tourism and its associated negative externalities thereby led to the origin of the concept of Green Hotels wherein the hotels make use of varied green practices to ensure that the negative environmental impact is mitigated. It gave rise to rising consumer awareness along with rating of hotels based on their use of green practices. It is a perfect example of how technology can produce results that are pro-environment (Yadav et al., 2019).

3.2. Community-based Tourism

Tourism as discussed earlier has the potential to generate income and other externalities for its stakeholders. A contemporary addition to the tourism industry has been the community-based tourism that helps in developing the local communities (Lee & Jan, 2019). Places like Malaysia, New Zealand, Vietnam, and Philippines have seen success because of community-based tourism. By showcasing the unique culture, traditions and natural landscapes of a destination, CBT initiatives contribute to diversifying tourism offerings. This reduces dependency on mass tourism and helps spread the benefits of tourism across different regions and communities.

3.3. Responsible EcoTourism

Responsible tourism plays a crucial role in fostering sustainable tourism by advocating for methods that reduce harmful effects on the environment, culture, and communities, while maximising the advantages for all parties involved. Responsible tourism, with its focus on environmental protection, effectively mitigates pollution, preserves natural resources, and safeguards biodiversity, thereby guaranteeing the sustainability of destinations in the long run. Furthermore, it promotes cultural reverence and conservation by motivating travellers to interact with local people in a considerate way, safeguarding cultural identities and history. Responsible tourism places high importance on ensuring that economic gains are distributed fairly, strengthening local communities by providing job opportunities and supporting small businesses. Responsible tourism employs education and awareness campaigns to educate both travellers and local populations about sustainable practices. It also promotes a feeling of responsibility towards taking care of the destinations. By adhering to the principles of responsible tourism, stakeholders together strive to develop tourist destinations that are commercially viable, socially inclusive, and ecologically sustainable.

The above points highlight how green hotels, community-led tourism, and responsible tourism have promoted sustainable tourism across various places. Other than these on a general front educating the masses and spreading awareness about the practice of following sustainable tourism practice is going to go a long way in producing significantly positive results for the environment and the global economy as a whole (Sezerel & Karagoz, 2023).

4. Bibliometric Analysis of Sustainable Tourism

Sustainable tourism has gained prominence in policymaking, research, and popular awareness throughout the years, as shown by a thorough examination of previous literature in the field. The modern Bibliometric analysis, described in this portion of the chapter, offers a systematic and comprehensive overview of all the research conducted throughout the years using a scientific method (Pahlevan Sharif et al., 2019). For extraction of the studies related to sustainable tourism,

SCOPUS was used and the following string and inclusion and exclusion criteria were applied to extract the final list for the bibliometric analysis:

> *TITLE-ABS-KEY (("sustainab* Tourism") OR (ecotourism)) AND PUBYEAR > 1999 AND PUBYEAR < 2025 AND (LIMIT-TO (SUBJAREA, "ENVI") OR LIMIT-TO (SUBJAREA, "BUSI") OR LIMIT-TO (SUBJAREA, "ECON")) AND (LIMIT-TO (DOCTYPE, "ar")) AND (LIMIT-TO (LANGUAGE, "English")) AND (LIMIT-TO (PUBSTAGE, "final")) AND (LIMIT-TO (EXACTKEYWORD, "Ecotourism") OR LIMIT-TO (EXACTKEYWORD, "Sustainable Tourism") OR LIMIT-TO (EXACTKEYWORD, "Tourism Development"))*

A total of 6,612 documents were examined using Biblioshiny by Bibliometrix-R, based on the specified search criteria. The preliminary examination focuses on the expansion of research in the domain of sustainable tourism. Fig. 1.3 illustrates the increasing interest in this field, which gained momentum around the year 2000. Subsequently, there was a significant surge in studies around 2008 and again in 2014. By the end of 2019, the number of studies reached a peak of approximately 650.

Another important area of analysis in the bibliometrics is the most relevant source which is provided in Fig. 1.4 highlighting the prominence of a few journals like – *Sustainability, Journal of Sustainable Tourism, Journal of Ecotourism, Tourism Management,* and *Current Issues in Tourism*. An understanding of the most relevant sources gives an idea about the places wherein academic work related to sustainable tourism and its implications and future prospects can be looked for. A detail about the sources' specific parameters related to h_index, and citations along with a total number of publications is provided in Table 1.1.

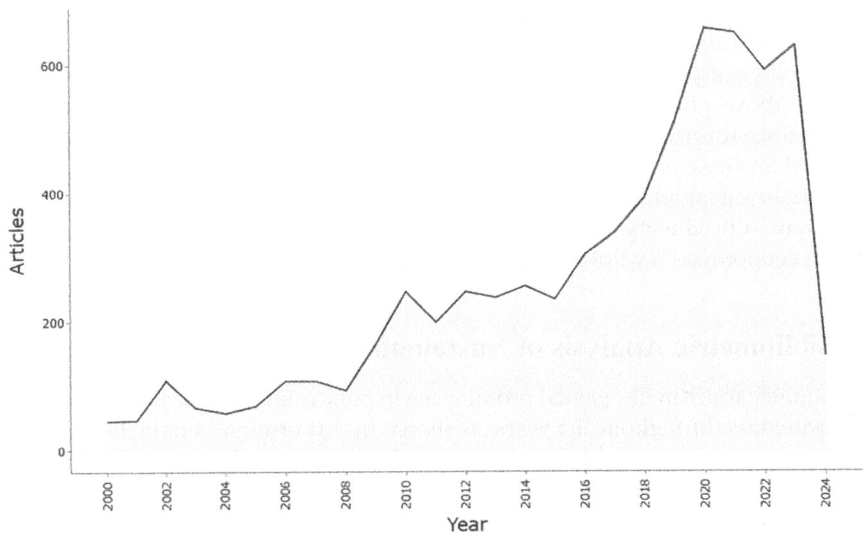

Fig. 1.3. Annual Scientific Production. *Source*: Scopus.

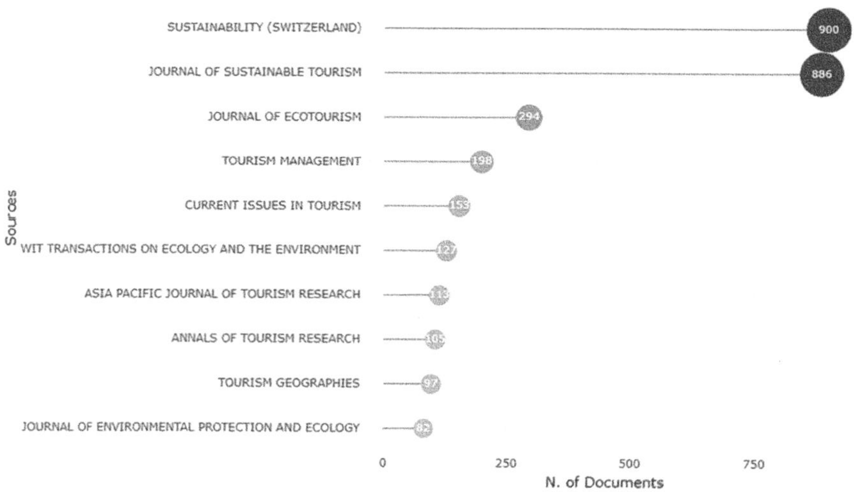

Fig. 1.4. Most Relevant Sources. *Source*: Biblioshiny.

Table 1.1. Source Local Impact.

Element	h_index	TC	NP	PY_start
Journal of Sustainable Tourism	110	50629	886	2000
Tourism Management	88	22151	198	2000
Sustainability (Switzerland)	53	15384	900	2013
Annals of Tourism Research	51	9479	105	2000
Journal of Ecotourism	44	7550	294	2002
Current Issues in Tourism	43	6773	153	2001
Tourism Geographies	38	5092	97	2000
Journal of Travel Research	35	3957	63	2000
Biological Conservation	32	3242	47	2000
Asia Pacific Journal of Tourism Research	30	2709	113	2003

Source: Biblioshiny.

A brief about the most relevant authors is provided in Fig. 1.5 which highlights the most significant contribution in the area of sustainable tourism based on the total number of publications. The knowledge of important authors in the area is crucial to identify as it can help in policy formulation and sustainable solutions to an emergent issue.

An analysis of countries involved in sustainable tourism reveals that the primary contributors in this sector are developed nations such as Australia, the

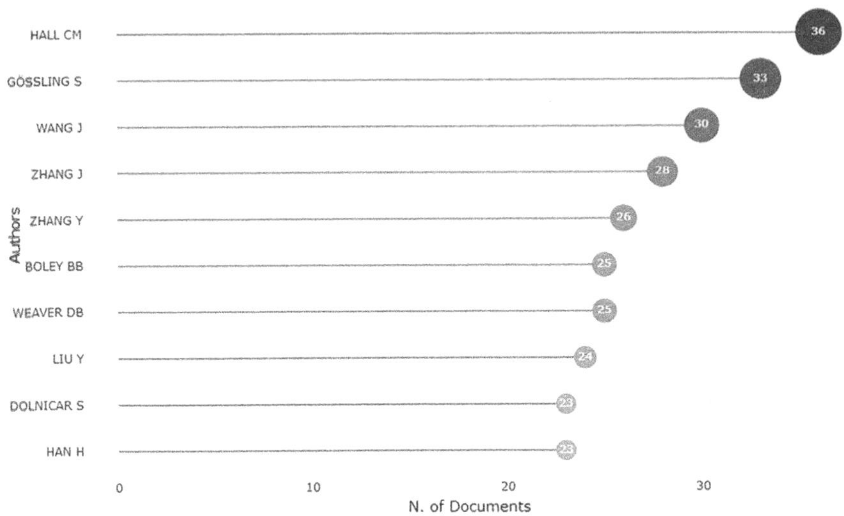

Fig. 1.5. Most Relevant Authors. *Source*: Biblioshiny.

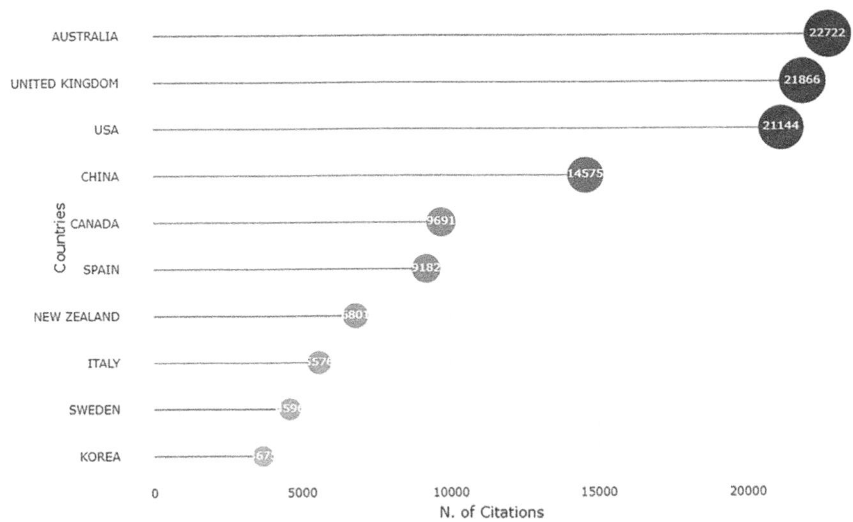

Fig. 1.6. Most Cited Countries. *Source*: Biblioshiny.

United States, and the United Kingdom. Following closely behind are the growing countries of China, Canada, Spain, and others. The rising interest of different countries in the domain of sustainable tourism highlights how the situation is prevalent across nations. The countries producing most work in the area of sustainable tourism are presented in Figs. 1.6 and 1.7 showing how a majority of the countries are involved in the research as the issue is increasingly recognised as

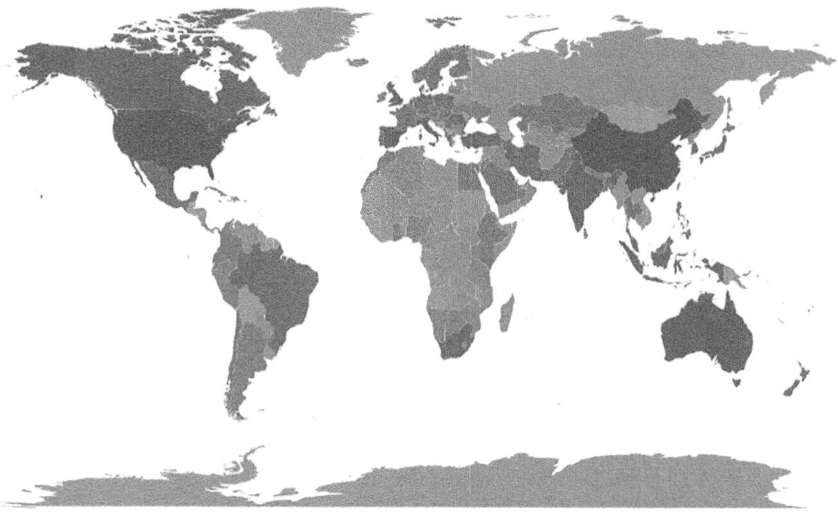

Fig. 1.7. Country Scientific Production. *Source*: Biblioshiny.

Fig. 1.8. Most Frequent Words. *Source*: Biblioshiny.

critical for the preservation of natural resources and cultural heritage, as well as for fostering socio-economic development.

Research in every field follows multiple avenues as ideas develop and their new aspects are revealed. Fig. 1.8 provides a concise overview of the most often-used words that emphasise key study themes in sustainable tourism. The figure emphasises terms such as eco-tourism, sustainable development, tourist development, sustainability, and others. The notion is reinforced by the trend themes (Fig. 1.9), which provide a thorough representation of the significant change in trajectory over the years.

In conclusion, bibliometric analysis provides invaluable insights into the research landscape surrounding sustainable development, highlighting trends, gaps, and emerging areas of focus. As we move forward, leveraging these findings

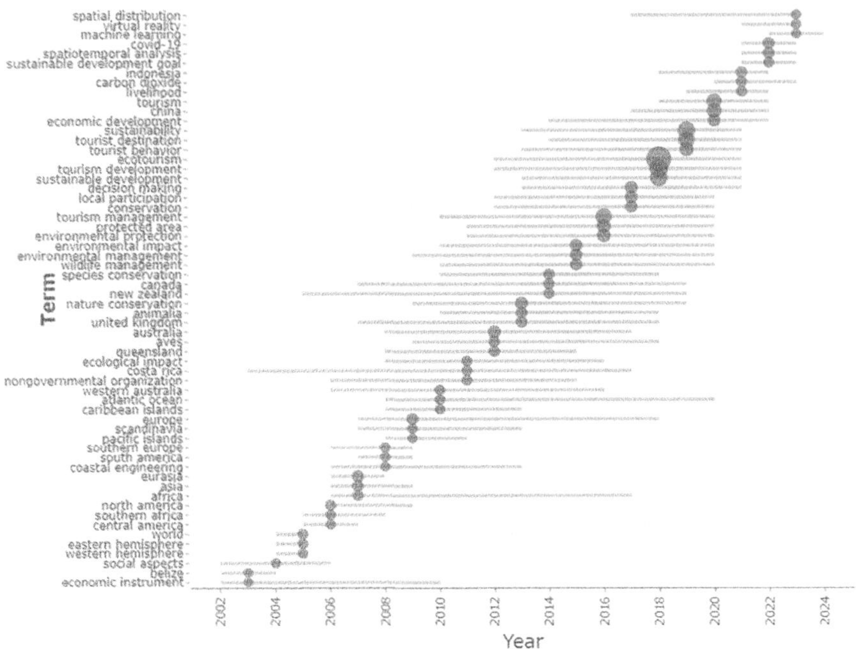

Fig. 1.9. Trend Topics. *Source*: Biblioshiny.

can guide policymakers, researchers, and practitioners in shaping more effective strategies and interventions to address global sustainability challenges. By harnessing the power of bibliometrics, we can pave the way for a more informed and impactful approach towards achieving a sustainable future for generations to come.

5. Future Prospects and Conclusion

Undoubtedly, travel and tourism are essential in today's context as they not only disrupt the repetitive routine but also contribute to the growth of global prosperity. However, these advantages are not without cost and are often linked to an increase in carbon emissions, which in turn disrupts the delicate environmental equilibrium. This discovery led to the development of the idea of sustainable tourism, which aims to minimise the negative aspects while fully realising the benefits of tourism, without compromising the ability of future generations to enjoy the same opportunities. The tourism sector no doubt is a critical component of the overall global growth. However, the exponential growth that the sector has witnessed over the years has led to severe environmental issues and hence it requires mediation to ensure that the negative impacts are minimised and the true benefits of travelling can be ensured (Luo et al., 2021). Adoption of sustainable tourism practices is not just a choice but a need for protecting our world and ensuring a more environmentally friendly future for future generations.

The crux of the issue is the urgent need to recognise and tackle the environmental consequences of worldwide tourism. Given that tourism is responsible for a substantial amount of carbon emissions on a global scale, the sector finds itself at a critical juncture where it has the potential to either worsen or alleviate the effects of climate change. The core idea of sustainable tourism is responsible travel, which requires travellers, companies, and legislators to give priority to environmental protection, social equality, and cultural preservation. Promoting increased recognition and admiration of the environment and cultural legacy may promote a more profound feeling of responsibility among tourists, motivating them to be mindful and make a beneficial contribution wherever they go. Nevertheless, the pursuit of sustainable tourism is accompanied by other obstacles. Stakeholders encounter several challenges, ranging from insufficient infrastructure to economic pressures, as they strive for sustainability. Through the use of technology, entrepreneurship, and policy change, we may surmount obstacles and stimulate a shift towards more sustainable practices in the tourist industry. There is a vast amount of untapped potential for creating good change in the sector, whether it is by developing environmentally friendly mobility options, promoting community-based tourism projects, or implementing carbon offset programmes. In conclusion, amidst the current confluence of environmental accountability and global economic prosperity, the imperative for sustainable tourism emerges as an important concept to transcend towards a greener and sustainable future.

References

Balsalobre-Lorente, D., Abbas, J., He, C., Pilař, L., & Shah, S. A. R. (2023). Tourism, urbanization and natural resources rents matter for environmental sustainability: The leading role of AI and ICT on sustainable development goals in the digital era. *Resources Policy*, *82*(March), 103445. https://doi.org/10.1016/j.resourpol.2023.103445

Boháč, A., & Drápela, E. (2022). Overtourism hotspots: Both a threat and opportunity for rural tourism. *European Countryside*, *14*(1), 157–179. https://doi.org/10.2478/euco-2022-0009

Fauzel, S., Seetanah, B., Sannassee, R., & Nunkoo, R. (n.d.). *Investigating the impact of climate change on the tourism sector of SIDS* (pp. 1–15).

Khan, A., Chenggang, Y., Hussain, J., Bano, S., & Nawaz, Aa. A. (2020). Natural resources, tourism development, and energy-growth-CO2 emission nexus: A simultaneity modeling analysis of BRI countries. *Resources Policy*, *68*(May), 101751. https://doi.org/10.1016/j.resourpol.2020.101751

Kryczka, M. (2019). Overtourism vs. sustainable development of tourism. Attempts to handle overtourism following the example of Venice. *Studia Periegetica*, *2*(26), 2019. https://doi.org/10.26349/st.per.0026.03

Lee, T. H., & Jan, F. H. (2019). Can community-based tourism contribute to sustainable development? Evidence from residents' perceptions of the sustainability. *Tourism Management*, *70*(September 2018), 368–380. https://doi.org/10.1016/j.tourman.2018.09.003

Lenzen, M., Sun, Y. Y., Faturay, F., Ting, Y. P., Geschke, A., & Malik, A. (2018). The carbon footprint of global tourism. *Nature Climate Change*, *8*(6), 522–528. https://doi.org/10.1038/s41558-018-0141-x

León-Gómez, A., Ruiz-Palomo, D., Fernández-Gámez, M. A., & García-Revilla, M. R. (2021). Sustainable tourism development and economic growth: Bibliometric review and analysis. *Sustainability (Switzerland)*, *13*(4), 1–20. https://doi.org/10.3390/su13042270

Luo, J. M., Chau, K. Y., Fan, Y., & Chen, H. (2021). Barriers to the implementation of green practices in the integrated resort sector. *SAGE Open*, *11*(3). https://doi.org/10.1177/21582440211030277

Merli, R., Preziosi, M., Acampora, A., & Ali, F. (2019). Why should hotels go green? Insights from guests experience in green hotels. *International Journal of Hospitality Management*, *81*(November 2018), 169–179. https://doi.org/10.1016/j.ijhm.2019.04.022

Pahlevan Sharif, S., Mura, P., & Wijesinghe, S. N. R. (2019). Systematic reviews in Asia: Introducing the "PRISMA." In S. Rezaei (Ed.), *Protocol to tourism and hospitality scholars* (pp 13–33). Springer. https://doi.org/10.1007/978-981-13-2463-5_2

Ratten, V. (2023). Gender and climate change: A sustainable tourism perspective. *Thunderbird International Business Review*, *65*(6), 553–566. https://doi.org/10.1002/tie.22351

Safaa, L., Atalay, A., Makutėnienė, D., Perkumienė, D., & Bouazzaoui, I. El. (2023). Assessment of carbon footprint negative effects for nature in international traveling. *Sustainability (Switzerland)*, *15*(16), 1–14. https://doi.org/10.3390/su151612510

Sezerel, H., & Karagoz, D. (2023). The challenges of sustainable tourism development in special environmental protected areas: Local resident perceptions in Datça-Bozburun. *Sustainability (Switzerland)*, *15*(4), 3364. https://doi.org/10.3390/su15043364

Weaver, D. (2011). Can sustainable tourism survive climate change? *Journal of Sustainable Tourism*, *19*(1), 5–15. https://doi.org/10.1080/09669582.2010.536242

Yadav, R., Balaji, M. S., & Jebarajakirthy, C. (2019). How psychological and contextual factors contribute to travelers' propensity to choose green hotels? *International Journal of Hospitality Management*, *77*(April), 385–395. https://doi.org/10.1016/j.ijhm.2018.08.002

Weblinks

https://www.statista.com/statistics/1099933/travel-and-tourism-share-of-gdp/

https://sustainabletravel.org/issues/carbon-footprint-tourism/#:~:text=Tourism%20is%20responsible%20for%20roughly,the%20top%20of%20the%20list

https://www.science.org/content/article/tourism-four-times-worse-planet-previously-believed

Chapter 2

Tourism in North East India: A Case Study of Sustainable Tourism with Successes and Lessons Learned

Bhartrihari Pandiya

School of Management Studies, National Forensics Sciences University, Gandhinagar, India

Abstract

The North East India region is known for its rich cultural diversity and pristine landscapes, and has witnessed a growing tourism industry but has maintained its sustainability. This research explores the paradigm of sustainable tourism of five tourist destinations in North East India. Through an extensive literature review and comprehensive case study method, the research discusses the successes and lessons learned from the implementation of sustainable tourism practices in North East India, shedding light on the region's unique socio-economic and environmental challenges. It also compares the sustainability model of North East India and the rest of India. The study identifies successful models through five cases that have contributed to the growth of responsible tourism while minimising negative impacts. The research also evaluates the effectiveness of sustainable tourism strategies in preserving local cultures, protecting biodiversity, and fostering community engagement. Upon comparison of the sustainable development model of North East India with the rest of India, it was found that North East India is better on the aspects of cultural preservation, environmental conservation, community involvement, mass tourism and infrastructure, government policies, biodiversity, and unique attractions. The lessons learned from challenges encountered during the implementation of sustainable tourism provide insights for policymakers,

The Need for Sustainable Tourism in an Era of Global Climate Change: Pathway to a Greener Future, 13–33

Published under exclusive licence by Emerald Publishing Limited
doi:10.1108/978-1-83608-668-020241006

practitioners, and local communities. This research thus offers valuable lessons and implications for similar regions grappling with balancing tourism growth and environmental conservation.

Keywords: North East India; sustainable tourism; case study; community engagement; implications

1. Introduction

The travel and tourism sector plays a vital role in the global economy due to its capacity to create jobs and serve as a spur for economic growth (Croes et al., 2021; Lasisi et al., 2023). According to the United Nations World Tourism Organisation (2018), the industry saw a growth rate of 7.1% in tourist arrivals and roughly 1,322 million and is predicted to see a growth rate of 4–5%. Even after the pandemic, there is a huge opportunity and growth in the tourism sector (Lew et al., 2020; Nepal, 2021). The concept of sustainable tourism has gained prominence globally as destinations grapple with the challenges of balancing economic development with environmental and cultural preservation. Scholars such as Bramwell (2015), Zolfani et al. (2015), and McGehee et al. (2018) have contributed to the theoretical framework of sustainable tourism, emphasising the need for a holistic and long-term approach that considers the triple bottom line – economic, environmental, and socio-cultural aspects.

India's numerous cultures and customs, ecosystems, and varied physiography make it an intriguing travel destination (Das & Hussain, 2016). Particularly, tourism in North East India has become an increasingly captivating subject for scholarly exploration due to the region's exceptional cultural diversity, abundant biodiversity, and a growing emphasis on sustainable tourism practices (Cabral & Dhar, 2020; Sharma & Sarmah, 2021). The North East region comprises of eight states – Arunachal Pradesh, Assam, Manipur, Meghalaya, Mizoram, Nagaland, Sikkim, and Tripura. Known as the 'Paradise Unexplored', and these states have a great deal of tourist potential since they offer a diverse range of indigenous cultures, lively customs, and breathtaking landscapes (Banik, 2017; Bhattacharjee, 2019). North East India is becoming more and more popular as a travel destination because of its breathtaking scenery, unique cultural variety, and potential for sustainable tourism. With its own traditions, dialects, and customs, each state in the region adds to the region's cultural richness. The native tribes contribute to the rich tapestry of cultures by providing visitors with chances to engage in colourful customs, regional celebrations, traditional foods, and gracious hospitality (Brooks, 2008). North East India boasts unmatched natural beauty with its grassy hills, thick forests, waterfalls, and majestic rivers. The region's varied landscapes, which include the calm lakes of Manipur, the tea gardens of Assam, and the Himalayan mountains of Sikkim, make it a sanctuary for ecotourism, adventure seekers, and peace-seekers. Due to the geography, there are many options for adventure tourism, such as paragliding, trekking, river rafting, and cave exploration (Dikshit & Dikshit, 2014; Das, 2017). With national parks like Assamese's Kaziranga and Manas, a UNESCO World Heritage Site, drawing tourists with their varied flora and fauna, North East India is also a wildlife enthusiast's dream (Dikshit & Dikshit, 2014; Das, 2017). The region is a melting pot of spiritual and religious diversity, featuring sites like Tawang Monastery in Arunachal Pradesh, Kamakhya Temple in Assam, and the ancient ruins of Unakoti in Tripura, etc. (Dkhar & Tiwari, 2020; Singh, 2022). Recognised as a pioneer in sustainable tourism practices, North East India actively promotes responsible tourism, waste management, and eco-friendly initiatives. Communities in places like Mawlynnong in Meghalaya exemplify this commitment to balancing tourism growth with environmental and cultural preservation. The region's vibrant festivals, such as Bihu in Assam, Hornbill Festival in Nagaland, and Ziro Festival of Music in Arunachal Pradesh, attract tourists, providing a glimpse into the traditional music, dance, and art forms of North East India. Researchers like Mitra and Khan (2017), Zou and Prasain (2017), Choudhury et al. (2018), and Sati (2019) have discussed the region's potential as a sustainable tourism destination, highlighting its diverse landscapes, vibrant cultures, and untapped tourism opportunities. While North East India has emerged as a multifaceted tourism phenomenon with its unique blend of cultural diversity, natural splendour, and historical significance, the surge in visitor numbers raises concerns about potential negative impacts on the environment, cultural heritage, and local communities (Baruah, 2020; Paul et al., 2019). Challenges such as connectivity issues, limited infrastructure, and inadequate promotion have historically hindered tourism growth in the region, but ongoing initiatives aim to address these issues and promote North East India as a desirable tourist destination (Chong, 2018; Haokip, 2015).

In Indian philosophy, there is a deep-rooted belief in the interconnectedness of people with the natural world, as reflected in Vedic mythology. The Vedic tradition highlights the reverence that humans hold for natural phenomena and objects perceived as divine (Baindur, 2015). The central tenet in Vedic philosophy is the notion that all natural entities are fundamentally unified. A deep and persistent belief in the interdependence of humans and the natural world is woven into the very fabric of Indian philosophy. The ancient texts and stories of Vedic mythology beautifully reflect this symbiotic interdependence, offering a philosophical basis that is timeless. The idea that people are inextricably connected to the natural events around them is fundamental to Indian philosophy and is masterfully expressed in the Vedic scriptures. The intricate network of relationships between people and their environment is reflected in Vedic literature, which is a massive corpus of ancient literature that includes the Rigveda, Samaveda, Yajurveda, and Atharvaveda (Agarwal, 2013; Goel, 2003). These texts' poems and songs express awe for the natural world, endowing objects, scenery, and heavenly bodies with divine attributes. According to this perspective, nature is a living, breathing thing that people have a close, mutually beneficial relationship with, rather than just a setting for human existence. Instead of being seen as distinct entities, rivers, mountains, forests, and animals are seen as essential parts of a harmonious and holistic cosmic order. The components of the natural world are frequently personified as deities or divine manifestations because they are treasured as sacred.

Joshi et al. (2023) discuss how Indian philosophy sees humans as being inextricably linked to the natural world, which goes beyond a mere appreciation of the surroundings. By highlighting the interconnectedness of all living things and the realisation that nature and humans are both infused with a global consciousness, it highlights a profound spiritual connection. By encouraging people to live in peace with nature and take care of the environment, this holistic viewpoint promotes a sense of responsibility. Essentially, the idea that humans and the natural world are intertwined, which has its roots in Indian philosophy and is represented in Vedic mythology, goes beyond simple thinking. It creates a perspective that combines ethical duty, spiritual awe, and ecological awareness, providing a timeless roadmap for people looking to live in harmony with the environment and comprehend their place in it better (Alter, 2015; Baindur, 2015).

The sustainable tourism industry in North East India has attained managerial importance for various stakeholders. Sustainable tourism aims to delicately balance economic development, environmental preservation, and social responsibility. It underscores the importance of adopting responsible travel practices that minimise ecological impact, show respect for local cultures, and promote community engagement. Applying the principles of sustainable tourism in the North East is not merely a theoretical concept but a practical necessity, given the region's ecological vulnerability and the imperative to safeguard its unique cultural identity. The North East, distinguished by its geographical remoteness and historical marginalisation, offers a distinct context for examining sustainable and medical tourism (Biswas & Rai, 2022). The region's pristine landscapes, home to a wealth of flora and fauna, are juxtaposed with the challenges of infrastructure development, accessibility, and socio-economic disparities. Against this backdrop, the successes and lessons learned from sustainable tourism initiatives in the North East carry broader implications for similar regions globally, grappling with the dual imperative of tourism development and environmental conservation. The introduction of sustainable tourism practices in North East India is not a recent phenomenon but has evolved over time in response to the changing dynamics of tourism and the recognition of its potential impact on the region's fragile ecosystems (De, 2013). Nature-based tourism has taken centre stage in the northeastern India. Many tourists visit this region each year primarily to take in the breathtaking scenery waterfalls, limestone caves, trekking through hilly jungles, forest biodiversity, and indigenous cultures. Tourists are prepared to invest a substantial sum of money and time in order to appreciate and help preserve such natural beauty (De & Devi, 2011). To enhance the authentic experience for tourists, many regions in the North East promote homestays and community-based tourism. This not only provides visitors with a chance to experience local lifestyles firsthand but also ensures that the economic benefits of tourism directly reach local communities. The study will trace the historical trajectory of tourism in the North East, examining key inflection points that led to the realisation of the need for sustainability. This historical lens will provide valuable insights into the evolution of policies, regulations, and community-led initiatives aimed at fostering sustainable tourism practices. It is essential to acknowledge the diverse stakeholders involved in the tourism ecosystem of North East India – from government bodies and non-governmental organisations to local communities and tourists themselves. Their collective actions and interactions shape the narrative of sustainable tourism in the region. Through an interdisciplinary perspective, this study seeks to integrate perspectives from environmental science, cultural studies, economics and tourism management, etc. to offer a holistic understanding of the successes and challenges encountered in the pursuit of sustainable tourism in North East India.

This comprehensive study aims to unravel the complexities of sustainable tourism in North East India through a detailed case study approach. Based on this background, the objectives of the study are as follows.

1.1. Objectives

(i) To compare sustainable tourism models of North East India with other tourism regions in India.
(ii) To explore the sustainable tourism model implemented in North East India and portray the successes and lessons learned in this context.

2. Methodology

The present investigation complies with Torraco's (2005) integrative literature review requirements. Many academics have used this literature review approach (Amui et al., 2017; Jabbour, 2013; Seuring & Muller, 2008).

2.1. Literature Review

For the first objective, a thorough literature scan leads to distinct sustainability models for North East India. A comparative study is also done with respect to the sustainability tourism models of North East and the rest of India. According to Barbhuiya and Chatterjee (2020), the vulnerability and resilience in sustainability of tourism models can vary significantly across different regions in India. While North East India faces its own set of challenges, it exhibits certain characteristics that contribute to a relatively more sustainable tourism model compared to some other popular tourist destinations in the country. North East India is renowned for its rich biodiversity, with lush forests, diverse flora, and fauna. The region's commitment to preserving its unique ecosystems has led to the establishment of numerous national parks and wildlife sanctuaries. Strict conservation measures and community involvement in these efforts contribute to the sustainable protection of the environment. In contrast, some other tourist destinations in India may face challenges related to overexploitation of natural resources and inadequate conservation measures. The cultural sensitivity and community involvement lead to sustainable tourism in North East India as it often places a strong emphasis on respecting and preserving local cultures. Many tourism initiatives in the region actively involve indigenous communities, promoting cultural exchange while ensuring that economic benefits are distributed equitably. This community-centric approach fosters a sense of ownership among local residents, mitigating the negative social impacts often associated with tourism. On the other side, problems like inadequate community involvement and cultural commodification plague some of India's most well-known tourist sites. Due to its remote location, North East India has not seen the same level of commercialisation and mass tourism as some other regions of the country. This has made it possible for tourism to flourish in a more managed and sustainable way. The lack of an excessive amount of visitors contributes to the preservation of the natural environment, avoids crowding, and lessens the burden on the available resources in the area. Conversely, more popular tourist spots in India could experience problems with traffic, pollution, and infrastructure stress. The governments of various nations have increasingly recognised the importance of sustainable tourism and have implemented policies to regulate and guide the industry (Edgell, 2019; Fletcher et al., 2019) and this holds true even in the case of North East India. From promoting eco-friendly practices to supporting community-based tourism initiatives, these policy measures contribute to a more sustainable tourism model. In contrast, some other tourist destinations in India may have been slower to adopt and enforce comprehensive sustainable tourism policies. North East India has seen a growing emphasis on responsible tourism practices. Initiatives encouraging responsible behaviour among tourists, such as waste reduction, eco-friendly accommodations, and the promotion of local handicrafts, contribute to a more sustainable tourism model. This contrasts with some other tourist destinations in India where irresponsible tourism practices, such as littering and damage to cultural sites, may be more prevalent. While North East India presents a more sustainable tourism model in certain aspects, it is important to note that challenges persist, including issues related to infrastructure development, tourism infrastructure, accessibility from within and from the rest of India, and overall economic disparities.

The direct comparison of the sustainability model of North East India and the rest of India can be done on these aspects.

(i) *Cultural Preservation*: the North East boasts diverse indigenous cultures, and many communities actively engage in preserving their heritage through sustainable tourism practices. In the case of the rest of India, while there are initiatives to promote cultural tourism in various parts of the country, challenges such as over-commercialisation and cultural commodification can be more pronounced in popular tourist destinations.

(ii) *Environmental Conservation*: the North East India is characterised by lush landscapes and diverse ecosystems. Sustainable practices, such as community-based conservation and eco-friendly tourism initiatives, are often integrated into the tourism model. For the rest of India, some popular tourist destinations face challenges related to environmental degradation, pollution, and strain on natural resources due to high visitor numbers.

(iii) *Community Involvement*: many tourism initiatives in the North East actively involve local communities, fostering a sense of ownership and ensuring that economic benefits are distributed more equitably whereas, in the rest of India, in certain areas, the tourism industry may not be as integrated with local communities, leading to issues of economic disparity and limited community engagement.

(iv) *Mass Tourism and Infrastructure*: the North East region has not experienced the same level of mass tourism and commercialisation as some other parts of India. This has allowed for more controlled and sustainable growth. Popular tourist destinations in other parts of India often grapple with issues like overcrowding, inadequate infrastructure, and the strain on public services.

(v) *Government Policies*: some North Eastern states have actively embraced sustainable tourism in their policy frameworks, promoting responsible travel and conservation. While there are national guidelines for sustainable tourism, enforcement and implementation can vary across states, and some areas may face challenges in aligning policies with sustainable practices.

(vi) *Biodiversity and Unique Attractions*: the North East Indian region is known for its rich biodiversity and unique attractions, contributing to a distinctive and often less-explored tourist experience. In case of the rest of India, while there are diverse ecosystems across the country, some heavily visited areas may struggle to balance tourism with the preservation of unique ecological features.

It's important to note that the comparison is based on an extensive literature review and there are exceptions on both sides. Moreover, the notion of 'better' can be subjective and depends on specific criteria and perspectives. Sustainable tourism involves a complex interplay of cultural, environmental, and economic factors, and ongoing efforts are required in all regions to address challenges and enhance sustainability. For the better understanding, the research recommends empirical studies to be conducted to validate the tourism statistics specific to the regions in question.

2.2. Case Studies

For the second objective, the research relies on the deep understanding of the five tourist places visited by the researcher during the period 2015–2019. The observation analysis coupled with facts from various literature led to the overall understanding of the sustainable tourism models in North East India. This region, renowned for its cultural diversity and pristine landscapes, has become a focal point for sustainable tourism initiatives. These five case studies delve into specific examples that highlight the region's commitment to a sustainable tourism model, showcasing successes and innovative approaches that address environmental, cultural, and community considerations.

(i) *Community-Based Tourism in Mawlynnong, Meghalaya* – Mawlynnong, often referred to as 'Asia's Cleanest Village' (Fig. 2.1) exemplifies a community-led sustainable tourism model. Local communities actively participate in waste management initiatives, maintaining the village's cleanliness and preserving its natural beauty. The establishment of community-based homestays ensures that economic benefits directly reach the villagers. By actively taking part in

Fig. 2.1. Photograph of Entry to Mawlynnong. *Source*: Author's own original photographic work.

traditions like weaving and listening to local music, travellers participate in cultural exchanges that foster mutual understanding. The world is familiar with Mawlynnong, a quaint village in the East Khasi Hills of Meghalaya, for its locally driven approach to sustainable tourism. The 'Cleanest Village in Asia', Mawlynnong, is a shining example of how locally focused tourism can advance economic development while also preserving the environment and local way of life. The village's superior waste management practices are significant in this regard. Mawlynnong's commitment to cleanliness is the cornerstone of its sustainable tourism strategy. Every family is responsible for maintaining the cleanliness of their area, and the neighbourhood actively participates in garbage management activities. Recycling is widespread, and the hamlet has made a name for itself as a pioneer in effective waste disposal. A crucial element of Mawlynnong's sustainable tourism strategy is the use of community homestays. Locals welcome travellers inside their homes in order to provide them with a true experience of Khasi culture. With this approach, the economic benefits of tourism have a direct impact on the well-being of community members. It is advised that tourists engage in cultural exchanges by making bamboo, weaving, and listening to traditional music while in Mawlynnong. These interactions show off the rich cultural heritage of the Khasi people while also promoting understanding between tourists and the local populace. The community-based tourism sector in Mawlynnong is highly dependent on how well the local government runs. In order to safeguard the community's interests and guarantee that development decisions adhere to sustainable principles, local leaders take an active role in shaping tourism policies. The village takes great pride in its robust system of decision-making. For both tourists and residents of Mawlynnong, education is of utmost importance. Presentations are made to visitors regarding the significance of eco-friendly travel, local customs, and environmental conservation efforts. In addition to improving the tourism experience, this educational element makes tourists feel more accountable. Mawlynnong's residents have a strong bond with the surrounding environment. The goal is to preserve the pristine environment by encouraging responsible tourism activities, such as planting trees, keeping green areas, and limiting the use of non-biodegradable products. Mawlynnong will continue to be a desirable travel destination for sustainable tourism because of its dedication to environmental conservation. Although there is remarkable success in the community-based tourism model, there are various challenges continuing to trouble. Balancing the influx of tourists with the need to preserve the village's tranquillity and cultural integrity requires ongoing efforts. Additionally, infrastructure development and access to resources remain areas that need attention for sustainable growth. As Mawlynnong continues to navigate the complexities of tourism, the village stands as a beacon for other communities seeking to embrace sustainable tourism. The lessons learned from Mawlynnong emphasise the importance of community involvement, cultural preservation, and environmental stewardship in creating a holistic and enduring model for tourism development.

(ii) *Kaziranga National Park, Assam* – Kaziranga, a UNESCO World Heritage Site (Fig. 2.2) showcases a successful blend of wildlife conservation and sustainable tourism. The park's robust anti-poaching measures protect its iconic one-horned rhinoceros population, while responsible tourism practices, such as guided safaris on designated paths, minimise disturbances to the ecosystem. Revenue generated from entry fees and eco-tourism initiatives

Fig. 2.2. A Picturesque of Kaziranga National Park. *Source*: Author's own original photographic work.

directly contributes to conservation efforts and supports local communities. Kaziranga National Park, situated in the Indian state of Assam, is a UNESCO World Heritage Site and a global biodiversity hotspot. Known for its population of the iconic one-horned rhinoceros, Kaziranga presents a unique case study of sustainable tourism, successfully integrating wildlife conservation with responsible visitor management. The foundation of sustainable tourism at Kaziranga lies in its steadfast dedication to wildlife preservation. Rigorous anti-poaching measures, incorporating advanced technologies like drones and infrared cameras, have been deployed to safeguard the endangered one-horned rhinoceros and other wildlife species. The park stands as a testament to the successful cohabitation of humans and wildlife. Noteworthy aspects include visitor management and guided safaris. Kaziranga has established a well-regulated visitor management system to mitigate the impact of tourism on the ecosystem. Guided safaris along designated paths ensure that tourists can observe wildlife in their natural habitat without causing undue disturbance. This methodical technique reduces stress on the park's wildlife and keeps vulnerable regions from deteriorating. The park has embraced eco-tourism initiatives in an effort to give conservation efforts a steady stream of income. These projects include educational programmes, bird-watching tours, and guided nature walks. The money made from eco-tourism directly helps to maintain the park and the surrounding communities. Local communities actively participate in and gain from tourism-related activities, emphasising benefit sharing and community involvement. Locals are given the opportunity to work, especially in positions pertaining to ecotourism and park management. There is an assurance that the locals benefit economically from tourism, this strategy promotes a feeling of community ownership and conservation support. To handle the increasing number of tourists, buffer zones have been established around the main wildlife areas in Kaziranga. By preventing congestion in the central areas, this tactic lessens its negative effects on animals and preserves the delicate balance between tourist and conservation. Additionally, it permits the construction of infrastructure to serve tourists without invading vital habitats. Despite its successes, Kaziranga faces ongoing challenges. The park is susceptible to natural disasters like floods, which can impact both wildlife and tourism infrastructure. Additionally, managing the delicate equilibrium between tourism and conservation necessitates continuous adaptation to changing conditions. The future prospects for sustainable tourism in Kaziranga involve a continued focus on technological innovations for wildlife protection, enhanced community engagement, and the development of eco-friendly tourism infrastructure. Striking a balance between increasing tourist numbers and preserving the park's ecological integrity remains a priority. Kaziranga National Park's sustainable tourism model serves as a testament to the possibility of harmonising wildlife conservation with responsible tourism. The successful integration of community involvement, guided visitor experiences, and eco-tourism initiatives provides valuable insights for other protected areas globally seeking to balance conservation imperatives with the economic benefits of tourism. Kaziranga stands as a beacon for sustainable wildlife tourism, showcasing the potential for coexistence between humans and the natural world.

(iii) *Dzukou Valley, Nagaland-Manipur Border* – The Dzukou Valley presents a unique case of sustainable trekking tourism. Local communities, in collaboration with government agencies, have implemented measures to preserve the valley's delicate ecosystem. Restricted entry, waste management systems, and community-led initiatives ensure that the influx of trekkers does not compromise the valley's ecological integrity. The money raised helps local infrastructure and goes towards maintaining hiking trails. Situated on the boundary between Nagaland and Manipur in Northeast India, Dzukou Valley is a pristine and ecologically sensitive region that is well-known for its stunning views, diverse flora, and remarkable wildlife. It is now seen as an example for sustainable tourism, emphasising the need for responsible behaviour to safeguard the fragile ecology. Controlled tourism and restricted access are prioritised, with measures implemented to restrict tourist numbers, prevent overcrowding, and mitigate adverse impacts on the valley's delicate ecosystem. Entry permits are needed, and there is a visitation cap in place to preserve the valley's natural beauty and biodiversity as well as the ecological balance of the area. The surrounding communities of Dzukou Valley are actively involved in conservation work. Governmental organisations and community-based organisations work together to manage the valley sustainably. This entails developing strategies for garbage management, hiking pathways, and protections for the region's endangered animal and plant species. As a popular destination for trekking enthusiasts, Dzukou Valley integrates sustainable tourism practices into trekking activities. Visitors are encouraged to adhere to designated trails to minimise soil erosion and disturbance to local fauna. Eco-friendly trekking practices, such as carrying back all non-biodegradable waste, are promoted to conserve the valley's pristine environment. Local communities play a pivotal role in shaping Dzukou Valley's sustainable tourism model. Actively engaging with visitors, they provide information about the region's cultural and environmental significance. Ecotourism initiatives, including homestays and cultural experiences, contribute to the local economy, instilling a sense of responsibility and pride among community members. Recognised for its unique biodiversity, including rare and endemic plant species, Dzukou Valley implements sustainable tourism practices that enforce guidelines against the collection of flora and fauna. This ensures the long-term survival of the valley's ecological treasures. While Dzukou Valley's sustainable tourism model has achieved success, challenges persist, particularly in infrastructural development, waste management during peak seasons, and the need for continuous community involvement. Striking a balance between visitor numbers and conservation efforts remains a critical consideration as the valley's popularity grows. Future prospects involve the implementation of advanced sustainable tourism strategies, improved waste management infrastructure, and increased community education. Dzukou Valley serves as a beacon, demonstrating how sustainable tourism can flourish in ecologically sensitive areas while maintaining ecological integrity and offering a responsible tourist experience. By embracing restricted entry, community-led conservation initiatives, and eco-friendly trekking practices, Dzukou Valley exemplifies a harmonious relationship between tourism and nature. This case study offers valuable insights into the successful implementation of sustainable tourism practices in a region celebrated for its natural beauty and biodiversity.

(iv) *Khonoma Village, Nagaland*– Khonoma Village has gained recognition for its community-driven conservation efforts. The village has established a biodiversity conservation zone, restricting hunting and logging. Tourists who visit Khonoma actively engage in eco-friendly activities such as bird-watching and nature walks, contributing to the community's income. The success of Khonoma's sustainable tourism model is rooted in the harmonious integration of cultural preservation and environmental conservation. Nestled in the hills of Nagaland, Khonoma Village serves as a pioneering example of sustainable tourism in Northeast India. Renowned for its rich cultural heritage and dedication to environmental conservation, Khonoma has successfully woven tourism into community-led initiatives, safeguarding both its cultural identity and natural surroundings. A key feature of Khonoma's sustainable tourism model is the creation of a Biodiversity Conservation Zone. This initiative designates areas where hunting and logging are restricted, allowing the natural ecosystem to thrive. The community has taken proactive measures to protect the diverse flora and fauna of the region, including endangered species such as the Blyth's tragopan. Khonoma offers visitors an authentic cultural experience through community-based tourism. Homestays are a prevalent accommodation option enabling tourists to reside with local families, partake in daily activities, and engage in cultural exchanges. This not only brings economic benefits to the community but also ensures the preservation of traditional practices and rituals. Khonoma actively educates tourists about the village's cultural and environmental significance. Guided tours and interpretative programs provide insights into the history, customs, and the importance of sustainable living. Tourists are sensitised to the necessity for responsible behaviour, fostering respect and appreciation for the local way of life. Sustainable agriculture practices are integral to Khonoma's tourism model, evident in its embrace of traditional farming methods promoting biodiversity and conservation. Organic farming, terrace cultivation, and traditional irrigation systems contribute to the sustainability of agricultural activities, maintaining fertile and resilient land. The success of Khonoma's sustainable tourism is closely tied to its robust community governance. The village council, in collaboration with local organisations, plays a pivotal role in decision-making related to tourism activities, ensuring alignment with principles of sustainability and prioritising the community's interests. Despite significant strides in sustainable tourism, Khonoma Village faces challenges, including the need for improved infrastructure, waste management solutions, and considerations regarding the potential impact of increased visitor numbers. Addressing these challenges requires a holistic and adaptive approach to sustainable tourism management. The future prospects for Khonoma involve a continued focus on community engagement, the implementation of eco-friendly practices, and the development of sustainable tourism infrastructure. Balancing the economic benefits of tourism with the preservation of cultural and environmental assets remains a key consideration for Khonoma's sustainable tourism journey. Khonoma Village stands as a beacon for sustainable tourism, demonstrating that economic development can go hand in hand with cultural preservation and environmental conservation. By prioritising biodiversity, engaging in community-led initiatives, and fostering responsible tourism practices, Khonoma provides a blueprint for other regions seeking to create a sustainable and harmonious relationship between tourism and local communities.

(v) *Tawang Monastery, Arunachal Pradesh* – Tawang Monastery, a cultural and spiritual hub, exemplifies responsible tourism within a heritage context. The monastery administration, in collaboration with local authorities, regulates visitor numbers and enforces respectful conduct within the premises. Revenue from entrance fees supports the maintenance of the monastery and the welfare of the local monks. Educational programs for tourists enhance cultural awareness, contributing to a harmonious interaction between visitors and the religious community. Tawang Monastery is a prominent spiritual centre and a model of sustainable tourism, situated at an elevation of about 3,000 metres in the tranquil landscapes of Arunachal Pradesh (Fig. 2.3). The monastery's tourism strategy skilfully combines cultural preservation, religious practices, and appropriate tourist management. Tawang Monastery has implemented protocols to control the quantity of guests and mandate decorum on its grounds, guaranteeing peace of mind.

Fig. 2.3. A View of Tawang Monastery. *Source*: Author's own original photographic work.

It is recommended that pilgrims and visitors alike maintain a sombre and thoughtful manner, cultivating an atmosphere of respect and cultural admiration that preserves the site's holy holiness. Tawang Monastery actively participates in maintaining its rich cultural legacy by providing guests with educational programmes that shed light on Buddhist customs, ceremonies, and the history of the monastery. The focus on cultural sensitivity makes sure that tourism-related activities help people recognise and comprehend the region's distinctive history. The upkeep of the monastery and the welfare of the resident monks (Fig. 2.4) are directly supported by the money received from contributions and admission fees. This financial support contributes to the self-sufficiency of the monastery by providing a means of subsistence for the monastic community as well as ensuring the maintenance of the sacred location. Tawang Monastery provides educational programmes and guided tours to visitors in order to improve their experience. These offerings give visitors a deeper understanding of the monastery's historical significance, architectural wonders, and religious practises. By promoting cultural awareness, this educational component encourages visitors to interact with the site in a responsible and knowledgeable manner. The monastery proactively advocates for conscientious tourism behaviours, encompassing waste minimisation, environmentally conscious endeavours, and sustainable modes of transportation. In keeping with the monastery's dedication to environmental care in the ecologically delicate Himalayan region, pilgrims and visitors are urged to reduce their environmental impact. Tawang Monastery is successful in fostering sustainable tourism, yet it still has to deal with issues like trash management, infrastructure development, and the possible effects of growing tourist numbers. Finding a careful balance between protecting

Fig. 2.4. A Photograph with Resident Child Monks. *Source*: Author's own original photographic work.

the sacredness of cultures and religions and benefiting economically from tourism is still a difficult task. Tawang Monastery's future prospects include building eco-friendly infrastructure, fostering greater community involvement, and carrying out sustainable tourism practises. Working together with community stakeholders and local authorities will be essential to addressing issues and guaranteeing the sustainability of the tourism model. Tawang Monastery serves as an exemplary model for sustainable tourism in religious and cultural heritage sites. By integrating responsible tourism practices with cultural preservation and economic sustainability, Tawang demonstrates the potential for harmonious coexistence between tourism and sacred sites. The case of Tawang Monastery provides valuable insights for regions seeking to balance the spiritual, cultural, and economic dimensions of tourism in a sustainable manner.

The case study of sustainable tourism in North East India illustrates the region's proactive efforts to balance tourism development with environmental conservation, cultural preservation, and community well-being. These instances not only exemplify effective models but also offer valuable lessons for other regions aspiring to adopt sustainable tourism practices. As North East India grapples with the challenges of tourism expansion, the knowledge gleaned from these initiatives contributes to a more nuanced and encompassing approach to sustainable tourism on a larger scale.

3. Conclusion

The research concludes the exploration of sustainable tourism in North East India, a region renowned for its cultural diversity, unspoiled landscapes, and distinct identity; it becomes apparent that this part of the country embraces a unique and commendable approach to balancing tourism growth with environmental conservation, cultural safeguarding, and community well-being. A thorough analysis of case studies from various states in the North East, including Mawlynnong, Kaziranga National Park, Dzukou Valley, Khonoma Village, and Tawang Monastery, has provided insights into the intricate dynamics of sustainable tourism in the region, focusing on cultural preservation and identity. North East India has a top-notch position in preserving its gorgeous cultural heritage and the efforts are commendable. In places like Khonoma Village and Mawlynnong, community-based tourism initiatives not only yield economic benefits but also serve as platforms for cultural exchange. Homestays, traditional activities, and cultural experiences become integral elements of the tourist experience, fostering mutual understanding between visitors and local communities. The emphasis on cultural sensitivity and the active involvement of indigenous populations play crucial roles in safeguarding unique traditions, ensuring that tourism enhances rather than erodes cultural identities. The North East's commitment to environmental conservation is evident in its sustainable tourism practices. Kaziranga National Park and Dzukou Valley stand as a testament to the region's dedication to protecting its natural heritage. From strict anti-poaching measures and regulated visitor management in Kaziranga to restricted entry and eco-friendly trekking practices in Dzukou Valley, these initiatives underscore the importance of balancing tourism with the preservation of fragile ecosystems. The emphasis on biodiversity conservation, the establishment of conservation zones, and community-led efforts in these areas exemplify a forward-looking approach that places environmental sustainability at the forefront. The success of sustainable tourism in North East India is intricately linked to community engagement and inclusivity. In Khonoma Village, the establishment of a Biodiversity Conservation Zone involves the active participation of local communities, signalling a collaborative approach to environmental protection. Similarly, the Tawang Monastery and Mawlynnong showcase how economic benefits from tourism directly support local communities. Through homestays, cultural initiatives, and educational programs, tourists are not merely spectators but active participants in the life and well-being of these communities. While celebrating the successes of sustainable tourism in North East India, it is crucial to acknowledge the persistent challenges and consider the path forward. Infrastructure development, waste management, and the potential impact of increasing visitor numbers emerge as common challenges. Balancing the need for economic growth with environmental and cultural preservation remains an ongoing endeavour. The North East will need to navigate these challenges by fostering collaborative efforts, integrating technological solutions, and continuously adapting its sustainable tourism model. As discussed earlier, the comparative analysis shows the comparison of North East India with the rest of India. North East India stands out for its more controlled and community-centric approach to tourism. The region's lower exposure to mass tourism, active community involvement, and a focus on responsible practices contribute to a more sustainable model. In contrast, popular tourist destinations in other parts of India may grapple with issues related to overcrowding, pollution, and cultural commodification. The role of government initiatives and policies in shaping sustainable tourism cannot be overstated. North East Indian states have increasingly recognised the importance of sustainable tourism, evident in policies promoting eco-friendly practices, community involvement, and conservation. While efforts are made at the national level, the implementation and enforcement of sustainable tourism policies may vary across different states, impacting the overall sustainability of tourism in the country.

As we look ahead, the case studies from North East India provide valuable lessons and inspiration for regions globally. The region's commitment to balancing economic growth with environmental and cultural preservation exemplifies a holistic approach to sustainable tourism. The people who live in North East are equally as important to the region's success as its breathtaking scenery. When cultural identity, environmental protection, and economic sustainability are combined, tourism can be used to promote change rather than endanger the environment and local customs. The pursuit of sustainable tourism necessitates constant adaptation and cooperation. With its varied cultures and landscapes, North East India has set the standard for using tourism as a vehicle for constructive change. The North East's experiences offer a model for a more peaceful and sustainable future where a destination's charm lies not only in its natural wonders but also in the communities that protect and nurture them, while other regions struggle with the complexities of tourism development. The knowledge gathered from North East India emphasises the interdependence of people, culture, and environment, and suggests that there is room for a global move towards more responsible and sustainable tourist practices.

4. Implications

The implications and lessons learned from North East India's sustainable tourism are substantial. The sustainable tourism practices of the region have significant ramifications and can teach other regions in the world important lessons. With its well-balanced approach to cultural preservation, environmental stewardship, and community engagement, North East India's tourism development model is a model for others hoping to make the tourism industry more sustainable and responsible in the long run.

Firstly, North East India's sustainable tourism has been remarkably successful because of its all-encompassing and holistic development strategy. Rather than viewing tourism solely through an economic lens, the region recognises the interconnectedness of cultural, environmental, and economic aspects. This holistic perspective emphasises the importance of developing a sustainable tourism model that prioritises the well-being of local communities, the preservation of cultural heritage, and the protection of natural ecosystems. Secondly, a critical lesson from North East India is the significance of community empowerment and inclusivity in tourism development. Community-based tourism initiatives, exemplified in places like Mawlynnong and Khonoma Village, demonstrate the transformative power of including local communities in the decision-making process. When communities actively participate in and benefit from tourism, it not only enhances economic opportunities but also fosters a sense of ownership and pride in preserving cultural identities. Thirdly, when it comes to sustainable tourism, North East India gives environmental conservation top importance. The region offers important lessons for other locations because of its dedication to preserving biodiversity, which is demonstrated by places like Kaziranga National Park and Dzukou Valley. A proactive approach to conservation is necessary for sustainable tourism, and this includes community-led conservation efforts, eco-friendly practises, and entry restrictions for the preservation of fragile ecosystems. Fourth, the protection of cultural heritage is the cornerstone of sustainable tourism in Northeast India. There are lessons to be learned for destinations trying to strike a balance between responsible tourism practices and cultural promotion from the region's emphasis on cultural sensitivity, visitor education programmes, and community-led initiatives to preserve traditions. It emphasises how crucial cultural preservation is to sustainable tourism as a vital and essential element. In addition, North East India's sustainable tourism industry depends on ethical visitor behaviour. The success of the area depends on visitors acting responsibly, supporting programmes like eco-friendly activities, waste reduction, and responsible tourism. This focus on responsible behaviour promotes a more thoughtful and polite exchange between travellers and the places they visit, thereby reducing the negative effects of tourism. Last but not least, it is impossible to overestimate the influence of governmental assistance and policies for sustainable tourism. States in North East India have shown their dedication to sustainable tourism by developing and enforcing laws that support eco-friendly behaviour, community engagement, and conservation initiatives. This emphasises the necessity of a national and regional policy framework that is supportive in order to guarantee the long-term viability of tourism-related initiatives.

Beyond its boundaries, North East India's sustainable tourism model has important lessons to teach other areas struggling with the challenges of developing their tourism industries. The prosperous histories of Mawlynnong, Kaziranga, Dzukou Valley, Khonoma, and Tawang Monastery offer a model for a more conscientious and peaceful coexistence between tourism and the local populace. One of the most important lessons from North East India is the fine line that separates economic expansion from conservation. Although local businesses benefit from sustainable tourism, the area shows that conservation efforts and economic expansion can coexist. Conservation zones, access restrictions, and community-led projects are examples of a well-balanced strategy that protects the environment and promotes economic growth. The success of sustainable tourism initiatives in North East India is largely dependent on cooperation between many stakeholders, such as local people, governmental agencies, and non-governmental organisations. The adoption of community-led projects and the active participation of communities in decision-making processes highlight the necessity of teamwork in developing an inclusive and sustainable tourism model. Sustainable tourism is a dynamic endeavour that calls for ongoing learning and adaptation. The experiences of North East India emphasise how crucial it is to adapt quickly to changing conditions, whether they have to do with the amount of tourists visiting the region, environmental issues, or community needs. Global destinations may learn a lot from the region's dedication to ongoing development. The ability of North East India to present distinctive and genuine experiences is essential to its success in sustainable tourism. Travellers who are responsible and looking for real interactions are drawn to the area by its unique customs, gorgeous scenery, and different civilisations. This emphasises the allure of travel locations that put originality and individuality ahead of mass tourism. The difficulties North East India faces, such as those pertaining to waste management, infrastructure, and the possible effects of a rise in tourists, highlight the resilience needed for sustainable tourism. For locations negotiating the complexity of tourism development, the region's capacity to confront problems head-on, put solutions in place, and adjust to changing conditions offers a resilience lesson. The sustainable tourism journey of North East India focuses on the necessity for an integrated and thoughtful strategy, providing a global roadmap for places to follow. Lessons from North East India become more pertinent as travel destinations struggle to maintain their distinct identities while promoting economic growth. The region serves as an example of how tourism may be used as a catalyst for positive change by placing a high priority on community engagement, environmental conservation, and cultural preservation. North East India serves as a monument to the transforming power of well-executed sustainable tourism, as the global tourist landscape continues to change. It emphasises the notion that, when handled sensibly, tourism may promote economic empowerment, cross-cultural exchange, and environmental preservation. The various stakeholders involved in the holistic process of tourism in North East India have these points to be learned by others, special the ones where tourism is causing serious damage to the ecosystem. The implications and insights learned from North East India show the way for other travel destinations elsewhere to set out on a path towards a more inclusive, sustainable, and peaceful tourism future.

References

Agarwal, M. K. (2013). *The vedic core of human history: And truth will be the savior*. iUniverse.

Alter, J. S. (2015). Nature cure and Ayurveda: Nationalism, viscerality and bio-ecology in India. *Body & Society*, *21*(1), 3–28.

Amui, L. B. L., Jabbour, C. J. C., de Sousa Jabbour, A. B. L., & Kannan, D. (2017). Sustainability as a dynamic organizational capability: A systematic review and a future agenda toward a sustainable transition. *Journal of Cleaner Production*, *142*, 308–322.

Baindur, M. (2015). *Nature in Indian philosophy and cultural traditions*. Springer.

Banik, S. (2017). Stereotyping India's North East: Examining the "Paradise Unexplored" in tourism discourse. *Rupkatha Journal on Interdisciplinary Studies in Humanities*, *9*(3), 1–7.

Barbhuiya, M. R., & Chatterjee, D. (2020). Vulnerability and resilience of the tourism sector in India: Effects of natural disasters and internal conflict. *Tourism Management Perspectives*, *33*(2), 100616.

Baruah, S. (2020). *In the name of the nation: India and its Northeast*. Stanford University Press.

Biswas, T., & Rai, A. (2022). Medical travel from north–east India: an assessment of domestic medical tourists' travel profile and experience. *Asian Geographer*, *41*(1), 1–25.

Bhattacharjee, B. J. (2019). Tourism in North East India's state of Tripura: Paradise unexplored. *ITIHAS-The Journal of Indian Management*, *9*(4), 44–54.

Bramwell, B. (2015). Theoretical activity in sustainable tourism research. *Annals of Tourism Research*, *54*, 204–218.

Brooks, L. T. (2008). *The common pot: The recovery of native space in the Northeast* (Vol. 7). University of Minnesota Press.

Cabral, C., & Dhar, R. L. (2020). Ecotourism research in India: From an integrative literature review to a future research framework. *Journal of Ecotourism*, *19*(1), 23–49.

Chong, B. (2018). Pursuing development through connectivity: An analysis of India's northeast region. *Centre on Asia and Globalization, Lee Kaun Yew School of Public Policy. National University of Singapore*.

Choudhury, K., Dutta, P., & Patgiri, S. (2018). Rural tourism of North East India: Prospects and challenges. *IOSR Journal of Humanities and Social Science*, *23*(2), 69–74.

Croes, R., Ridderstaat, J., Bąk, M., & Zientara, P. (2021). Tourism specialization, economic growth, human development and transition economies: The case of Poland. *Tourism Management*, *82*, 104181.

Das, D. (2017). Park, people and biodiversity conservation in Kaziranga National Park, India. *Space and Culture, India*, *5*(1), 36–48.

Das, D., & Hussain, I. (2016). Does ecotourism affect economic welfare? Evidence from Kaziranga National Park, India. *Journal of Ecotourism*, *15*(3), 241–260.

De, U. K. (2013). Sustainable nature-based tourism, involvement of indigenous women and development: A case of North-East India. *Tourism Recreation Research*, *38*(3), 311–324.

De, U.K., & Devi, A. (2011). Valuing recreational and conservational benefits of a natural tourist site: Case of Cherrapunjee. *Journal of Quantitative Economics*, *9*(2), 154–172.

Dikshit, K. R., & Dikshit, J. K. (2014). Natural vegetation: Forests and grasslands of North-East India. In *North-East India: Land, people and economy* (Advances in Asian Human-Environmental Research, pp. 213–255). Springer.

Dkhar, M., & Tiwari, B. K. (2020). Traditional ecological knowledge of tribal communities of North East India. *Biodiversitas Journal of Biological Diversity*, *21*(7), 3209–3224.

Edgell, D. L., Sr. (2019). *Managing sustainable tourism: A legacy for the future*. Routledge.

Fletcher, C., Pforr, C., & Brueckner, M. (2019). Factors influencing Indigenous engagement in tourism development: An international perspective. In *Sustainable tourism and Indigenous peoples* (pp. 34–54). eBook ISSN - 9781315112053.

Goel, A. (2003). *Environment and ancient Sanskrit literature*. Deep and Deep Publications.

Haokip, T. (2015). India's look east policy: Prospects and challenges for Northeast India. *Studies in Indian Politics*, *3*(2), 198–211.

Jabbour, C. J. C. (2013). Environmental training in organisations: From a literature review to a framework for future research. *Resources, Conservation and Recycling*, *74*, 144–155.

Joshi, A., Roy, S., Manik, R. K., & Sahoo, S. K. (2023). Scientific philosophy: Exploring existential, metaphysical, and ethical research philosophy behind the question "WHO AM I?". *Journal of Pharmaceutical Negative Results*, *14*(3), 1648–1671.

Lasisi, T. T., Odei, S. A., & Eluwole, K. K. (2023). Smart destination competitiveness: Underscoring its impact on economic growth. *Journal of Tourism Futures*. https://doi.org/10.1108/JTF-09-2022-0243

Lew, A. A., Cheer, J. M., Haywood, M., Brouder, P., & Salazar, N. B. (2020). Visions of travel and tourism after the global COVID-19 transformation of 2020. *Tourism Geographies*, *22*(3), 455–466.

McGehee, N. G., Knollenberg, W., & Komorowski, A. (2018). The central role of leadership in rural tourism development: A theoretical framework and case studies. In *Rural tourism* (pp. 145–165). Routledge.

Mitra, A., & Khan, A. (2017). Green tourism management in India 'A 3D study of the seven sisters states of north-east with special reference to eco-tourism'. *International Journal of Innovative Research in Science, Engineering and Technology*, *6*(4), 6923–6932.

Nepal, S. K. (2021). Adventure travel and tourism after COVID-19–business as usual or opportunity to reset? In A. A. Lew, J. M. Cheer, P. Brouder, & M. Mostafanezhad (Eds.), *Global tourism and COVID-19* (pp. 192–196). Routledge.

Paul, A., Deka, J., Gujre, N., Rangan, L., & Mitra, S. (2019). Does nature of livelihood regulate the urban community's vulnerability to climate change? Guwahati city, a case study from North East India. *Journal of Environmental Management*, *251*, 109591.

Sati, V. P. (2019). Potential and forms of sustainable village tourism in Mizoram, northeast India. *Journal of Multidisciplinary Academic Tourism*, *4*(1), 49–62.

Seuring, S., & Muller, M. (2008). From a literature review to a conceptual framework for sustainable supply chain management. *Journal of Cleaner Production*, *16*, 1699–1710.

Sharma, N., & Sarmah, B. (2021). Consumer engagement in village eco-tourism: A case of the cleanest village in Asia–Mawlynnong. In S. K. Dixit, K.-H. Lee, Poh, & T. Loo (Eds.), *Consumer behaviour in hospitality and tourism* (pp. 96–113). Routledge.

Singh, R. S. (2022). Geographical patterns of indigenous religious belief systems in north-east India: A case study. In R. S. Singh, B. Dahiya, A. K. Singh, & P. C. Poudel (Eds.), *Practising cultural geographies: Essays in honour of Rana PB Singh* (pp. 241–254). Singapore.

Torraco, R. J. (2005). Writing integrative literature reviews: Guidelines and examples. *Human Resource Development Review*, *4*(3), 356–367.

United Nations World Tourism Organisation. (2018). *2017 International tourism results: The highest in seven years*. http://media.unwto.org/press-release/2018-01-15/2017-international-tourism-results-highest-seven-years

Zolfani, S. H., Sedaghat, M., Maknoon, R., & Zavadskas, E. K. (2015). Sustainable tourism: A comprehensive literature review on frameworks and applications. *Economic research-Ekonomska istraživanja*, *28*(1), 1–30.

Zou, J. M., & Prasain, G. P. (2017). North-eastern states tourism: Exploring the unexplored on earth. *IOSR Journal of Humanities and Social Science*, *22*(8), 35–40.

Chapter 3

Evaluating Service Quality and Sustainability at the Largest São João in the World in 2023: A Tourist Perception Analysis

Gabriela Dione Florêncio de Lima[a], *Kettrin Farias Bem Maracajá*[a] *and Tiago Savi Mondo*[b]

[a]*Federal University of Campina Grande (UFCG), Brazil*
[b]*Federal Institute of Santa Catarina (IFSC), Brazil*

Abstract

The significance of event tourism in destination development underscores the importance of assessing the quality of events and their sustainability for competitiveness. This study investigates tourists' perceptions of service quality and sustainability at the "Biggest São João in the World" event in 2023. Employing a quantitative approach, we conducted field research using questionnaires based on the Tourqual model, integrating sustainability indicators to address environmental impacts associated with such events. Data collection took place between June and July 2023, with a non-probabilistic convenience sample of 709 visitors. Descriptive statistics were applied to the 2022 survey data (397 respondents), on the 2023 data (709 respondents). Results reveal that the event's diverse activities received the highest average rating at 4.03, while the use of sustainable transport during the event garnered the lowest rating at 1.93. This emphasizes the need for event organizers to implement continuous improvement plans, focusing particularly on areas with critical evaluations to enhance visitor satisfaction. This study's originality and innovative approach contribute

The Need for Sustainable Tourism in an Era of Global Climate Change: Pathway to a Greener Future, 35–52

doi:10.1108/978-1-83608-668-020241007

valuable insights for event managers and researchers and promote both local development and environmental awareness.

Keywords: Event tourism; service quality; sustainability; tourists' perceptions; tourqual

1. Introduction

The event tourism sector demonstrates remarkable potential, significantly supporting local economic development and promoting growth. This positive impact arises from the increase in tourist flow and the new employment opportunities these events provide. However, it is crucial to highlight that effective management of these events is essential to guarantee positive impacts on tourist destinations and visitors (Vico et al., 2019). Tourist events emerge as crucial strategies to combat seasonality, attracting visitors at different times of the year and increasing the visibility of destinations (Almeida et al., 2019; Marujo, 2014).

With the constant increase in tourism resulting from these events, it is imperative to address environmental concerns to minimize ecological impacts. Therefore, holding tourist events requires a sustainable approach, considering economic, social, and environmental aspects (Getz & Page, 2016). Promoting sustainable events involves adopting responsible practices in managing waste, energy, transport, water, and other resources (Mokhtar et al., 2020), incorporating sustainability into strategic decisions.

The offering of events must consider environmental impacts, such as noise pollution, solid waste production, water disposal, increased traffic, and urban congestion. Therefore, implementing strategies to reduce these impacts is essential to align event promotion with sustainability (Avigo, 2008; Lima et al., 2021). Sustainable events contribute to economic prosperity and create local opportunities, achieving financially viable objectives and mitigating adverse impacts (de Assumpção & Neto, 2020).

Given the importance of events in the region, expanding the audience becomes a challenge for managers. Tourist services play a crucial role in visitors' experience, requiring careful analysis to develop appropriate destination strategies (da Silva Santos & Forte, 2015). The quality of tourist services is directly related to competitiveness, profitability, and satisfaction (Moretti et al., 2016). From the 1980s onwards, models emerged to measure the quality of services, aiming for more accurate analyses and superior experiences for visitors (Parasuraman et al., 1985).

However, considering the importance of event tourism, offering quality services is essential to encourage repeat visits and recommendations from tourists (Chang et al., 2017). The Tourqual model emerges as an alternative to evaluate the quality of tourist services, considering visitors' perceptions (Mondo, 2014). Understanding tourist perception benefits local development, meeting the desires and needs of visitors (Kotler & Armstrong, 2023).

Given this context, the following problem question arises: what is tourists' perception of the quality of services and sustainability of the "Biggest São João in the World in 2023"? This study aims to analyze tourists' perceptions regarding the quality of services and the event's sustainability in 2023. To this end, the Tourqual Model evaluates the quality of tourist services at the event based in Campina Grande, Paraíba, contributing to research development in this area.

Furthermore, the analyzed tourist destination management can benefit from the knowledge obtained, identifying areas that require improvement based on the visitors' perspective to attract more tourists to the region, promoting a positive image of the destination. The next section will discuss the theoretical framework, addressing tourist events, sustainability, and evaluation of tourist services. The subsequent section will describe the methodology used, followed by the discussion of the results in Section 4 and the conclusions in Section 5, including theoretical and practical implications, limitations, and suggestions for future research.

2. Theoretical Foundation

2.1. Sustainable Tourism in Events

Event tourism has experienced notable growth in recent decades, consolidating itself as one of the main segments of tourism and contributing significantly to the increase in the flow of tourists to destinations (Alho, 2023). These events are crucial in generating an economic return, strategically mitigating destinations' seasonality by attracting tourists even during low-season periods (Henz et al., 2014). From an economic perspective, attracting tourists to events substantially impacts the economy and local development, standing out as an undeniably relevant factor for the region (Bem & Assis, 2010).

These events significantly attract tourists, acting as strategic elements to promote the destination's positive image and boost regional development (Getz & Page, 2019). Marketing strategies attract visitors, and tourist events enable a higher revisit and recommendation rate (Singh et al., 2020). They also stimulate development and urban renewal (Ferrari & Guala, 2017) and, in the case of mega-events, promote the international image of the destination, bringing global prestige and opportunities for social inclusion (Chen et al., 2020; Getz, 2008; Nadotti & Vannoni, 2019).

Event tourism is recognized as a driver of long-term sustainable development, generating economic benefits for the community (UNWTO, 2020) and playing a crucial role in creating jobs and income (Santamaria & Filis, 2019). In this context, developing programs and policies aimed at event tourism is essential and can be used as stimuli for the sustainable development of the region (Serrão & Costa, 2020). The growing concern about environmental issues related to event tourism in the 21st century highlights the need for consumption considering sustainable aspects (Dias, 2021).

The relevance of sustainability in environmental preservation highlights the crucial role of consumers, who value actions committed to sustainable practices (Akatu, 2011; da Silva Gonçalves et al., 2017). The creation of ISO 20121, in response to concerns about the environmental impacts of events, highlights the importance of integrated event management solutions and sustainable development (Maciel & Damke, 2021; Ranzan et al., 2016). Sustainable events emerge as strategies to promote environmental, social, and cultural actions, encouraging social interaction and adopting policies to reduce environmental impact in their planning and execution (de Oliveira & Santos, 2019).

These events, characterized by their ability to generate economic impact and profit while committing to environmental preservation, contribute to a positive image of the destination and benefit the community (Campos, 2020; Getz, 2008). The search for event sustainability requires responsible planning that minimizes negative environmental impacts, resulting in continuous improvements (Lopes et al., 2020). The ecological actions adopted by the event management influence the visitors' experience, impacting their behavior and practices regarding the environment (Skandalis et al., 2019; Henche et al., 2020).

Applying knowledge about sustainability and the Sustainable Development Goals (SDGs) of the 2030 Agenda is fundamental for planning tourist events. The SDGs aim to promote a dignified life, reduce poverty, and strengthen world peace, establishing specific goals for each national context (Maracajá & Fraga, 2023). These objectives are global commitments to be fulfilled by governments, organizations, and civil society, addressing issues relevant to the entire planet (Eisenmenger et al., 2020). Considering the environment and adopting sustainable practices are crucial to promoting change and achieving greater social equality and economic development (da Silva & Maracajá, 2021; Schwab & Malleret, 2020).

Sustainability, directly impacting the local community, must be developed through continuous improvement of sustainable practices, promoting economic, social, and environmental viability. Sustainable events satisfy the three pillars of sustainability (Buathong & Lai, 2019), raising awareness about consumption in tourist destinations and the importance of environmental preservation. The theme of sustainable events is gaining prominence, opportunities to promote best practices (da Silva & Sander, 2017), bringing business opportunities, credibility, promotion, and prestige to the tourist destination (Maciel & Damke, 2021; Trigo & Senna, 2016).

Sustainability in tourist events, covering the social, economic, and environmental pillars, promotes equal access to quality goods and services, values local culture, and drives technological and social development, directly impacting the economy (Leme et al., 2014). Recognized as a relevant topic, sustainability in tourist events is challenging for managers (Moreira et al., 2022), requiring incorporating sustainable actions into planning, implementation, and execution to achieve continuous improvements.

Sustainable events promote green marketing strategies for tourist destinations, offering competitive solutions, greater yield, environmental benefits, and greater competitiveness, improving the destination's image for customers (Tomasetto & Brandalise, 2018). From an economic perspective, the competitiveness of sustainable tourism events attracts more visitors and provides quality services (de Chaves & Secchi, 2023). Therefore, it is essential to analyze the services offered at tourist events and how they influence the visitors' experience. Strategic planning is necessary in tourist destinations, where concern is intense due to high competitiveness, making the analysis of the tourist services offered essential (da Silva Santos & Forte, 2015; Scalabrini et al., 2023).

2.2. Tourqual Model

The Tourqual Model, developed by Tiago Savi Mondo in 2014, represents a significant advance in assessing the quality of tourist services, filling a notable gap in specific tools in this sector. Its origins date back to bibliometric research aimed at identifying models in the literature to evaluate the quality of services (Mondo, 2014). This analysis identified 36 models, classified based on scope, indicator, dimension, form of presentation, and scale, totaling 211 indicators.

Through rigorous qualitative analysis, Mondo formulated a new model comprising 35 fundamental indicators for the tourism sector. This model makes it possible to evaluate the quality of tourist services, providing a competitive advantage, management benefits, and reduced errors.

In the next development phase, the Tourqual Model collected 6 thousand tourist comments on TripAdvisor and analyzed them with the T-LAB software. This analysis resulted in the exclusion of 12 indicators, the maintenance of 23, and the creation of 3 new ones, totaling 26 specific indicators for evaluating the quality of services in tourism (Mondo et al., 2018).

The Tourqual Model was used in a study with 476 tourists in Florianópolis, Santa Catarina, representing an innovative extension for tourist attractions. It stands out as the only validated model for analyzing the quality of tourist services (Mondo, 2014). Since then, it has been applied in several studies (de Santana et al., 2022; Mondo et al., 2020; Reges et al., 2021;), providing assertive diagnoses and assisting managers in strategic decisions in planning and management of tourist services.

Tourqual has six categories of analysis: access, environment, security, experience, human element, and technical quality, in which the 26 indicators. These categories and indicators form the basis for measuring and analyzing the quality of tourist services.

The flexible model can be applied in different contexts and adapted to specific services, such as hotels, restaurants, events, breweries, wineries, destinations, and attractions. This flexibility allows quality levels to be assessed based on tourists' perceptions, contributing to effective planning and management (Mondo, 2014) and, ultimately, promoting the sustainable development of tourist destinations.

3. Methodology

3.1. Type of Search

This research adopts a quantitative approach based on the quantification, measurement, and rigorous control of the data collected. This methodology allows the researcher to achieve descriptive values through analysis and calculations (Knechtel, 2014). The central characteristic of this approach is the researcher's neutrality in investigating the facts, contributing to informed decision-making. The choice for descriptive research is justified by its basis in literature and consolidated theoretical models, aiming to analyze the quality of services in tourist events. The main focus of descriptive research is the description of the characteristics of a sample, with an emphasis on collecting opinions, beliefs, and attitudes of the population in question (Heerdt & Leonel, 2022).

The field research method was applied, using 709 questionnaires during June and July, in person, at the "Biggest São João in the World" event in Campina Grande, Paraíba. The questionnaires, previously structured and validated by Prof. Dr. Tiago Mondo, were used to obtain a more in-depth understanding of the particularities of the observed reality involving on-site studies (Mondo, 2014). The application of the questionnaires also extended to an online survey, a comprehensive method to evaluate specific aspects of the sample, often using questionnaires as a research instrument (Wohlin et al., 2012).

The analysis of the perceptions of quality perceived by visitors to the "Biggest São João in the World" in Campina Grande, Paraíba, on the categories and indicators of the Tourqual Model (Mondo, 2014). Furthermore, indicators related to sustainable development were incorporated, considering the potential adverse impacts of events on the environment (Nascimento, 2022). The research process followed a comprehensive cycle, including review, treatment, and analysis of the data obtained (Flick, 2012).

3.2. Contextualization of the Study Object

This study focuses on the city of Campina Grande, located in Paraíba, Brazil, and specifically on the tourist event known as the "Biggest São João in the World." This celebration, which began in 1983, became a significant milestone by rescuing and exalting the rich northeastern culture over 30 consecutive days in June, consolidating itself as the main event in the state (Pbtur, 2022).

In five years, the "Biggest São João in the World" rose to position itself among the most prominent tourist destinations, standing out for the prestige achieved and its lasting tradition (Jornal da Paraíba, 2023). Over this period, the celebration evolved, becoming an attractive spectacle for visitors from all over Brazil and abroad, characterized by its June square dance performances, typical corn-based cuisine, forró rhythms, and various other manifestations. Cultural. The event expanded across different areas of the city, incorporating innovations such as the "Collective Wedding," the "Museu do Biggest São João in the World," the "Eldorado Casino," the "Cathedral and the National Telegraph," and the creation of a scenographic city that replicates the City Council buildings. In addition, the event offers "Fogueira Gigante," "Forró na Praça," "Ilhas de Forró" and a "Symphony Orchestra," with "Parque do Povo" as the epicenter of presentations and shows (de Andrade Lima, 2020).

The "Biggest São João in the World" brings together people around northeastern culture and June tradition, with the shows being one of the highlights. On the main stage, renowned artists of national and international reach, such as Elba Ramalho, Alceu Valença, Wesley Safadão, and Alok, provide outstanding performances (Jornal da Paraíba, 2023). With free entry, the event is crucial in preserving the heritage of June culture, rescuing and sharing monuments, works of art, parties, music, and dances with the public (Moreira, 2018; Ferreira Filho et al., 2022).

It is one of the most significant events in Brazil (Embratur, 2019); the "Biggest São João in the World" not only significantly boosts the flow of tourists in the region but is also recognized as a catalyst for local economic development. The event emerges as a source of profitability for several sectors involved, promoting tourism in the region and allowing visitors to appropriate elements of popular culture and Campinas traditions (Andrade Lima, 2020). It is worth mentioning that, according to information from the state secretariat for Tourism and Economic Development, Rosália Lucas (Portal TV Cariri, 2023), the "Biggest São João in the World" in Campina Grande moves more than R$500 million with a circulation of 2.5 million tourists.

3.3. Sample Selection and Size

This study covers a sample of individuals who participated in the "Biggest São João in the World" event in Campina Grande, Paraíba. The approach adopted is non-probabilistic for accessibility. In this context, non-probability sampling on criteria of typicality or accessibility, in which the researcher selects elements that he considers representative of the target audience based on his experience (Kraus et al., 2019; Marconi & Lakatos, 2008; Vergara, 2006). Accessibility is a representative and significant contribution to advancing the study (Prodanov & De Freitas, 2013). At the end of data collection, we obtained 709 responses.

3.4. Data Collection Procedures and Research Instrument

Data collection of questionnaires was online and in person. We used the Questionpro platform to develop the questionnaire, based on the "Tourqual Model" developed by Tiago Mondo in 2014. The research instrument was formally structured, adopting an easy-to-understand language and avoiding including extensive and tiring texts to facilitate participants' understanding.

The questionnaire has four distinct sections. In the first section, we describe the research entitled "Analysis of the Quality of Tourist Services in Biggest São João in the World – PB." We then provided the Free and Informed Consent Form (TCLE) to guarantee the confidentiality of the participants. The second section focused on collecting demographic information, seeking to outline the profile of visitors to "Biggest São João in the World" – PB, including an indicator of general satisfaction evaluated on a scale of 0 to 10, asking participants to assess their overall level of satisfaction during the destination experience.

The third section of the questionnaire encompassed indicators assessed using a five-point Likert Scale (ranging from 1 to 5), with options including (1) very poor, (2) bad, (3) medium, (4) good, and (5) excellent. Additionally, we have included a sixth alternative, "none of the options," for attendees who did not purchase tickets to the event. The "Tourqual Model" encompasses six categories of analysis: (1) access, (2) environment, (3) human element, (4) experience, (5) security, and (6) technical quality. Given the relevance of analyzing environmental impacts, we introduced a new category, "sustainable development," composed of six indicators related to environmental preservation (Nascimento, 2022), as detailed in Table 3.1.

Table 3.1. Questionnaire Categories and Indicators.

Categories	Indicators
Access	Waiting for service (queues)
Access	PCD Accessibility
Access	Accessible Bathrooms
Access	Ease of purchase
Access	Accessibility/Location
Environment	Temperature/Acoustics
Environment	Comfortable and Inviting Place
Human Element	Service
Experience	Aesthetics
Security	Security
Technical Quality	Cleaning
Technical Quality	Maintenance (equipment and infrastructure)
Technical Quality	Variety of Activities
Technical Quality	Signaling
Technical Quality	Technology
Technical Quality	Price
Sustainable Development	Air quality at the event
Sustainable Development	Noise level at the event
Sustainable Development	Water quality in areas close to the event
Sustainable Development	Waste management at the event
Sustainable Development	Use of renewable energy at the event
Sustainable Development	Use of sustainable transport during the event

Source: Created by the authors, based on Mondo (2014) and Nascimento (2022).

3.5. Data Analysis Procedures

After completing data collection, the information obtained through the questionnaire and analyzed using SPSS software. The analysis included the application of Mann-Whitney mean comparison tests. The statistical indicators evaluated include:

Frequency of Responses: This analysis aims to describe the sample, focusing on responses related to demographic questions.

Mean: Used to measure central tendency and calculated by adding all response values and dividing by the total number of responses.

Standard Deviation: This indicator quantifies the uniformity of data within a set, providing insights into the dispersion of responses.

Additionally, the research explored the possibility of structuring the indicators into factors through an exploratory factor analysis. The objective was to investigate whether quality indicators influenced the overall satisfaction of participants in "Biggest São João in the World" – PB. Subsequently, we conducted a confirmatory factor analysis, employing principal component extraction without applying rotation. This analysis aimed to confirm the factors identified in the exploratory factor analysis.

3.6. Collection Period and Disclosure Strategies

Quantitative data collection began on 06/05/2023 and ended on 07/28/2023. The questionnaire was made available online through various platforms, such as WhatsApp, Instagram, LinkedIn, Twitter, and Facebook. Furthermore, the questionnaires apply in person during the "Biggest São João in the World" event – PB. After the initial dissemination, participants were encouraged to share the questionnaire with others who also attended the event, significantly increasing the responses collected and enabling a more comprehensive analysis of the perceptions under study. Closing of the Questionnaire and Prize Draw

In the last section of the questionnaire, participants asked about the possibility of recommending the event to friends and relatives. To conclude, participants expressed their interest in participating in a draw, the prize of which consisted of a PIX[1] worth U$20.00. Interested parties provided their contact details, such as telephone number or email address. The result of the research was published on the Instagram page of the Postgraduate Program in Administration (PPGA) at the Federal University of Campina Grande (UFCG), and the winner received the PIX prize.

4. Analysis and Discussion

4.1. Sample Characteristics

In analyzing the variables that outline the profile of the event's participants, the significant majority, corresponding to 92.5% of the sample, consists of residents of Paraíba. On the other hand, tourists represent 7.5% of the total number of participants. Concerning gender, 65.3% of the sample, equivalent to 463 participants, comprises male individuals, while 34.6%, totaling 245 participants, are

[1]PIX is a real-time electronic payment system introduced by the Central Bank of Brazil. It stands for "Pix," which is derived from the Portuguese word "pixar," meaning to make a quick and small cut. PIX was officially launched in November 2020 as a means to enhance the efficiency and speed of financial transactions in Brazil.

female. When examining the level of education, we found that 0.3% of respondents completed primary education, 8% completed secondary education, 4.4% have technical training, 52.3% have a degree, and 35% of participants have some post-graduate education level.

Regarding the age group, there is a predominance of young adults. The age group from 25 to 35 stands out, covering 49.1% of the sample, with 348 respondents. The subsequent age group, 18 to 24 years old, comprises 30.5% of the sample, with 216 respondents. For the age group from 36 to 50 years old, 16.1% of the sample is recorded, with 114 respondents, while the age group over 51 represents 4.4%, with 31 respondents.

4.2. Assessment of the Perception of Service Quality and Sustainability in the "Biggest São João in the World" – PB

This section evaluates the quality indicators of the Tourqual model based on the perceptions of visitors who participated in the event and responded to the questionnaire. The participants react according to their perceptions about the event, recognizing no correct or incorrect answers.

The indicator (Table 3.2) related to queues at the event presented an average of 2.44, indicating a deterioration in the classification compared to the previous year. Event management must prioritize minimizing long and tiring queues at the tourist destination, ensuring clear signage for entry and exit to the event, in addition to reinforcing monitoring at the attraction's entrances. Keeping queues organized is crucial for perceived quality, directly impacting the development of the tourist destination, visitor preferences, and revisits (Campos & Maia, 2015). The increase in space for boxes and VIP areas in 2023 is highlighted as contributing to the rise in queues, considering the reduction in space for free entry to the event (Folha de São Paulo, 2023).

The indicator related to evaluating the prices of tickets and event products obtained an average of 2.43, indicating that visitors consider the prices high. In this context, event management must pay attention to supplier management, financial control, reduction of unnecessary costs, and adequate pricing of products and services. Authors who explored queue management at events or customer service efficiency at tourist events can offer valuable insights to understand changes in this indicator (Andrade & Veiga, 2012). Analyzing visitors' perceptions is essential to generate positive impacts in the studied area (Berselli & Zucco, 2019).

In the scope of sustainability, the noise level assessment at the event obtained an average of 3.26, classifying it as regular. Although visitors perceive the noise level as acceptable, improvement actions should be implemented, especially in areas closer to the stage, where noise is more pronounced. This indicator is in line with authors who address sustainability in tourist events, highlighting practices to reduce environmental impact, including noise pollution, which must be incorporated by event management (de Oliveira & Santos, 2019).

The air quality assessment at the event, with an average of 3.12, is also classified as regular, according to visitors' perception. Therefore, management must monitor and control air, analyze environmental risks, and develop improvement

Table 3.2. Assessment of Tourqual Events Indicators.

Evaluation items in the attraction	Mean	*N*	Standard Deviation
Assessment of the variety of activities at the event	4.03	707	0.975
Event service assessment	3.8	706	0.882
Assessment of event infrastructure	3.76	708	0.996
Evaluating the ease of purchasing tickets	3.69	381	1.005
Assessment of signage to get to the event	3.67	699	0.951
Assessment of comfort and well-being at the event	3.51	707	1.128
Perception of security at the event	3.49	708	0.962
Facilities cleanliness assessment	3.43	707	0.999
Security assessment in campina grande	3.13	705	1.034
Transport assessment to the event	3.12	697	1.144
Evaluation of the technology offer at the event	2.82	676	1.192
Assessment of accessibility for people with disabilities at the event	2.71	617	1.101
Assessment of bathroom availability and cleanliness	2.63	675	1.133
Evaluation of queues at the event	2.44	702	1.125
Evaluation of the price of tickets and products for the Event	2.43	646	1.086

Source: Primary Data (2023).

Good Rating Regular Classification Bad Rating.

strategies. Scholars such as Lopes et al. (2020) and Campos (2020) extensively explore the correlation between tourist satisfaction and sustainable practices in event tourism. They emphasize that they must implement sustainability to reduce environmental impacts and promote continuous improvement.

The analysis of waste management during the event revealed an average of 3.05, indicating a regular participant assessment. Although there is a waste collection and treatment station on site, there is room for improvement in raising awareness among visitors about the importance of proper waste disposal. Given the magnitude of the event and the high number of participants, efficiency in waste management is crucial. Can implement Marketing strategies to raise awareness among visitors, promoting the adoption of more conscious practices about waste disposal. Environmental actions, such as strategic waste collection, applied as part of the event's management strategies, have a direct influence on consumer behavior, impacting their practices about the environment (Henche et al., 2020; Skandalis et al., 2019). You can find additional details about evaluating sustainability indicators at the event in Table 3.3.

The last three indicators revealed unfavorable evaluations, considered critical points that require implementing corrective measures to increase satisfaction in the tourist destination. The water quality assessment in areas adjacent to the event recorded an average of 2.94. Recognizing the importance of water quality for improving quality of life and preventing diseases (Bittencourt & De Paula, 2014), must adopt control strategies. Such strategies can enhance the quality of life and generate economic returns for the city while promoting sustainable practices to preserve the environment (Maciel & Damke, 2021).

The evaluation of the use of renewable energy during the event obtained an average of 2.30, reflecting a negative review by the participants. The finding that there was no use of renewable energy at the event highlights the need to

Table 3.3. Assessment of Sustainability Indicators.

Assessment Items at the Attraction	***N***	**Mean**	**Standard Deviation**
Assessment of the noise level at the event (refers to the impact of the sound of loud music, construction noise, and vehicle traffic)	695	3.26	1.032
Assessment of air quality at the event (refers to odor, smoke, and pollution)	701	3.12	1.085
Assessment of waste management at the event (refers to the availability of bins, the separation of recyclable and non-recyclable waste, and the treatment of organic waste)	683	3.05	1.137
Assessment of water quality in areas close to the event (rivers, beaches, lakes, dams), as well as the supply of drinking water at the event	640	2.94	1.107
Assessment of the use of renewable energy at the event (refers to the use of solar panels and wind turbines)	520	2.30	1.160
Assessment of the use of sustainable transport during the event (shared bicycles, electric buses, or other options that reduce the emission of polluting gases)	578	1.93	1.075

Source: Primary Data (2023).

implement improvement actions. These actions contribute to the population's quality of life, promoting the reduction of pollutants and waste that are harmful to health, and also play an essential role in raising awareness about environmental preservation, becoming increasingly relevant in the context of sustainability in events (Verplanken & Roy, 2016).

The evaluation of the use of sustainable transport during the event presented an average of 1.93, receiving an unfavorable review from visitors. The event does not encourage the use of sustainable transport. Given the significant flow of people at the destination and the congestion on roads close to the event, promoting this type of transport benefits the population and reduces the emission of polluting gases. Adopting sustainable transport provides advantages for local development and the environment, corroborating authors who highlight the fundamental role of sustainability in developing tourist destinations (Rosa, 2018; Ruschel, 2008; UNWTO, 2020).

As a conclusion to the survey, participants would recommend the event to friends and family. The results revealed that 88.7% (629) responded affirmatively, indicating a significant willingness to share the event experience. On the other hand, 11.3% (80 people) responded negatively. These numbers reflect visitors' perception of the event experience and their willingness to recommend it, highlighting the importance of improving critical aspects, such as sustainable transport, to increase satisfaction and promote positive word-of-mouth recommendations.

5. Conclusion

Given the purpose of this study, which consists of analyzing tourists' perceptions about the quality of services and sustainability in the Biggest São João do Mundo in 2023, the best-evaluated indicators of the event stand out. The areas that received positive evaluations were the variety of activities (4.03), service (3.80), and event infrastructure (3.76). This event shows that factors such as quality service, kindness, education, diversity of attractions, maintenance/equipment, and quality services are fundamental to visitor satisfaction in the Biggest São João do Mundo. The most extensive set of indicators is classified as regular, covering aspects such as ease of purchasing tickets (3.69), signage for the event (3.67), comfort and well-being (3.51), safety at the event (3.49), cleanliness of the facilities (3.43), noise level (3.26), security in Campina Grande (3.13), and transportation to the event (3.12), providing opportunities for improvement to managers, who they must pay attention to changes in the assessments of these aspects.

The last group of indicators, classified as poor, includes aspects such as technology supply (2.82), accessibility for people with disabilities (2.71), availability and cleanliness of bathrooms (2.63), queues at the event (2. 44), and ticket and product prices (2.43). These require special attention from managers, negatively impacting visitor satisfaction. Although there was an improvement in the technology supply in 2023, the indicator continues to have a low rating, suggesting the need for a technological improvement plan, including the provision of free internet during the event. Prices are perceived as high, highlighting the importance of careful financial management and a more effective pricing strategy.

As for sustainability indicators, three are in a critical situation:

- Addressing water quality in areas close to the event (2.94).
- Using renewable energy at the event (2.30).
- Using sustainable transport during the event.

Managers must implement Event Strategies (1.93) to improve tourist satisfaction rates and promote the reduction of environmental impacts. Actions such as reducing greenhouse gas emissions, saving energy, and monitoring water quality are crucial, and investing in disseminating socio-environmental practices can increase visitors' awareness of the importance of environmental preservation.

Event managers should focus on indicators in critical situations since most can structurally improve and promote continuous improvement. The analysis of perceptions about the quality of services and sustainability proves to be fundamental to increasing the destination's competitiveness and visitor satisfaction, constituting a crucial strategy in the event tourism sector.

As limitations of the study, the lack of internet signal in several event locations and the long queues stand out, which sometimes resulted in the gates being closed due to the large number of people, making it difficult for visitors to respond. The study contributes to a deeper understanding of the quality of services and sustainability practices in event tourism, benefiting the environment, researchers in the field, event managers, and visitors to the destination, promoting local development.

For future research in event tourism in "Biggest São João in the World," specific themes such as deepening sustainability practices are recommended. This entails investigating more targeted strategies to enhance the event's sustainability, including implementing waste management practices, initiatives to reduce carbon emissions, and increased local community involvement in sustainable actions. Additionally, it would be valuable to examine how community participation and engagement can be enhanced to strengthen collaboration between residents and event managers, benefiting both residents and visitors. These research focuses can significantly contribute to promoting sustainability, fostering community relationships, and enhancing the overall experience of event tourism in the region.

Implementing Event Strategies is crucial to improving tourist satisfaction and reducing environmental impacts. The study's unique focus on service quality and sustainability and its detailed categorization of indicators present a valuable framework for event managers to enhance destination competitiveness strategically. Despite limitations such as internet signal issues, the study contributes significantly to understanding and improving event tourism practices, benefiting the environment, researchers, managers, and visitors, ultimately fostering local development. This study's implications extend to theoretical insights and practical strategies for future event planning and management.

References

Akatu, O. (2011). dia "D" do consumo, 2008. Instituto Akatu. Andrade, Daniel Caixeta; Romeiro, Ademar Ribeiro. Degradação ambiental e teoria econômica: algumas reflexões sobre uma "Economia dos Ecossistemas". *Economia, 12*(1), 3–26.

Alho, L. D. A. (2023). *Sazonalidade e desenvolvimento turístico sustentável: da avaliação dos impactos à proposta para a sua mitigação-O caso de Porto Covo* [Master's thesis, Universidade de Évora].

Almeida, A., Teixeira, S. J., & Franco, M. (2019). Uncovering the factors impacting visitor's satisfaction: Evidence from a portfolio of events. *International Journal of Event and Festival Management, 10*(3), 217–247.

Andrade, P. P., & Veiga, H. M. D. S. (2012). Avaliação dos trabalhadores acerca de um programa de qualidade de vida no trabalho: validação de escala e análise qualitativa. *Psicologia: Ciência e profissão, 32*, 304–319.

Avigo, R. A. (2008). A aplicação dos princípios da sustentabilidade em eventos corporativos. In *XVIII Congresso de Ciências da Comunicação na Região Sudeste*.

Bem, K., & Assis, M. A. (2010). A importância dos clubes de Caicó para o desenvolvimento do turismo de eventos local. *Caderno Virtual de Turismo, 10*(2), 91–101.

Berselli, C., & Zucco, F. D. (2019). A imagem da Festa Nacional do Doce–Fenadoce em Pelotas (RS), Brasil, perante a percepção dos residentes. *Applied Tourism, 4*(1), 41–56.

Bittencourt, C., & De Paula, M. A. S. (2014). *Tratamento de água e efluentes Fundamentos de saneamento ambiental e gestão de recursos hídricos*. Saraiva Educação SA.

Buathong, K., & Lai, P. C. (2019). Event sustainable development in Thailand: A qualitative investigation. *Journal of Hospitality, Leisure, Sport & Tourism Education, 24*, 110–119.

Campos, A. G. D. M. (2020). Modelo de gestão de projetos de eventos: uma abordagem ambientalmente sustentável para a entrega de um serviço de qualidade.

Campos, D. F., & Maia, (2015). Qualidade dos serviços turísticos no destino Pipa/Brasil: Um estudo sob a ótica de uma análise de cluster. *Revista Brasileira de Pesquisa em Turismo, 9*(2), 258–277.

Chen, Z., Suntikul, W., & King, B. (2020). Constructing an intangible cultural heritage experiences cape: The case of the Feast of the Drunken Dragon (Macau). *Tourism Management Perspectives, 34*, 100659.

da Silva, I. C. A., & Maracajá, K. F. B. (2021). Analisando os objetivos de desenvolvimento sustentável (ODS) na gastronomia do Sertão Paraibano. *Research, Society and Development, 10*(7), e12510716501.

da Silva, E. Y. K., & Sander, A. Os impactos ambientais que um grande evento ocasiona na região em que é realizado e as ações sustentáveis para minimizá-los, na visão de um gestor de eventos. *Revista Metodista de Administração do Sul, 3*(3), 143–184.

da Silva Gonçalves, M., Barata, A. J. T. S. S., & Alves, R. R. (2017). *Eventos sustentáveis: Caracterização das ações ambientais promovidas no rock in rio*.

da Silva Santos, E., & Fortes, M. (2015). Performance of Brazilian state capitals as tourism destinations. *Revista Brasileira de Pesquisa em Turismo, 9*(1), 22–39.

de Andrade Lima, E. C. (2020). A festa de São João e a invenção da cultura popular. *Revista Internacional de Folkcomunicação, 11*(23), 13–29.

de Assumpção, M. R., & Neto, M. P. M. (2020). State-of-the-art practices being reported by the PRME champions group: A reference to advance education for sustainable development. *The International Journal of Management Education, 18*(2), 100369.

de Chaves, T. J., & Secchi, L. (2023). Uma análise comparativa dos indicadores de competitividade turística regional com base na teoria de crouch e Ritchie (1999). *Revista de Turismo Contemporâneo, 11*(1).

de Oliveira, M. T. C., & Santos, É. (2019). Responsabilidade social empresarial: estudose proposições para festivais de inverno do vale Paraíba Paulista. *Revista H-Tec Humanidades e Tecnologia*, *3*(1), 30–46.

de Santana, J. C., Maracajá, K. F. B., de Araújo Machado, P., & Mondo, T. S. (2022). *Perceptions of tourist comments through the semantic network of a cultural tourism attraction in Paraíba*, Brazil.

Dias, S. M. A. D. J. (2021). *Eventos de turismo desportivo de pequena escala e desenvolvimento local sustentável: o caso do trail de Conímbriga Terras de Sicó 2019* [Doctoral dissertation].

Eisenmenger, N., Pichler, M., Krenmayr, N., Noll, D., Plank, B., Schalmann, E., Wandl, M.-T., & Gingrich, S. (2020). The sustainable development goals prioritize economic growth over sustainable resource use: a critical reflection on the SDGs from a socio-ecological perspective. *Sustainability Science*, *15*, 1101–1110.

Embratur. (2019). Instituto Brasileiro de Turismo. Eventos realizados no Brasil. https://embratur.com.br/2023/06/20/festas-de-sao-joao-sao-apresentadas-para-operadores-de-turismo-internacional/

Ferrari, S., & Guala, C. (2017). Mega-events and their legacy: Image and tourism in Genoa, Turin and Milan. *Leisure Studies*, *36*(1), 119–137.

Ferreira Filho, J. V., do Nascimento, A. M., Bezerra, A. B., de Sousa, B. F. L., & de Arruda Silva, R (2022). O "Biggest São João in the World" e o caso do museu condenado à morte: apontamentos iniciais sobre uma pesquisa em andamento na Comunicação.

Flick, U. (2012). *Introdução à metodologia de pesquisa: um guia para iniciantes*. Penso Editora.

Folha de São Paulo. (2023). São João vive a 'camarotização' no Nordeste, com áreas vip e estrutura de festival em 2023. https://www1.folha.uol.com.br/cotidiano/2023/06/sao-joao-vive-camarotizacao-no-nordeste-com-areas-vip-e-estrutura-de-festival.shtml

Getz, D. (2008). Event tourism: Definition, evolution, and research. *Tourism*, *29*(3), 403–428.

Getz, D., & Page, S. J. (2016). Progress and prospects for event tourism research. *Tourism Management*, *52*, 593–631.

Getz, D., & Page, S. J. (2019). *Event studies: Theory, research and policy for planned events*. Routledge.

Heerdt, M. L., & Leonel, V. (2022). *Metodologia científica e da pesquisa* (rev. e atual). Unisul Virtual.

Henche, B. G., Salvaj, E., & Cuesta-Valiño, P. (2020). Um modelo de gestão sustentável para ecossistemas de turismo cultural criativo. *Sustentabilidade*, *12*(22), 9554.

Henz, A. P., Leite, F. D. L., & Ruiz, T. C. D. (2014). Um ensaio teórico sobre sazonalidade e turismo. *Anais VIII do Fórum Internacional de Turismo do Iguassu*.

Jornal da Paraíba (2023). Tadeu, Emanuel. Parque do Povo: a casa do Biggest São João in the World. https://jornaldaparaiba.com.br/guia-qualeaboa/parque-do-povo-a-casa-do-maior-sao-joao-do-mundo

Knechtel, M. D. R. (2014). *Metodologia da pesquisa em educação: uma abordagem teórico-prática dialogada* (p. 106). Intersaberes.

Kotler, P., & Armstrong, G. (2023). *Princípios de marketing*. Bookman Editora.

Kraus, C. B., Kraus, I. B., dos Santos, T., da Silva Flores, L. C., & Zucco, F. D (2019). *Festival Brasileiro da Cerveja: uma avaliação do nível de satisfação do evento e das fontes de informação segundo os participantes. Revista Acadêmica Observatório De Inovação Do Turismo*, (S.l), 25–46.

Leme, P. C. S., Mortean, A. F., & Brandão, M. S. (2014). *Sustentabilidade em eventos acadêmicos: guia prático para instituições de educação superior*. EESC-USP

Lima, D. R., Simões, A. F., Mercedes, S. S., & Jacino, R. (2021). A (in) sustentabilidade dos megaeventos: regulação, autorregulação e injustiça ambiental da reciclagem de resíduos nas grandes festas brasileiras. *Revista Geográfica de América Central*, *1*(66), 439–477.

Lopes, V., Moreno Pires, S., & Costa, R. (2020). Uma estratégia para um desenvolvimento turístico sustentável da ilha grega de Chios. *Tourism: An International Interdisciplinary Journal*, *68* (3), 243–260.

Maciel, A. L. T., & Damke, E. J. (2021). Gestão sustentável de eventos: Uma revisão sistemática por meio da metodologia proknow-c. *Revista Metropolitana de Sustentabilidade (ISSN 2318-3233)*, *11*(1), 257–285.

Maracajá, K. F. B., & Fraga, C. C. L. (2023). Turismo e eventos pela ótica da Agenda 2030. *Cenário: Revista Interdisciplinar em Turismo e Território*, *10*(2), 270–282.

Marconi, M. D. A., & Lakatos, E. M. (2008). Metodologia qualitativa e quantitativa. *Marconi, MA; Lakatos, em Metodologia Científica*, *5*, 269–288.

Marujo, N. (2014). Os eventos turísticos como campo de estudo académico. *Turydes-Revista Turismo y Desarrollo*, *17*(7), 1–11.

Mokhtar, M. R., Yazid, M. F. M., & Shamsudin, M. F. (2020). Sustainability of tourism industry in Malaysia. *Journal of Postgraduate Current Business Research*, *1*(1), 1–3.

Mondo, T. S. (2014). TOURQUAL: Proposta de um modelo de avaliação da qualidade de serviços em atrativos turísticos.

Mondo, T. S., da Silva, F. V. C., & Vidor, A. L. B. (2018). Qualidade de serviços em eventos: aplicação do TOURQUAL na 16ª Fenaostra–Florianópolis-Brasil. *Caderno Virtual de Turismo*, *18*(2).

Mondo, T. S., Marques, O. R. B., & Gândara, J. M. G. (2020). Rio 2016 Olympics: analysis of the tourist perception of quality using TOURQUAL. *Tourism & Management Studies*, *16*(2), 26–34.

Moreira, A., Encarnação, T., Viseu, J., & Sousa, M. J. (2022). Job crafting and job performance: The mediating effect of engagement. *Sustainability*, *14*(22), 14909.

Moreira (2018). São João, d. e. l. r. e. i. Fernanda Aparecida Fonseca Moreira.

Moretti, S. L. D. A., Zucco, F. D., & Pozo, H. (2016). Motivações para visitar a Oktoberfest de Blumenau (BR) e Munique (RFA): uma análise comparativa utilizando a técnica Partial Least Squares. *Revista Brasileira de Pesquisa em Turismo*, *10*, 374–393.

Nadotti, L., & Vannoni, V. (2019). Cultural and event tourism: An interpretative key for impact assessment. *Eastern Journal of European Studies*, *10*(1), 115.

Nascimento, J. L. F. (2022). *Repensando o turismo de eventos sob a perspectiva triple bottom line e a biossegurança*.

Parasuraman, A., Zeithaml, V. A., & Berry, L. L. (1985). A conceptual model of service quality and its implications for future research. *Journal of Marketing*, *49*(4), 41–50.

Pbtur. (2022). *Empresa Paraibana de Turismo S/A*. https://www.pbtur.pb.gov.br/2022/06/09/pbtur-traz-influenciadora-digital-para-conhecer-e-divulgar-maior-sao-joao-do-mundo-cabaceiras-e-cidades-do-brejo/

Portal TV Cariri. (2023). Governo da Paraíba anuncia mais de R$9 milhões em investimentos no São João de Campina Grande 2023. https://portaltvcariri.com.br/noticia/87425/governo-da-paraiba-anuncia-mais-de-r-9-milhoes-em-investimentos-no-sao-joao-de-campina-grande-2023

Prodanov, C. C., & De Freitas, E. C. (2013). *Metodologia do trabalho científico: métodos e técnicas da pesquisa e do trabalho acadêmico* (2nd ed.). Editora Feevale.

Ranzan, E. M., Mauricio, A. L., & de Sousa, R. P. L. (2016). A aplicação da NBR ISO 20121 na gestão da sustentabilidade em eventos: o fortalecimento da marca por meio da produção de eventos mais sustentáveis em Santa Catarina. *Criar Educação*.

Reges, K. E. A., de Paz, T. K. F., Aragão, S. G., & Maracajá, K. F. B. (2021). Avaliação da qualidade de serviço através da reputação online em hotéis de Campina Grande utilizando o modelo Tourqual. *Cenário: Revista Interdisciplinar em Turismo e Território*, *9*(2), 223–239.

Rosa, D. (2018). *Contributos para a implementação do conceito de evento sustentável à feira de maio de Leiria* [Doctoral dissertation].

Ruschel, D. (2008). A bicicleta como meio de transporte sustentável no turismo. *Monografia, Pontifícia Universidade Católica do Rio Grande do Sul.*

Santamaria, D., & Filis, G. (2019). Tourism demand and economic growth in Spain: New insights based on the yield curve. *Tourism Management*, *75*, 447–459.

Scalabrini, E., Correia, A. I., Nunes, A., Veloso, C. M., Vieira, E., Ferreira, F. A., Silva, G., Fonseca, M., Carrança, P., Santos, S., & Fernandes, P. O. (2023). The use of mobile technology in outdoor tourism: A systematic review. In A. Mesquita, A. Abreu, & J. V. Carvalho (Eds.), *Perspectives and trends in education and technology* (pp. 667–675). Springer Singapore.

Schwab, K., & Malleret, T. (2020, March). The great reset. In *World economic forum, Geneva* (vol. 22).

Serrão, S., & Costa, R. (2020). Os Eventos Culturais como promotores do desenvolvimento local: O Caso da Festa do Vinho Madeira. *Revista Turismo & Desenvolvimento*, (34), 149–164.

Singh, N., Shalender, K., & Joan Su, C. H. (2020). Developing impacts and indicators for sustainable event management using a triple bottom line approach: A study of auto expo. *Event Management*, *24*(1), 1–16.

Skandalis, A., Byrom, J., & Banister, E. (2019). Experiential marketing and the changing nature of extraordinary experiences in post-postmodern consumer culture. *Journal of Business Research*, *97*, 43–50.

Tomasetto, S. R., & Brandalise, L. T. (2018). Percepção ambiental dos usuários de bandeira em relação ao ciclo de vida do produto. *Revista de Gestão Ambiental e Sustentabilidade: GeAs*, *7*(1), 23–42.

Trigo, A. G. M., & Senna, J. D. (2016). Sustentabilidade em eventos: características, motivações e análise de eventos sustentáveis. In *XII Congresso Nacional de Excelência em Gestão & III Inovarse–Responsabilidade Social Aplicada* (No. 12).

United Nations World Tourism Organization (UNWTO). (2020). *World Tourism Barometer*, *18*(1).

Vergara, S. C. (2006). Projetos e relatórios de pesquisa. *São Paulo: Atlas*, *34*, 38.

Verplanken, B., & Roy, D. (2016). Empowering interventions to promote sustainable lifestyles: Testing the habit discontinuity hypothesis in a field experiment. *Journal of Environmental Psychology*, *45*, 127–134.

Vico, R. P., Uvinha, R. R., & Gustavo, N. (2019). Sports mega-events in the perception of the local community: The case of Itaquera region in São Paulo at the 2014 FIFA World Cup Brazil. *Soccer & Society*, *20*(6), 810–823.

Virgínio, D. F., Moreira, Y. S., de Lima, P. M. D. P., & Chacon, L. S. (2019). Interface entre municípios, Turismo e eventos: Um olhar para a microrregião do Litoral Sul Potiguar e impactos possíveis. *Revista de Turismo Contemporâneo*, *7*(1), 49–70.

Wohlin, C., Mendes, E., Felizardo, K. R., & Kalinowski, M. (2020). Guidelines for the search strategy to update systematic literature reviews in software engineering. *Information and Software Technology*, *127*, 106366.

Chapter 4

Sustainable Tourism: A Catalyst for the Green Economy

Shehnaz Tehseen, Kim Yoke Ong, Syed Monirul Hossain and Elizabeth Andrews

Department of Management, Sunway Business School (SBS), Sunway University, Malaysia

Abstract

This chapter provides an in-depth examination of the role that sustainable tourism plays in fostering a more environmentally and economically sustainable global economy. It delves into the advantages, disadvantages, and potential of sustainable tourism, emphasizing environmental conservation, societal inclusion, and economic progress. The shift towards sustainable tourism underscores the importance of minimizing negative impacts while maximizing positive outcomes for the environment, local communities, and cultural heritage. It underscores the necessity of balancing economic development, environmental preservation, and social well-being. Sustainable tourism encourages environmentally friendly practices, promotes renewable energy, and supports local businesses, thereby driving both environmental protection and economic growth. Challenges and opportunities are identified, necessitating innovative solutions for a sustainable tourism model. Effective policies and collaboration among stakeholders can address issues like excessive tourism impacts and ensure equitable distribution of benefits to local populations. Technology and community involvement play crucial roles in fostering sustainability and responsible travel. The chapter emphasizes energy efficiency, waste management, and transportation enhancements, while recognizing digital platforms' role in promoting informed travel decisions. Future trends focus on climate adaptation, responsible tourism, and circular economies. Thus, a triple bottom-line approach integrating economic, social, and

The Need for Sustainable Tourism in an Era of Global Climate Change: Pathway to a Greener Future, 53–67

Published under exclusive licence by Emerald Publishing Limited
doi:10.1108/978-1-83608-668-020241009

environmental considerations, alongside resource management, technology, community engagement, and cultural preservation, emerges as essential for advancing sustainable tourism practices.

Keywords: Sustainable tourism; environmental preservation; social inclusivity; economic development; green economy

1. Introduction

Tourism has been considered as one of the most important economic sectors in most of the world's countries (Streimikiene et al., 2021). Sustainable tourism stands as a beacon of hope in the realm of travel and leisure, transcending traditional paradigms to offer a transformative approach that prioritizes environmental preservation, social inclusivity, and economic development. As the world grapples with the challenges posed by climate change, biodiversity loss, and socioeconomic inequalities, the importance of adopting sustainable practices in the tourism industry becomes increasingly evident. At its core, sustainable tourism embodies a commitment to minimizing the negative impacts of travel on the environment, fostering social inclusivity, and promoting economic benefits for local communities. To comprehend the intricacies of sustainable tourism, it is crucial to explore its fundamental principles. These principles revolve around responsible resource management, community engagement, and the promotion of cultural integrity. Sustainable tourism strives to strike a delicate balance between fulfilling the desires of travelers and safeguarding the destinations they visit. One of the defining features of sustainable tourism is its adherence to the triple bottom line – a framework that evaluates the success of an endeavor based on its environmental, social, and economic impacts (Butler, 2020). In the context of sustainable tourism, this means assessing the ecological footprint, social equity, and economic viability of travel activities. By examining each of these dimensions, stakeholders can gain a comprehensive understanding of the true cost and benefits associated with tourism. This interplay between the triple bottom line framework and its application for successful sustainable tourism practices are further substantiated by previous studies (Gupta et al., 2023; Yeap & Liow, 2023) which validates the need for such frameworks by policymakers and tourism businesses to put into practice.

A key focus of sustainable tourism is environmental preservation, recognizing that the natural beauty and biodiversity of destinations are invaluable assets that must be conserved. This entails adopting eco-friendly practices such as minimizing carbon emissions, reducing waste, and protecting fragile ecosystems. Through initiatives like eco-tourism and wildlife conservation programs, sustainable tourism seeks to not only minimize harm but actively contribute to the preservation of the planet's natural wonders. Sustainable tourism places a strong emphasis on social inclusivity, aiming to empower and benefit local communities (Streimikiene et al., 2021). This involves fostering positive interactions between tourists and residents, respecting local cultures, and ensuring that the economic benefits generated by tourism reach the grassroots level. Community-based tourism initiatives, homestays, and local entrepreneurship programs are integral components of this approach, creating a symbiotic relationship between tourists and the communities they visit.

Beyond its environmental and social dimensions, sustainable tourism serves as a powerful catalyst for economic development. By channeling resources into local economies, tourism becomes a force for positive change. Job creation, skill development, and the establishment of sustainable enterprises contribute to the economic resilience of host communities.

The symbiotic relationship between economic development and environmental preservation lies at the heart of sustainable tourism. This chapter delves into the intricate balance that sustainable tourism strikes, elucidating how it harmonizes economic progress with environmental and social well-being. Through prioritizing eco-friendly practices, adopting renewable energy sources, and supporting local enterprises, sustainable tourism emerges as a powerful catalyst for economic growth while safeguarding the delicate ecosystems that host travellers. This chapter seeks to delve into the pivotal role of sustainable tourism as a catalyst for the green economy, shedding light on its fundamental principles, significant impacts, and the promising opportunities it holds for shaping a more sustainable and resilient global economy. To gauge the effectiveness of sustainable tourism in fostering a green economy, it is essential to analyze its impacts. Positive outcomes, such as the conservation of natural resources, empowerment of local communities, and economic growth, are evident. However, challenges and potential negative consequences must also be addressed. Striking a balance between the desire for increased tourism and the need for sustainability is a delicate task that requires careful planning and collaboration among various stakeholders, consumers, the community and policy makers.

2. Literature Review

2.1. Economic Benefits of Sustainable Tourism

Sustainable tourism, by its very nature, goes beyond traditional economic paradigms, integrating environmental and social considerations into its framework (León-Gómez et al., 2021). The economic benefits are manifold, stemming from a conscious effort to create a positive impact on both local and global scales. For instance, sustainable tourism is a driver of employment, creating opportunities that extend far beyond traditional sectors. Local communities, often marginalized in conventional tourism models, find avenues for employment and skill development. From eco-guides and conservationists to artisans crafting sustainable souvenirs, sustainable tourism fosters a diverse array of job opportunities. Moreover, the economic benefits of sustainable tourism are not confined to large corporations; they permeate down to the grassroots level. Local enterprises, including family-owned accommodations, community-based tourism initiatives, and sustainable agriculture, receive direct financial support. This economic empowerment enhances the overall resilience of communities, reducing dependency on volatile industries. Likewise, the sustainable tourism places a premium on preserving cultural heritage. This, in turn, becomes an economic asset as tourists seek authentic experiences. Local communities are encouraged to showcase their traditions, crafts, and cuisine, creating sustainable revenue streams. By valuing and preserving cultural identity, sustainable tourism transforms heritage into an economic driver. Unlike conventional tourism, which often relies on a narrow range of attractions, sustainable tourism encourages the diversification of experiences. This diversification mitigates the risks associated with dependency on a single tourism hotspot. By exploring and promoting lesser-known destinations, sustainable tourism contributes to a more evenly distributed distribution of economic benefits.

2.2. Environmental Benefits of Sustainable Tourism

The environmental benefits of sustainable tourism are equally compelling, representing a commitment to responsible stewardship of natural resources and ecosystems (Rasoolimanesh et al., 2023). For instance, sustainable tourism places a premium on minimizing carbon footprints. This involves adopting eco-friendly transportation options, promoting energy-efficient practices, and encouraging travellers to choose sustainable modes of commuting. Through such measures, the sector actively contributes to the global effort to combat climate change. By its very essence, sustainable tourism seeks to protect biodiversity and fragile ecosystems. Initiatives such as wildlife conservation programs, marine protection projects, and habitat restoration efforts become integral components of sustainable tourism practices. The result is the preservation of biodiversity, ensuring that future generations can continue to marvel at the wonders of the natural world. Moreover, sustainable tourism aims to minimize waste generation and promote recycling. Accommodations, restaurants, and tour operators adopt practices that reduce single-use plastics, encourage responsible waste disposal, and engage in recycling programs. This commitment to waste reduction contributes to cleaner environments and healthier ecosystems. Sustainable tourism serves as a beacon for the wider adoption of sustainable practices. By showcasing successful models of eco-friendly operations, it inspires other industries to follow suit. The ripple effect of sustainable tourism extends beyond its immediate sphere, influencing businesses and individuals to embrace environmentally responsible practices.

2.3. Sustainable Tourism and the Green Economy

The OECD's publication "Towards Green Growth" articulates the concept of green growth as the pursuit of economic growth and development in tandem with the preservation of natural assets, ensuring the continued provision of resources and environmental services essential for human well-being. Achieving this balance necessitates a concerted effort to promote investment and innovation. By fostering a climate conducive to increased investment and encouraging innovative practices, the objective is to establish the foundations for sustained economic growth (Ahmad et al., 2023). Simultaneously, these initiatives are expected to engender novel economic opportunities, thereby aligning economic progress with environmental sustainability (Satta et al., 2019).

A green economy is conceptualized as an economic framework characterized by low carbon emissions, resource efficiency, and social inclusivity (Ali et al., 2023). Within the context of a green economy, the augmentation of income and employment is envisioned to arise from strategic investments (Chaaben et al., 2022). These investments are directed towards activities that mitigate carbon emissions, minimize pollution, improve energy and resource efficiency, and safeguard biodiversity and ecosystem services. The underlying premise is to foster economic development that is not only financially robust but also environmentally sustainable and socially equitable.

The shift towards a greener economy poses a prolonged and formidable challenge for the tourism industry, as it does for numerous other sectors. Given its transversal nature and interactions with various industries and services, tourism stands as a substantial contributor to the broader movement towards more sustainable, cleaner, and low-carbon economic growth. Entrepreneurs and policymakers are increasingly recognizing the pivotal role of innovation in enhancing environmental performance and realizing sustainable objectives within the tourism sector (Momayez et al., 2023). The concept of innovation is integral to the refinement of existing products and the development of tourism offerings that align with sustainability principles (Baiocco et al., 2023).

The OECD's "Towards Green Growth" emphasizes that innovation is central to the transformative processes of an economy, contributing to the establishment of new markets and the creation of employment opportunities (Kanianska, 2017). In the journey towards a greener economy, green innovation in tourism services takes the spotlight. For the purposes of this project, green innovation, or eco-innovation, is conceptualized as innovation leading to a deliberate or unintentional reduction in environmental impact. The OECD's publication, "Eco-Innovation in Industry: Enabling Green Growth" (2009), underscores that eco-innovation has the potential to drive more radical improvements compared to traditional forms of innovation. It is characterized by its multidimensional nature, encompassing targets, mechanisms, and impacts. The report also suggests that improved benchmarking and enhanced indicators would foster a deeper understanding and make the advancements achieved through eco-innovation more apparent to both industry stakeholders and customers.

2.4. Role of Technology and Innovation

Technological advancements are pivotal in transforming tourism into a more sustainable industry (Abdelfattah et al., 2023). Innovations in energy efficiency, waste management, and transportation contribute to reducing the environmental impact of travel (Lv & Shang, 2023). Moreover, the use of digital platforms and data analytics enables more informed and responsible travel choices, facilitating the promotion of sustainable practices. The integration of technology and innovation stands as a cornerstone in the quest to transition the tourism industry towards sustainability, aligning with the principles of a green economy (Wu et al., 2024). Technological advancements have the potential to reshape various facets of tourism, ranging from energy consumption and waste management to transportation efficiency (Lv & Shang, 2023).

Innovation is delineated as the introduction of a novel or significantly enhanced product, process, or method (Baccini et al., 2023). It encompasses a broad spectrum of activities, ranging from those that are ground breaking for a particular firm to those that are pioneering for the market or the world (Felicetti et al., 2024). Hence, it involves both the successful adoption of processes or business methods developed by other entities and being the first to introduce a new product to the market. Within the sphere of innovation, Green innovation emerges as a nuanced concept with a discernible positive impact on the environment. It can be specifically characterized as innovation leading to a reduction in environmental impact and/or optimizing the utilization of resources throughout the entire lifecycle of associated activities (Xing & Liu, 2023).

2.5. The Importance of Green Growth Strategies

The imperative to re-evaluate the paradigm of economic growth has gained prominence due to the imbalances generated by the environmental impacts of economic activities (Van Kien et al., 2023). Effectively steering an economy toward a green growth trajectory is contingent upon a nuanced interplay of policy and institutional frameworks, developmental stages, resource endowments, and specific environmental stress points (Henrysson & Nuur, 2021). The notion of a one-size-fits-all prescription for implementing a green growth strategy is dispelled, emphasizing the need for tailored approaches to each unique context. Common considerations, however, must underpin all strategies, with a pivotal role assigned to robust economic policy that extends beyond explicitly environmental policies. At the core of a green growth framework lies the recognition of natural capital as a fundamental factor of production, contributing to the enhancement of overall well-being (Khan et al., 2023). Rather than displacing existing priorities and areas of policy reform, green growth should be envisaged as a strategic complement, targeting domains where environmental and economic policies converge beneficially (Vazquez-Brust & Sarkis, 2012). Strategies should prioritize the identification of cost-effective measures for alleviating environmental pressures and initiating a transition toward novel growth patterns that circumvent critical environmental thresholds.

The perceived trade-offs between economic growth and environmental protection diminish when alternative metrics, including the quality and composition of growth, are employed alongside GDP to gauge progress. Green growth, integral to sustainable development, addresses these concerns by offering an operational policy agenda that facilitates tangible advancements at the nexus of the economy and the environment (Ekonomou & Halkos, 2023). Emphasizing a narrower scope than sustainable development, Green growth entails a focused policy agenda capable of effecting concrete and measurable progress. It necessitates specific attention to social issues and equity concerns arising from the green transition, focusing on the broader social pillar of sustainable development (Hasan et al., 2023).

Tourism, historically plagued by low productivity in developed economies, faces challenges in attracting capital investment and skilled personnel. The labor-intensive nature of the industry is a major factor contributing to this challenge. However, this paper contends that the future trajectory of businesses, industries, or economies within tourism is more contingent on their level of innovation than their current productivity (Gürlek et al., 2021). Historically, innovations in the transport sector have triggered transformative processes in tourism, ranging from the advent of railways to the mass production of automobiles and the global connectivity facilitated by modern jet airplanes. In the current era of globalization, process innovations become imperative for survival in the tourism sector. Process innovations can drive down costs, lead to quality enhancements, lower prices, and consequently, boost profitability (Vo et al., 2023).

2.6. Challenges for the Tourism Sector in Making Innovation Routine

Despite its critical importance, the innovation process is not yet ingrained as a routine in the tourism industry. As different segments of the tourism sector mature, pioneering companies become less prevalent. While business opportunities are explored in emerging sectors like eco-, green-, or sustainable tourism, overall entrepreneurship tends to decline (Turekulova et al., 2022). In such a landscape, the "spirit of enterprise" becomes a crucial resource in fostering innovation. The absence of radical innovations leads to a gradual, feedback-driven process where each innovation paves the way for incremental improvements in products and processes (Hendricks & Matthyssens, 2023).

Innovation in tourism follows a pattern where successful product-level innovations are swiftly imitated by competitors, while process-level innovations often get outsourced. Imitation and outsourcing are key mechanisms for disseminating innovation within the tourism industry (Kornilova et al., 2023). This holds true in the context of the increasing emphasis on sustainable and greener tourism. As the industry moves towards sustainability, imitation and outsourcing are expected to play pivotal roles in the adoption of green innovation products and practices by tourism businesses (Le et al., 2023).

2.7. Challenges and Spill-over Effects

The propensity for imitation may lead to spill-over effects, potentially depriving innovators of the full benefits of their efforts. This implies that the evolving landscape of green innovation in tourism necessitates strategic thinking and proactive measures to safeguard the interests of pioneering innovators. The tourism industry's future hinges not only on current productivity levels but significantly on its capacity for innovation. The complex interplay between imitation, outsourcing, and spill-over effects underscores the need for a nuanced understanding of the innovation dynamics within the tourism sector, particularly as it navigates towards a more sustainable and greener future (Kuo et al., 2022).

The concept of "Tourism in a Green Economy" aligns with the principles outlined by the United Nations Environment Programme (UNEP), which defines a green economy as a framework that enhances human well-being, fosters social equity, and concurrently mitigates environmental risks and scarcities. In its essence, a green economy is characterized by being low carbon, resource-efficient, and socially inclusive. In a green economy, the trajectory of growth in income and employment is steered by strategic investments, both from the public and private sectors. These investments are directed towards initiatives that actively reduce carbon emissions and pollution, enhance energy and resource efficiency, and prevent the depletion of biodiversity and ecosystem services. The pivotal goal is to create a sustainable economic model where prosperity is intertwined with environmental preservation and social inclusivity (Green, 2023).

For tourism to thrive within the framework of a green economy, it must adhere to the principles of sustainability. This involves a fundamental shift towards low-carbon practices, efficient resource utilization, and a commitment to social inclusivity. Investments in tourism infrastructure and services should prioritize environmental conservation, energy efficiency, and the preservation of local cultures and ecosystems (Nugroho et al., 2023, May).

The transformation towards tourism in a green economy necessitates coordinated efforts involving public and private stakeholders (Pan et al., 2018). Public expenditure plays a crucial role in catalyzing and supporting sustainable tourism initiatives. Policy reforms and regulatory changes become essential tools in steering the industry towards practices that align with the principles of a green economy. The paradigm of "Tourism in a Green Economy" embodies a holistic approach where economic prosperity, environmental stewardship, and social equity are interlinked. It calls for a departure from traditional models of tourism towards practices that are environmentally responsible, resource-efficient, and socially inclusive. By aligning tourism with the principles of a green economy, the industry can contribute significantly to the broader goals of sustainable development (Fauzi, 2023).

2.8. Challenges and Opportunities in Achieving Sustainable Tourism in a Green Economy

Tourism, more than any other sector, holds the potential to exert both positive and negative impacts on the resources it relies upon. The concept of tourism in a green economy revolves around activities that can be sustained indefinitely, encompassing social, economic, cultural, and environmental dimensions – termed sustainable tourism (Apine et al., 2023). Sustainable tourism involves policies, practices, and programs that not only consider tourists' expectations for responsible natural-resource management but also address the needs of communities supporting or affected by tourism initiatives and the environment at large. The industry confronts significant sustainability-related challenges, including energy and greenhouse gas emissions, water consumption, waste management, loss of biodiversity, and effective cultural heritage management (Liu, 2023). Despite these challenges, several promising opportunities exist to green tourism activities. Tourism emerges as a pivotal driver of global economic growth, making it crucial from a resource perspective. Even marginal shifts towards sustainability can have significant impacts due to the sector's size and reach. Tourism's intricate connections with various sectors mean that changes in its practices can stimulate broader transformations among diverse public and private actors (Annamalah et al., 2023). Furthermore, the tourism sector holds the potential to support development goals, particularly in new destinations, primarily in developing countries. The trend toward sustainable tourism aligns with changing consumer patterns, where tourists increasingly prioritize sustainability considerations, such as the demand for locally produced goods and services. Evidence suggests that tourists actively seeking environmentally and culturally differentiated experiences are willing to pay more for such encounters. Making tourism more sustainable fosters stronger linkages with local economies, contributing to local development and poverty reduction. The sector's labor-intensive nature, especially suited for women and disadvantaged groups, further enhances its potential for societal impact (Tabaeeian et al., 2023). The Global Sustainable Tourism Council (GSTC) has outlined baseline criteria for tourism businesses to protect and sustain the world's natural and cultural resources. These criteria focus on sustainability planning, maximizing social and economic benefits for local communities, enhancing cultural heritage, and reducing negative environmental impacts. While initially designed for accommodation and tour operations, the GSTC asserts the applicability of these criteria to the entire tourism industry (Harms, 2013). The challenges faced by the tourism industry in achieving sustainability are significant but not insurmountable. With concerted efforts, the sector can transform itself into a powerful force for positive economic, social, and environmental change, contributing to the broader goals of a green economy.

2.9. Community Engagement and Social Inclusivity

Sustainable tourism thrives on the engagement and empowerment of local communities. Recognizing the cultural heritage, involving indigenous communities, and ensuring their participation in decision-making processes are crucial elements in fostering a more inclusive and sustainable tourism industry. Social inclusivity remains a fundamental aspect in redefining the way tourism interacts with local populations. Sustainable tourism is intricately connected to community engagement and social inclusivity, embodying a paradigm shift from conventional tourism models. It places paramount importance on fostering the active involvement and empowerment of local communities to create a more resilient and sustainable industry. This commitment is underpinned by recognizing and respecting cultural heritage, actively involving indigenous communities, and ensuring their participation in decision-making processes (Buhalis et al., 2023). Central to the ethos of sustainable tourism is a departure from exploitative practices that disregard the cultural richness of destination communities. Instead, it seeks to establish a more symbiotic relationship where tourism becomes a force for positive change and community development. Cultural heritage is not viewed merely as an attraction but as a living, dynamic aspect that shapes the identity of a destination. Consequently, sustainable tourism initiatives aim to preserve and celebrate this heritage while ensuring that local communities share in the benefits (Fafouti et al., 2023).

Indigenous communities, often the stewards of unique cultural and natural resources, are recognized as key stakeholders in sustainable tourism. Their inclusion goes beyond token representation; it involves active participation in decision-making processes that impact their land, culture, and socio-economic conditions. Empowering indigenous communities economically and socially is integral to the sustainability agenda, ensuring that tourism becomes a vehicle for their prosperity rather than a threat to their way of life (Buhalis et al., 2023). The participatory approach in decision-making is a cornerstone of sustainable tourism practices. Local communities, including indigenous groups, are not passive recipients but active contributors to the planning and management of tourism activities. This engagement is vital for ensuring that the tourism experience aligns with the values, aspirations, and well-being of the local population (Rahmawati et al., 2023).

Social inclusivity, as a core tenet of sustainable tourism, transcends economic dimensions. It embraces a broader understanding of the impacts of tourism on the social fabric of communities. Sustainable tourism seeks to minimize negative social consequences such as cultural erosion, social stratification, and the disruption of traditional lifestyles. It strives to create tourism experiences that are respectful, enriching, and inclusive for both visitors and local communities (Page & Duignan, 2023). Community engagement and social inclusivity are not ancillary considerations but intrinsic to the very essence of sustainable tourism. By recognizing the agency of local communities, valuing cultural heritage, and fostering social inclusivity, sustainable tourism charts a course towards a more equitable and mutually beneficial coexistence of the tourism industry and the communities it serves. This approach envisions a future where tourism becomes a catalyst for positive social change, contributing to the well-being and empowerment of destination communities (Roodbari & Olya, 2024).

2.10. Future Trends and Strategies to Promote Sustainable Tourism

The evolving landscape of sustainable tourism encompasses future trends and strategies aimed at addressing emerging challenges. Climate change adaptation, the promotion of responsible and regenerative tourism practices, and the development of circular economies within tourism sectors are among the key strategies paving the way for a more sustainable future. The trajectory of sustainable tourism is shaped by a dynamic interplay of evolving trends and strategic responses aimed at navigating emerging challenges. This forward-looking approach is critical in ensuring the long-term viability of tourism as an industry while mitigating its environmental and socio-cultural impacts. Several key trends and strategies are instrumental in defining the future landscape of sustainable tourism (Morrison & Buhalis, 2023).

The looming spectra of climate change pose unprecedented challenges to the tourism industry. As destinations grapple with the intensification of extreme weather events, rising sea levels, and shifts in ecosystems, climate change adaptation becomes imperative. Sustainable tourism strategies are increasingly focused on enhancing resilience, implementing eco-friendly infrastructure, and developing destination-specific plans to mitigate the impacts of climate change (Mancini et al., 2022). Moreover, a paradigm shift towards responsible and regenerative tourism practices is gaining momentum. Beyond the conventional notions of sustainability, regenerative tourism seeks to leave a positive impact on destinations, going beyond merely minimizing harm. It involves actively contributing to the restoration of ecosystems, conservation of biodiversity, and the enhancement of local communities. Responsible tourism practices emphasize ethical considerations, cultural sensitivity, and the equitable distribution of benefits among stakeholders (Bellato et al., 2023). The concept of a circular economy, where resources are used sustainably and waste is minimized through recycling and regeneration, is gaining traction in the tourism sector. Circular economies within tourism involve reducing waste, promoting resource efficiency, and encouraging practices that contribute to the overall well-being of destinations. This approach requires rethinking traditional linear models of tourism development and adopting circular strategies that prioritize the longevity of resources (Gabor et al., 2023). Likewise, the integration of technology is a pervasive trend shaping the future of sustainable tourism. From blockchain applications for transparent supply chains to the use of big data for informed decision-making, technology plays a crucial role in enhancing the sustainability of tourism operations. Innovations in virtual reality and augmented reality also contribute to reducing the environmental footprint by offering alternative ways for travelers to experience destinations without physical travel (Consoli et al., 2023). The future of sustainable tourism envisions stronger community empowerment and local engagement. Strategies are being developed to ensure that local communities actively participate in the decision-making processes that affect them. Initiatives promoting community-based tourism, where locals are active partners and beneficiaries, are gaining prominence. This approach aligns with the principles of social inclusivity and ensures that the benefits of tourism are distributed equitably (Giampiccoli & Mtapuri, 2021). To ensure accountability and transparency in sustainable tourism practices, certification and standardization are emerging as critical strategies. Various global and regional certification programs assess and endorse businesses and destinations that adhere to specific sustainability criteria (Rodríguez-García et al., 2023). These standards provide a framework for continuous improvement and guide consumers in making responsible choices. The future trends and strategies in sustainable tourism underscore a commitment to resilience, responsibility, and innovation. As the industry grapples with unprecedented challenges, these strategies pave the way for a more sustainable and resilient future. Through the integration of climate adaptation, responsible practices, circular economies, technological innovations, community empowerment, and robust certification systems, sustainable tourism seeks not only to endure but to thrive in a rapidly changing global landscape.

2.11. Conclusion

In conclusion, sustainable tourism is both a trendy buzzword and a practical strategy for building a more just and resilient global economy. The fundamental principles of this method are to revolutionize the leisure and tourism sector by making environmental preservation, social inclusion, and economic growth their top priorities. A commitment to a triple-bottom-line approach, which evaluates success by considering the economic, social, and environmental impacts, is fundamental to sustainable tourism. By advocating for responsible resource management, community engagement, and cultural integrity, this travel paradigm seeks to bring tourists' desires into harmony with the necessity to preserve the places they visit. Using eco-friendly practices and trying to safeguard species are at the heart of sustainable tourism, which prioritizes environmental preservation. Tourism benefits local communities through fostering constructive exchanges, honoring cultural variety, and bolstering local economies from the ground up. Social inclusiveness ensures this. This approach is driven by more than just a desire to help others; it also helps the economy grow. The economic resilience of the communities that welcome sustainable tourists is bolstered because these visitors put money into their local economies, which in turn generates jobs, knowledge, and support for sustainable companies. There are challenges to sustainable tourism despite the fact that it has many advantages, such as helping to preserve resources and boosting economic development. Careful preparation and collaboration among all stakeholders are required to guarantee both the necessity of sustainability and the attraction of new tourists. This chapter argues that sustainable tourism is more than just a concept; it is a driving force behind the green economy movement. It presents promising opportunities to achieve economic growth while safeguarding the fragile ecosystems that sustain our world. Consistent evaluation, adjustment, and cooperation are key to realizing sustainable tourism's potential in fostering a green economy and making sure that current and future generations reap the benefits. Sustainable tourism can help build a more equitable, resilient, and environmentally friendly world economy if its practitioners' band together and adhere to its principles. Sustainable tourism holds immense potential as a catalyst for the green economy. Its ability to reconcile economic prosperity with environmental and social sustainability positions it as a fundamental driver of a more resilient and equitable global economy. To fully realize this potential, concerted efforts from governments, businesses, communities, and travellers are necessary. By embracing and fostering sustainable tourism, we set the stage for a more sustainable and prosperous future for generations to come. Therefore, it is vital to engage in multiple strategies which encompass provisions to enhance and sustain triple-bottom-line initiatives for the purpose of effective and long-term sustainable tourism practice. However, this requires due diligence, commitment, and constructive actions by relevant agencies, communities, and tourism bodies to ensure that sustainable tourism should prioritize planetary conservation throughout its value chain.

References

Abdelfattah, F., Al-Alawi, A., Abdullahi, M. S., & Salah, M. (2023). Embracing the industrial revolution: The impact of technological advancements and government policies on tourism development in Oman. *Journal of Policy Research in Tourism, Leisure and Events*, 1–25. https://doi.org/10.1080/19407963.2023.2294789

Ahmad, M., Peng, T., Awan, A., & Ahmed, Z. (2023). Policy framework considering resource curse, renewable energy transition, and institutional issues: Fostering sustainable development and sustainable natural resource consumption practices. *Resources Policy*, *86*, 104173.

Ali, A. (2023). Estimating the recreational value of mountain tourism to shape sustainable development in Gilgit-Baltistan, Pakistan. *Journal of Cleaner Production*, *426*, 138990.

Annamalah, S., Paraman, P., Ahmed, S., Dass, R., Sentosa, I., Pertheban, Raja, T., Shamsudin, F., Baharudin, K., Logeiswaran, A., Raman, M., Hoo, W., & Singh, P. (2023). The role of open innovation and a normalizing mechanism of social capital in the tourism industry. *Journal of Open Innovation: Technology, Market, and Complexity*, *9*(2), 100056.

Apine, E., Ramappa, P., Bhatta, R., Turner, L. M., & Rodwell, L. D. (2023). Challenges and opportunities in achieving sustainable mud crab aquaculture in tropical coastal regions. *Ocean & Coastal Management*, *242*, 106711.

Baccini, P., & Brunner, P. H. (2023). *Metabolism of the anthroposphere: Analysis, evaluation, design*. MIT Press.

Baiocco, S., Leoni, L., & Paniccia, P. M. A. (2023). Entrepreneurship for sustainable development: Co-evolutionary evidence from the tourism sector. *Journal of Small Business and Enterprise Development*, *30*(7), 1521–1546.

Bellato, L., Frantzeskaki, N., & Nygaard, C. A. (2023). Regenerative tourism: A conceptual framework leveraging theory and practice. *Tourism Geographies*, *25*(4), 1026–1046.

Buhalis, D., Leung, X. Y., Fan, D., Darcy, S., Chen, G., Xu, F., Tan, G., Nunkoo, R., & Farmaki, A. (2023). Tourism 2030 and the contribution to the sustainable development goals: The tourism review viewpoint. *Tourism Review*, *78*(2), 293–313.

Butler, R. W. (2020). Sustainable tourism: A state-of-the-art review. *Tourism and Sustainable Development Goals*, *1*, 11–29.

Chaaben, N., Elleuch, Z., Hamdi, B., & Kahouli, B. (2022). Green economy performance and sustainable development achievement: Empirical evidence from Saudi Arabia. *Environment, Development and Sustainability*, 1–16.

Consoli, T., Désiron, J., & Cattaneo, A. (2023). What is "technology integration" and how is it measured in K-12 education? A systematic review of survey instruments from 2010 to 2021. *Computers & Education*, *197*, 104742.

Ekonomou, G., & Halkos, G. (2023). Exploring the impact of economic growth on the environment: An overview of trends and developments. *Energies*, *16*(11), 4497.

Fafouti, A. E., Vythoulka, A., Delegou, E. T., Farmakidis, N., Ioannou, M., Perellis, K., Giannikouris, A., Kampanis, N. A., Alexandrakis, G., & Moropoulou, A. (2023). Designing cultural routes as a tool of responsible tourism and sustainable local development in isolated and less developed islands: The case of Symi Island in Greece. *Land*, *12*(8), 1590.

Fauzi, M. A. (2023). Sustainable tourism and sustainable development goals (SDGs): A state-of-the-art review of past, present, and future trends. *Environment, Development and Sustainability*, 1–22. https://doi.org/10.1007/s10668-023-04077-0

Felicetti, A. M., Corvello, V., & Ammirato, S. (2024). Digital innovation in entrepreneurial firms: A systematic literature review. *Review of Managerial Science*, *18*(2), 315–362.

Gabor, M. R., Panait, M., Bacoş, I. B., Naghi, L. E., & Oltean, F. D. (2023). Circular tourism economy in European Union between competitiveness, risk and sustainability. *Environmental Technology & Innovation*, *32*, 103407.

Giampiccoli, A., & Mtapuri, O. (2021). From conventional to alternative tourism: Rebalancing tourism towards a community-based tourism approach in Hanoi, Vietnam. *Social Sciences*, *10*(5), 176.

Green, J. (2023). Comparative capitalisms in the Anthropocene: A research agenda for green transition. *New Political Economy*, *28*(3), 329–346.

Gürlek, M., & Koseoglu, M. A. (2021). Green innovation research in the field of hospitality and tourism: The construct, antecedents, consequences, and future outlook. *The Service Industries Journal*, *41*(11–12), 734–766.

Gupta, S., Kumar, S., & Thakkar, H. (2023). Examining the triple bottom line benefits of ecotourism: A multi-faceted approach. *Multidisciplinary Science Journal*, *5*, 2023ss0210.

Harms, E. (2013). The global sustainable tourism council. In K. S. Bricker, R. Black, & S. Cottrell (Eds.), *Sustainable tourism and the millennium development goals: Effecting positive change* (pp. 297–306). Jones & Bartlett Publishers.

Hasan, M. A., Mahjabin, T., Hossain, M. A., Kibria, M. G., Haseeb, M., & Hossain, M. E. (2023). Towards green economy and sustainable development in Bangladesh: Assessing the role of social and environmental indicators. *Environmental Science and Pollution Research*, *30*, 1–16.

Hendricks, L., & Matthyssens, P. (2023). *A platform perspective on ecosystem development, value co-creation and upscaling* (Doctoral dissertation, University of Antwerp).

Henrysson, M., & Nuur, C. (2021). The role of institutions in creating circular economy pathways for regional development. *The Journal of Environment & Development*, *30*(2), 149–171.

Kanianska, R. (2017). *Green growth and green economy* (pp. 11–20). Banská Bystrica, Blianum.

Khan, Z., Hossain, M. R., Badeeb, R. A., & Zhang, C. (2023). Aggregate and disaggregate impact of natural resources on economic performance: Role of green growth and human capital. *Resources Policy*, *80*, 103103.

Kornilova, I., Horbas, I., Poltoratska, A. A., Netreba, I., Derkach, O., & Ovcharenko, T. (2023). Types of innovation outsourcing: A systemic perspective. *Scientific Papers of the University of Pardubice. Series D. Faculty of Economics and Administration*, *31*(2), 1709.

Kuo, F. I., Fang, W. T., & LePage, B. A. (2022). Proactive environmental strategies in the hotel industry: Eco-innovation, green competitive advantage, and green core competence. *Journal of Sustainable Tourism*, *30*(6), 1240–1261.

León-Gómez, A., Ruiz-Palomo, D., Fernández-Gámez, M. A., & García-Revilla, M. R. (2021). Sustainable tourism development and economic growth: Bibliometric review and analysis. *Sustainability*, *13*(4), 2270

Le, T. T., Tran, P. Q., Lam, N. P., Tra, M. N. L., & Uyen, P. H. P. (2024). Corporate social responsibility, green innovation, environment strategy and corporate sustainable development. *Operations Management Research*, *17*(1), 114–134.

Liu, X. (2023). Tourism development, environmental regulations, and natural resource management: Evidence from G20 countries. *Resources Policy*, *86*, 104224.

Lv, Z., & Shang, W. (2023). Impacts of intelligent transportation systems on energy conservation and emission reduction of transport systems: A comprehensive review. *Green Technologies and Sustainability*, *1*(1), 100002.

Mancini, M. S., Barioni, D., Danelutti, C., Barnias, A., Bračanov, V., Pisce, G. C., Chappaz, G., Đuković, B., Guarneri, D., Lang, M., Martín, I., Reverté, S., Morell, I., Peçulaj, A., Prvan, M., Randone, M., Sampson, J., Santarossa, L., Santini, F., & Galli, A. (2022). Ecological Footprint and tourism: Development and sustainability monitoring of ecotourism packages in Mediterranean Protected Areas. *Journal of Outdoor Recreation and Tourism*, *38*, 100513.

Momayez, A., Rasouli, N., Alimohammadirokni, M., & Rasoolimanesh, S. M. (2023). Green entrepreneurship orientation, green innovation and hotel performance: The moderating role of managerial environmental concern. *Journal of Hospitality Marketing & Management*, *32*(8), 981–1004.

Morrison, A. M., & Buhalis, D. (Eds.). (2023). *Routledge handbook of trends and issues in global tourism supply and demand*. Taylor & Francis.

Nugroho, L. D., Melati, S. R., Wahyuliana, I., Pawestri, A., & Kurniawan, L. F. (2023, May). Legal Policy of Implementation Green Economy in the Tourism Sector to Realize Sustainable Tourism and Environment. *IOP Conference Series: Earth and Environmental Science* (vol. 1181, no. 1, p. 012018). IOP Publishing.

Page, S. J., & Duignan, M. (2023). Progress in tourism management: Is urban tourism a paradoxical research domain? Progress since 2011 and prospects for the future. *Tourism Management, 98*, 104737.

Pan, S. Y., Gao, M., Kim, H., Shah, K. J., Pei, S. L., & Chiang, P. C. (2018). Advances and challenges in sustainable tourism toward a green economy. *Science of the Total Environment, 635*, 452–469.

Rahmawati, R., Prayitno, G., Firdausiyah, N., Dinanti, D., Hayat, A., Efendi, A., & Roskruge, M. (2023). Harnessing social capital for fostering non-tourism actor involvement in sustainable tourism: A case study of an Indonesian village. *Journal of Urban Development and Management 2*(2), 69–83.

Rodríguez-García, R., Ferrero-Ferrero, I., & Fernández-Izquierdo, M. Á. (2023). Analysis of integration of sustainability in sustainability certifications in the hotel industry. *Frontiers in Sustainability, 4*, 1116359.

Roodbari, H., & Olya, H. (2024). An integrative framework to evaluate impacts of complex tourism change initiatives. *Tourism Management, 100*, 104829.

Rasoolimanesh, S. M., Ramakrishna, S., Hall, C. M., Esfandiar, K., & Seyfi, S. (2023). A systematic scoping review of sustainable tourism indicators in relation to the sustainable development goals. *Journal of Sustainable Tourism, 31*(7), 1497–1517.

Streimikiene, D., Svagzdiene, B., Jasinskas, E., & Simanavicius, A. (2021). Sustainable tourism development and competitiveness: The systematic literature review. *Sustainable Development, 29*(1), 259–271.

Satta, G., Spinelli, R., & Parola, F. (2019). Is tourism going green? A literature review on green innovation for sustainable tourism. *Tourism Analysis, 24*(3), 265–280.

Tabaeeian, R. A., Yazdi, A., Mokhtari, N., & Khoshfetrat, A. (2023). Host-tourist interaction, revisit intention and memorable tourism experience through relationship quality and perceived service quality in ecotourism. *Journal of Ecotourism, 22*(3), 406–429.

Turekulova, D., Beisengaliyev, B., Valiyeva, S., Kurmankulova, N., & Saimagambetova, G. (2022). Analysis of sustainable development of SMEs and factors influencing to the ecotourism industry. *Journal of Environmental Management & Tourism, 13*(1), 211–222.

Van Kien, P., Ou, J. P., Sadiq, M., Nguyen, T. T. H., Huy, P. Q., & Tran, T. K. (2023). What role financial inclusion, green trade and natural resources utilization play in ASEAN economic growth: Evidence from post COVID era. *Resources Policy, 85*, 103884.

Vazquez-Brust, D. A., & Sarkis, J. (2012). *Green growth: managing the transition to sustainable economies* (pp. 1–25). Springer Netherlands.

Vo, T. H. C., Hong-Hue, T. H., Bui, T. D., & Tran, M. L. (2023). Cross-level assessment of process innovation practices: An empirical approach to small and medium manufacturing enterprises in Southeast Asia. *Journal of Small Business & Entrepreneurship*, 1–31. https://doi.org/10.1080/08276331.2023.2243577

Wu, H., Feng, Z., Sun, T., Li, R., & Zhao, H. (2024). Efficiency, sustainability, and resilience a trifecta for a green economic recovery through natural resource markets. *Resources Policy, 88*, 104435.

Xing, Y., & Liu, Y. (2023). Integrating product-service innovation into green supply chain management from a life cycle perspective: A systematic review and future research directions. *Technovation, 126*, 102825.

Yeap, P. F., & Liow, M. L. S. (2023). Tourist walkability and sustainable community-based tourism: conceptual framework and strategic model. *International Journal of Tourism Cities, 10*(1), 78–104.

Chapter 5

Sustainable Tourism in a Changing Climate: Balancing Growth and Environmental Responsibility

Shehnaz Tehseen, Syed Monirul Hossain, Kim Yoke Ong and Elizabeth Andrews

Department of Management, Sunway Business School (SBS), Sunway University, Malaysia

Abstract

This chapter explores the present state of the worldwide tourism sector and its complex phenomenological interconnection with climate change. This analysis emphasizes the significant environmental consequences resulting from the rise of tourism, which can be attributed to the exponential growth facilitated by enhanced connectivity and the concurrent increase in disposable incomes. It thoroughly analyses the primary factors that contribute to the environmental impact of the sector, which encompasses emissions from transportation, infrastructure development, waste generation, and resource usage. Furthermore, this analysis sheds light on the obstacles encountered in the process of shifting toward sustainability, including limited knowledge and understanding, deficiencies in policy implementation, and reluctance to embrace transformation within the sector. However, it also provides insights into numerous activities and solutions currently in progress to alleviate the environmental impact of the industry. The aforementioned approaches encompass the implementation of sustainable tourism practices, legislation revisions, and advancements in technology. This chapter concludes by highlighting the imperative need for a collaborative initiative including governmental bodies, industrial participants, and tourists to establish a harmonious equilibrium between the pursuit of economic expansion and the protection of the environment. The future trajectory

The Need for Sustainable Tourism in an Era of Global Climate Change: Pathway to a Greener Future, 69–94

Published under exclusive licence by Emerald Publishing Limited
doi:10.1108/978-1-83608-668-020241012

of global tourism is contingent upon the adoption of sustainable practices, which are crucial for fostering a healthy coexistence between the tourism industry and the environment.

Keywords: Sustainable tourism; climate change; environmental impact; global tourism; sustainability initiatives

1. Introduction

Tourism, a dynamic and ever-expanding industry, has played an important role in the global economy, contributing to employment, cultural exchange, and economic growth (Jatav, 2023). However, the rapid growth of the tourism sector has raised concerns regarding its impact on the environment, particularly its contribution to climate change. The significance of sustainable tourism has transcended mere environmental responsibility to become a critical imperative caused by climate change challenges. In the face of climate change challenges, the tourism industry is at a critical juncture, necessitating the adoption of practices aimed at not only preserving but also enhancing the environments upon which it relies (Sahni et al., 2023).

Global tourism is classified as the activity of people traveling to and staying in different countries or regions around the world for leisure, business, or other purposes. It involves both international tourism, where visitors travel from their home country to another country, and domestic tourism, where visitors travel within their own country (Nguyen et al., 2023). Global tourism is a significant industry, with millions of people traveling internationally each year. It can have a significant impact on the economies of both the host countries and the countries of origin of the tourists, as well as on the environment, society, and culture of the places visited Global tourism can be facilitated by a range of factors, including improvements in transportation, communication, and technology. It can also be influenced by geopolitical factors; such as changes in visa requirements or political instability in certain regions (Kapoor & Jain, 2023).

Sustainable tourism emphasizes the importance of adapting to climate change by promoting destinations that implement resilient infrastructure and disaster preparedness plans (Lopes et al., 2022). Sustainable tourism, by its very nature, seeks to minimize the ecological footprint of travel activities. In the face of climate change, this approach becomes a linchpin for the preservation of fragile ecosystems. From coral reefs to mountainous landscapes, tourist destinations worldwide face the threat of degradation due to climate-induced changes (Chauka & Nyangoko, 2023). Sustainable tourism practices, such as low-impact accommodations, responsible waste management, and eco-friendly transportation, aid in conserving biodiversity and promoting ecosystem resilience (Baloch et al., 2023). It places a strong emphasis on the preservation of cultural heritage, recognizing that climate change can exacerbate the vulnerability of historical sites. By promoting responsible tourism practices, which include visitor education, conservation funding, and climate-resilient architecture, sustainable tourism becomes a custodian of the world's cultural treasures (Sheller, 2021). Sustainable tourism seeks to empower local communities by ensuring that they benefit from tourism activities economically, socially, and culturally (Buhalis et al., 2023). Community-based tourism initiatives, fair wage practices, and the preservation of indigenous knowledge contribute to the resilience of communities in the face of climate-related challenges. This chapter aims to delve into the current status of sustainable global tourism and explore its multifaceted impacts on climate change that balance growth and environmental responsibility.

Over the past few decades, the tourism industry has experienced exponential growth, with international tourist arrivals reaching unprecedented numbers. Factors such as increased connectivity, technological advancements, and rising disposable incomes have fuelled this expansion (Morrison & Buhalis, 2023). In 2019, international tourist arrivals surpassed 1.5 billion, a testament to the sector's robust growth (Vu & Hartley, 2022). Moreover, tourism is not solely limited to international travel. Domestic tourism contributes significantly to the industry, further amplifying its environmental impact. Increased urbanization, the development of tourist infrastructure, and the rise of digital platforms promoting travel have all fuelled domestic tourism growth (Nicola-Gavrilă, 2023).

One of the primary drivers of global tourism growth is the evolution of transportation infrastructure. The development of efficient air travel networks, characterized by the expansion of international airports and the introduction of more affordable airfares, has significantly reduced the barriers to long-distance travel. This has made it easier for people to explore destinations that were once considered remote and inaccessible (Zhu & Zhu, 2023).

As economies prosper, disposable incomes rise, enabling more individuals to allocate funds for leisure and travel. The tourism industry, in turn, has become a vital component of many countries' economies, contributing significantly to GDP and creating job opportunities (Marcouiller, 2023). Developing nations often leverage tourism as a means to diversify their economies and reduce dependence on traditional sectors.

The advent of digital technology has revolutionized the way people plan, book, and experience their trips. The internet has democratized access to information, allowing travelers to research destinations, read reviews, and make bookings online. Social media platforms have further amplified the influence of word-of-mouth marketing, with users sharing their travel experiences and recommendations, thus inspiring others to explore new places (Pencarelli, 2020). Cultural exchange has been a notable outcome of the growth in global tourism. As people from different corners of the world interact, share experiences, and immerse themselves in diverse cultures, there is a mutual exchange of ideas, values, and perspectives. This cultural intermingling fosters understanding and tolerance, contributing to a more interconnected global community (Du Cros, 2023).

However, the surge in tourism also poses challenges. Overtourism, wherein popular destinations become overcrowded, leading to environmental degradation and cultural erosion, has become a pressing issue (Siegel et al., 2023). Managing the environmental impact of increased travel, preserving local cultures, and ensuring equitable distribution of economic benefits are essential considerations for sustainable tourism (Pan et al., 2018).

2. Literature Review

2.1. Environmental Impact of Tourism on Climate Change

One major contributor to the environmental impact of tourism on climate change is transportation. Air travel, in particular, is a major source of greenhouse gas emissions. Commercial airplanes burn vast amounts of fossil fuels, releasing carbon dioxide (CO_2), along with other pollutants, into the atmosphere. The growth in air travel, driven by the increasing popularity of international tourism, exacerbates the industry's carbon footprint (Kamb et al., 2021).

The construction and operation of tourism-related infrastructure, such as hotels, resorts, and entertainment facilities, also have environmental consequences. The clearing of land for such developments can lead to deforestation and habitat loss, impacting biodiversity. Additionally, the energy-intensive nature of these facilities, especially in regions where renewable energy sources are not widely adopted, contributes to increased carbon emissions (Baloch et al., 2023).

Tourism's impact on climate change is also evident in its influence on natural ecosystems. Activities like hiking, wildlife watching, and water sports can disrupt fragile ecosystems, causing habitat degradation and fragmentation. Coral reefs, for example, are highly vulnerable to the impacts of tourism, with activities like snorkelling and diving causing physical damage, and the release of sunscreen chemicals contributing to coral bleaching (Gössling et al., 2023).

Melting glaciers and thawing permafrost in popular tourist destinations, such as the Arctic and mountainous regions, are consequences of climate change that are exacerbated by the influx of visitors. The increased foot traffic accelerates the melting process, with potential long-term implications for local ecosystems (Gordon, 2023).

Tourism's impact on climate change is further compounded by its contribution to waste generation. The consumption of single-use plastics, packaging materials, and the improper disposal of waste in tourist areas contribute to pollution and environmental degradation. Plastic pollution, in particular, poses a severe threat to marine ecosystems and contributes to the larger issue of environmental sustainability (Buhalis et al., 2023).

Addressing the environmental impact of tourism on climate change requires concerted efforts from various stakeholders, including governments, the tourism industry, and travelers. Sustainable tourism practices, such as the use of eco-friendly accommodations, responsible wildlife tourism, and the promotion of low-carbon transportation options, are crucial for mitigating these impacts (Ranasinghe et al., 2021). Additionally, governments and regulatory bodies play a key role in enforcing environmental standards, implementing carbon offset programs, and promoting sustainable tourism practices. While tourism brings economic benefits and cultural exchange, its environmental impact on climate change is a significant concern. Sustainable practices and responsible tourism are essential for mitigating these impacts and ensuring that tourism contributes positively to both local economies and the global environment (Pan et al., 2018). Nevertheless, there are multiple sustainable tourism perspectives as described below:

2.1.1. Environmental Perspective

Traditional ecotourism emphasizes the conservation of natural ecosystems and biodiversity. It explores how tourism can be a tool for environmental protection, involving tourists in conservation efforts (Das & Chatterjee, 2015).

Climate change resilience: As climate change poses a significant threat, this perspective looks at how sustainable tourism can adapt to and mitigate these changes. It examines strategies for reducing carbon footprints and building resilience in the face of environmental challenges (Xiao et al., 2024).

2.1.2. Social and Cultural Perspective

Community-based tourism: Focusing on the empowerment of local communities, this perspective explores how tourism can benefit, rather than exploit, indigenous populations. It delves into community-based models that ensure local ownership and cultural preservation (Ruiz-Ballesteros, 2023).

Cultural heritage tourism: Examining the preservation of cultural heritage, this perspective looks at how sustainable tourism can contribute to safeguarding traditions, historical sites, and intangible cultural heritage (Katelieva et al., 2020).

2.1.3. Economic Perspective

Sustainable business models: This perspective explores how businesses in the tourism sector can adopt sustainable practices not only for environmental and social benefits but also for long-term economic viability (Lupoae et al., 2024).

Economic impact on local communities: Beyond profit generation, sustainable tourism should positively impact local economies. This perspective evaluates how economic benefits can be equitably distributed among local communities, preventing over-tourism (Rivera et al., 2024).

2.1.4. Policy and Governance Perspective

Government regulations and incentives: This perspective assesses the role of governments in shaping sustainable tourism through regulations, incentives, and policies. It explores how effective governance can balance economic interests with environmental and social considerations (Ma & Su, 2024).

Public–private partnerships: Investigating collaborative approaches, this perspective looks at how partnerships between governments, private enterprises, and local communities can foster sustainable tourism initiatives (Shrestha et al., 2024).

2.1.5. Tourist Behavior Perspective

Educational tourism: This perspective focuses on how tourist behavior can be influenced through education (Xu & Ho, 2024). It explores the role of awareness campaigns and educational programs in promoting responsible tourism practices (Wani et al., 2024).

Consumer preferences: Understanding what drives tourist choices is crucial. This perspective examines consumer preferences for sustainable and responsible travel, exploring market trends and demands (Ray et al., 2024).

2.1.6. Global Perspective

Cross-cultural collaboration: As tourism is a global phenomenon, this perspective explores how cross-cultural collaboration can contribute to sustainable tourism. It delves into international efforts and agreements aimed at promoting sustainable practices (Wang & Brown, 2024).

Tourism in developing countries: Examining the impact in developing nations, this perspective looks at how sustainable tourism can contribute to socio-economic development while avoiding negative consequences (Wani et al., 2024). By engaging with these diverse perspectives, scholars can paint a comprehensive picture of sustainable tourism that considers its multi-faceted nature and addresses the complex challenges it faces (Kefi et al., 2024). The scholarly analysis of sustainable tourism indeed benefits from engaging with a variety of perspectives within the literature, as this approach allows for a deeper understanding of the complexities surrounding this topic. Deliberately contrasting these viewpoints can illuminate rich debates and reveal existing gaps in current research.

One such argument revolves around the tension between economic development and environmental conservation in sustainable tourism. Some scholars believe that economic growth through tourism can lead to improved standards of living and poverty alleviation in local communities, thereby contributing to sustainable well-being (Dwyer, 2023). However, others contend that unchecked tourism development can result in environmental degradation, loss of cultural authenticity, and exploitation of local populations (Mariam et al., 2024). Another viewpoint centers on the role of government regulation versus voluntary industry initiatives in promoting sustainable tourism. Advocates for government intervention argue that strict regulations are necessary to enforce environmental standards and ensure equitable distribution of benefits. Conversely, proponents of voluntary initiatives emphasize the importance of industry self-regulation and market-driven approaches to sustainability (Elkhwesky et al., 2024). Furthermore, there is ongoing discussion regarding the effectiveness of different sustainability certification schemes and eco-labels in guiding consumer choices and incentivizing sustainable tourism practices. While some argue that certification schemes provide valuable information to tourists and encourage businesses to adopt environmentally responsible practices, others question their credibility, transparency, and impact on actual behavior (Waris et al., 2024).

Moreover, there are divergent perspectives on the concept of carrying capacity in sustainable tourism planning. While some scholars advocate for strict limits on tourist numbers and infrastructure development to prevent overcrowding and environmental degradation, others argue that adaptive management strategies, such as zoning and visitor management plans, can effectively balance tourism growth with environmental protection (Glyptou, 2024).

2.2. Factors of Sustainable Global Tourism: A Comprehensive Perspective

Sustainable global tourism is a multifaceted concept that involves the careful consideration of various factors to ensure that tourism activities contribute positively to economic development, environmental conservation, and social well-being. In this discussion, we will explore key factors contributing to sustainable global tourism, focusing on improved transportation, technology, economic development, cultural exchange, natural attractions, political stability, business travel, and marketing and promotion.

2.2.1. Improved Transportation

Sustainable global tourism places a significant emphasis on transportation systems that are efficient, environmentally friendly, and socially responsible. Improved transportation is a linchpin in the paradigm of sustainable global tourism, embodying principles of efficiency, environmental friendliness, and social responsibility. This facet underscores the interconnectedness of diverse factors crucial for the tourism sector to thrive while minimizing its ecological footprint. In essence, the elevation of transportation systems aligns with the ethos of sustainable global tourism. It is not merely about facilitating movement but creating an intricate tapestry where efficiency, environmental consciousness, and social responsibility converge. Through strategic improvements in transportation, the tourism industry can chart a course that enriches traveler experiences, fosters economic prosperity for local communities, and safeguards the planet for future generations (Li et al., 2023). They are as follows:

First, improved transportation infrastructure enhances the accessibility of destinations, allowing for better connectivity and a more efficient flow of tourists. The bedrock of sustainable transportation lies in enhancing efficiency and connectivity. Upgrading transportation infrastructure ensures seamless accessibility to destinations, fostering a more efficient flow of tourists. This includes the development of well-connected networks, modernized airports, and efficient ground transport systems. Such enhancements not only benefit tourists but also contribute to the economic development of host communities by facilitating easier access to markets (Tomej & Liburd, 2019).

Second, promoting the use of eco-friendly modes of transportation, such as electric vehicles and public transport, reduces the carbon footprint associated with travel. A cornerstone of sustainability is the promotion of eco-friendly modes of transportation. Encouraging the use of electric vehicles and supporting robust public transport systems significantly mitigates the carbon footprint associated with travel. The adoption of electric vehicles, powered by renewable energy sources, reduces reliance on fossil fuels, aligning with global efforts to combat climate change. This shift towards eco-friendly transportation is emblematic of a commitment to environmental stewardship within the tourism industry (Lv & Shang, 2023).

Third, sustainable tourism requires careful planning of transportation infrastructure to minimize negative impacts on local ecosystems and communities. Critical to sustainable tourism is meticulous planning of transportation infrastructure. The aim is to minimize adverse impacts on local ecosystems and communities. Infrastructure development should adhere to eco-sensitive practices, avoiding disruption to wildlife habitats and ensuring the preservation of cultural heritage. By integrating sustainable infrastructure planning, destinations can harmonize tourism growth with environmental conservation, creating a symbiotic relationship between tourism and the natural environment (Baloch et al., 2023).

2.2.2. Technological Advancements in Sustainable Tourism

In the contemporary landscape of global tourism, technological advancements emerge as pivotal catalysts in steering the industry toward sustainability. The integration of technology, spanning digital platforms, smart tourism initiatives, and virtual tourism, presents a multifaceted approach to instilling responsible practices and augmenting the overall tourist experience (Abdelfattah et al., 2023). They are as follows:

First, technology enables the creation of digital platforms that promote responsible travel choices, offer information on sustainable practices, and facilitate eco-friendly bookings. These platforms serve as hubs for disseminating information on responsible travel choices, eco-friendly accommodations, and sustainable practices. Travelers can access comprehensive databases that guide them toward environmentally conscious decisions, from choosing green accommodations to opting for low-impact activities. Furthermore, digital platforms facilitate eco-friendly bookings, empowering tourists to make choices aligned with sustainability principles (Nicola-Gavrilă, 2023).

Second, implementing smart technologies, such as Internet of Things (IoT) devices and data analytics, can help manage tourist flows, reduce resource consumption, and enhance the overall tourist experience. IoT devices and data analytics contribute to the efficient management of tourist flows, aiding destinations in optimizing resource utilization. Smart tourism initiatives enable real-time monitoring of visitor numbers, environmental conditions, and resource consumption. By harnessing these technologies, destinations can develop responsive strategies to prevent overcrowding, reduce energy consumption, and enhance the overall quality of the tourist experience (Shafiee et al., 2023).

Third, virtual reality and augmented reality technologies provide alternatives to physical travel, allowing individuals to experience destinations virtually, reducing the environmental impact. The advent of virtual reality (VR) and augmented reality (AR) technologies introduces an innovative dimension to sustainable tourism. This not only reduces the environmental impact associated with travel but also addresses concerns related to over-tourism. Through immersive virtual experiences, individuals can engage with diverse cultures, natural wonders, and historical sites without the need for extensive travel. Virtual tourism aligns with the principles of sustainability by offering a low-impact way for individuals to quench their wanderlust. In essence, technology emerges as a powerful enabler in the pursuit of sustainable global tourism. Its integration fosters a paradigm where responsible choices, resource optimization, and immersive experiences converge, reshaping the tourism landscape toward a more sustainable and resilient future (Hoang et al., 2023).

When examining the range of transportation solutions crucial to sustainable tourism, it is essential for researchers to engage in rich debates regarding their feasibility and scalability in various tourism contexts. While some advocate for innovative transportation technologies as a means to reduce carbon emissions and enhance visitor experiences, others raise concerns about their practicality and broader implications (Elkhwesky et al., 2024).

One debate revolves around the balance between promoting eco-friendly transportation options and ensuring accessibility for all tourists. While initiatives such as electric vehicles and public transport systems can contribute to sustainability goals, there are concerns about their affordability and availability in remote or less developed tourism destinations. Critics argue that prioritizing these solutions may inadvertently exclude certain groups of travelers or perpetuate inequalities in access to tourism opportunities (Ofélia de Queiroz et al., 2024).

Another point of contention is the environmental impact of technological advancements in transportation. While electric vehicles and other green technologies offer the promise of reducing carbon emissions, their production, operation, and disposal still pose environmental challenges (Sathiyan et al., 2022). Some scholars caution against the "greenwashing" of technology, emphasizing the need for comprehensive life cycle assessments to accurately evaluate their sustainability credentials (Lee et al., 2024). Additionally, questions arise regarding the sustainability of sourcing materials for these technologies and the potential for increased energy consumption in their manufacturing processes. Furthermore, the role of technology in advancing sustainable tourism extends beyond transportation to encompass broader aspects of visitor management and experience enhancement (Elkhwesky et al., 2024). While digital platforms and data analytics can optimize resource allocation, enhance visitor safety, and improve destination management, they also raise concerns about data privacy, digital surveillance, and cultural commodification (Latinovic & Chatterjee, 2024). Critics argue that the pervasive use of technology in tourism may lead to the homogenization of visitor experiences and the erosion of authentic cultural identities (Varnajot et al., 2024). In addition to these debates, there is a growing recognition of the need to address the digital divide in the context of sustainable tourism. While technology holds the potential to democratize access to information and empower local communities, disparities in internet connectivity and digital literacy may exacerbate existing inequalities (Radovanović, 2023). Scholars emphasize the importance of ensuring equitable access to technology and digital skills training to prevent marginalized groups from being left behind in the digital transformation of tourism (Elkhwesky et al., 2024).

Overall, a nuanced discussion of the role of technology in sustainable tourism requires researchers to critically evaluate both its potential benefits and drawbacks. By engaging in rich debates surrounding issues such as accessibility, environmental impact, cultural authenticity, and digital equity, scholars can contribute to a more comprehensive understanding of how technology can be effectively leveraged to advance sustainability goals in tourism (Lin et al., 2024).

2.2.3. Sustainable Tourism and Economic Development

Sustainable global tourism seeks to contribute positively to the economic development of host communities. The principles of sustainable tourism encompass various strategies to foster economic growth while prioritizing local empowerment and resilience (Baiocco et al., 2023). They are as follows:

First, one of the fundamental contributions of sustainable tourism to economic development lies in its capacity to generate employment opportunities for local communities. The tourism industry, when managed sustainably, becomes a significant source of jobs, spanning roles from hospitality and guiding to cultural interpretation. This approach not only injects vitality into local economies but also ensures that the benefits of tourism reach beyond the confines of major tourist hubs, promoting inclusivity and reducing economic disparities (Manzoor et al., 2019).

Second, encouraging the development of local enterprises not only diversifies the economic landscape but also enhances community resilience. Sustainable tourism emphasizes fostering local entrepreneurship, particularly through the establishment of small-scale businesses catering to the tourism sector. Local entrepreneurs, whether engaged in culinary endeavors, artisanal crafts, or guided tours, contribute to the authenticity of the destination experience while simultaneously stimulating economic growth (Hertel et al., 2021).

Thirdly, sustainable tourism initiatives should empower local communities to actively participate in and benefit from tourism-related activities. The active involvement of local communities in tourism-related activities ensures that they not only participate but also benefit tangibly from the industry. This empowerment is multifaceted, encompassing the preservation of cultural heritage, environmental conservation efforts, and active engagement in decision-making processes related to tourism development. By empowering communities, sustainable tourism endeavors contribute to the long-term well-being and self-determination of host populations (Gautam & Bhalla, 2024).

2.2.4. Sustainable Tourism and Cultural Exchange

Cultural exchange stands as a cornerstone in the realm of sustainable global tourism, serving as a powerful conduit for mutual understanding and respect. Sustainable tourism practices are intricately woven into the fabric of cultural preservation, responsible engagement, and community participation to ensure that tourism becomes a force for cultural enrichment rather than erosion (Susila et al., 2023). They are as follows:

First, sustainable tourism emphasizes the preservation of cultural heritage, ensuring that tourism activities do not lead to cultural erosion. This entails a commitment to safeguarding the rich tapestry of traditions, customs, and historical landmarks that define a destination. Sustainable practices prioritize the conservation of cultural assets, ensuring that the influx of tourists does not compromise the integrity of these invaluable elements. Preservation efforts extend to promoting awareness about the significance of cultural heritage, both among tourists and local communities, fostering a sense of collective responsibility (Xiao et al., 2024).

Second, encouraging tourists to engage respectfully with local cultures, traditions, and customs promotes positive cultural exchange. Tourists are encouraged to immerse themselves respectfully in the traditions and practices of the communities they visit. This involves an appreciation for local customs, etiquettes, and ways of life. Through cultural sensitivity training and awareness programs, tourists become active participants in the preservation of cultural identity, contributing to a more enriching and harmonious travel experience (Pawalluri, 2023).

Third, tourism models, community involvement is not a token gesture but a fundamental aspect of experience design. Local artisans, performers, and storytellers become integral contributors, adding layers of authenticity to the cultural encounters that tourists enjoy. This participatory approach ensures that cultural exchange is reciprocal and mutually beneficial (Liburd et al., 2023).

When delving into the economic dimensions of sustainable tourism, researchers are presented with a plethora of rich debates that merit exploration. One such disputation revolves around the potential for economic disparity inherent in sustainable tourism practices and their true economic viability for stakeholders at various levels of the tourism ecosystem (Silva et al., 2023). The proponents of sustainable tourism argue that it can foster economic development by generating revenue, creating job opportunities, and supporting local businesses, particularly in marginalized communities (Dwyer, 2023). By prioritizing environmentally and socially responsible practices, sustainable tourism endeavors to distribute economic benefits more equitably and minimize negative externalities such as environmental degradation and cultural exploitation (Jahanger et al., 2023). However, critics raise concerns about the feasibility and inclusivity of sustainable tourism models, particularly in resource-constrained or politically unstable regions. They argue that the upfront costs associated with adopting sustainable practices, such as investing in eco-friendly infrastructure or obtaining certification, may pose significant barriers to entry for small-scale operators or developing countries. Moreover, the reliance on niche markets and the potential for greenwashing within the industry raises questions about the scalability and long-term economic viability of sustainable tourism initiatives (Rani et al., 2024).

Another contentious issue pertains to the implementation of cultural exchange initiatives within the framework of sustainable tourism. While cultural exchange can promote mutual understanding, appreciation, and respect among tourists and host communities, there is a risk of cultural commodification or disruption if not managed thoughtfully (Wu & Ma, 2024). Critics argue that commercialization and superficial representations of indigenous cultures for tourist consumption can lead to the erosion of cultural authenticity and perpetuate harmful stereotypes (Kunst et al., 2024). Furthermore, debates arise regarding the power dynamics inherent in cultural exchange initiatives, particularly between tourists from affluent countries and marginalized indigenous communities (Shrestha et al., 2024).

Questions of ownership, consent, and benefit-sharing underscore the need for ethical frameworks and community-driven approaches to cultural tourism. Scholars emphasize the importance of centering indigenous voices and perspectives in the design and management of cultural exchange programs to ensure they genuinely enrich local cultures and empower communities (Elers & Dutta, 2024).

2.2.5. Sustainable Tourism and Natural Attractions

In the realm of sustainable global tourism, the appreciation and preservation of natural attractions take center stage as an essential pillar. Sustainable practices in this context revolve around conscientious conservation efforts, the promotion of ecotourism, and the instillation of responsible nature tourism behaviors. These strategies are designed not only to safeguard the intrinsic value of natural wonders but also to create a harmonious balance between tourism activities and the delicate ecosystems they inhabit (Khizar et al., 2023). They are as follows:

First, sustainable tourism involves implementing conservation practices to protect natural habitats, wildlife, and ecosystems. Conservation efforts include the establishment and maintenance of protected areas, wildlife corridors, and sustainable resource management. By integrating these practices, sustainable tourism seeks to minimize the ecological footprint associated with tourist activities (Baloch et al., 2023).

Second, a key facet of sustainable tourism is the promotion of ecotourism, a concept that encourages tourists to engage with nature in a responsible and conservation-focused manner. Ecotourism experiences are designed to foster an understanding of the natural environment, emphasizing education and appreciation. Whether it's exploring pristine rainforests, observing wildlife in their natural habitats, or participating in habitat restoration projects, ecotourism endeavors to create a symbiotic relationship between tourists and the ecosystems they visit (Moorman et al., 2024).

Third, tourists are encouraged to follow responsible practices, such as leaving no trace, to minimize their ecological footprint in natural areas. This entails a set of guidelines and principles that tourists are encouraged to follow to minimize their impact on natural areas. These guidelines may include practices such as leaving no trace, refraining from disturbing wildlife, and adhering to designated trails to avoid habitat disruption. By instilling responsible behaviors, sustainable tourism aims to ensure that the allure of natural attractions remains untarnished for future generations (Baloch et al., 2022).

2.2.6. Sustainable Tourism and Political Stability

In the intricate tapestry of sustainable global tourism, the thread of political stability emerges as a vital determinant of the industry's sustainability. This factor exerts a profound influence on various aspects, including the safety and security of tourists, the formulation of government policies, and the facilitation of international collaboration – all of which are important for the enduring success of sustainable tourism endeavors (Banso et al., 2023). They are as follows:

First, tourists are more likely to visit destinations with stable political environments, ensuring their safety and security. Tourists naturally gravitate toward places where they can enjoy their experiences without the spectre of political unrest or security concerns. A politically stable environment fosters a sense of assurance among tourists, encouraging them to explore and engage with the local culture and attractions (Altinay & Kozak, 2021).

Second, supportive government policies that prioritize tourism development and sustainability contribute to the overall success of the industry. Political stability enables governments to formulate and implement policies that prioritize tourism development while ensuring its sustainability. These policies may include regulations for environmental protection, community engagement, and the responsible management of tourist activities. A stable political climate allows for the long-term planning and execution of initiatives that align with the principles of sustainable tourism (Lasisi et al., 2023).

Third, political stability creates a conducive environment for international collaboration and partnerships in the realm of sustainable tourism. Nations with stable political landscapes are more likely to engage in collaborative efforts with other countries, international organizations, and non-governmental entities. These collaborations can take various forms, including knowledge exchange, joint conservation projects, and the establishment of sustainable tourism standards (Martins & Ribeiro, 2023).

2.2.7. Sustainable Tourism and Business Travel

Sustainable business travel involves minimizing the environmental impact of corporate travel activities. As corporations navigate the complex terrain of global interactions, the imperative to minimize the environmental impact of business travel activities becomes paramount. This involves a multifaceted approach, including the adoption of green practices in meetings and conferences, the promotion of telecommuting and remote work, and the overarching commitment to corporate responsibility (Baloch et al., 2023). They are as follows:

First, implementing green practices in meetings and conferences, such as reducing waste and offsetting carbon emissions, contributes to sustainability. This may encompass initiatives such as minimizing waste through efficient resource management, adopting eco-friendly materials, and offsetting carbon emissions through the support of sustainable projects. By infusing sustainability into the fabric of corporate gatherings, businesses contribute to the broader ethos of responsible tourism (Joshi et al., 2023).

Second, a transformative element in sustainable business travel is the promotion of telecommuting and remote work. Embracing these practices reduces the necessity for frequent corporate travel, mitigating its associated environmental impact. Technological advancements that facilitate seamless virtual collaboration empower employees to engage in professional responsibilities without the need for extensive travel (Roberto et al., 2023).

Third, companies adopting sustainable business travel policies demonstrate corporate responsibility and contribute to the overall sustainability of the tourism sector. This involves not only implementing eco-friendly measures but also actively participating in initiatives that contribute to the broader sustainability of the tourism sector. By assuming responsibility for the environmental consequences of their travel-related activities, corporations become agents of positive change in the global tourism landscape (Paraskevas et al., 2022).

The examination of sustainable business travel practices triggers a multitude of rich debates, especially when considering the interplay between corporate objectives and sustainability ambitions, particularly within the expansive reach of multinational corporations (MNCs). One significant argument revolves around the inherent conflict between the pursuit of profit and environmental sustainability within corporate travel strategies. Critics argue that while MNCs may espouse commitments to sustainability, their primary focus often remains on maximizing shareholder value and financial returns (Stoelhorst & Vishwanathan, 2024). This raises questions about the extent to which corporate travel policies genuinely prioritize environmental concerns over economic interests. Another contentious issue is the effectiveness of voluntary corporate sustainability initiatives versus regulatory measures in driving sustainable business travel practices (de Mello, 2024). Advocates of voluntary initiatives argue that they allow for flexibility and innovation, enabling companies to tailor strategies to their unique circumstances (Ruiz-Castillo et al., 2024). However, skeptics contend that without mandatory regulations, many corporations may prioritize short-term financial gains over long-term sustainability goals (Chewpreecha et al., 2023). Moreover, there is ongoing debate regarding the role of technology in promoting sustainable business travel. Proponents argue that technological innovations, such as video conferencing and telecommuting, can significantly reduce the need for physical travel, thereby mitigating carbon emissions and environmental impact (Tao et al., 2024). Conversely, critics caution against over-reliance on technology, highlighting potential drawbacks such as increased energy consumption from digital infrastructure and the exacerbation of the digital divide, particularly in developing regions (Wang & Guo, 2024). Additionally, there is discourse surrounding the ethical implications of corporate travel, particularly concerning social justice and equity (Büchs & Mattioli, 2024). Critics argue that frequent business travel often exacerbates inequalities by entitling privileged employees who have the resources and flexibility to travel, while disadvantaging those who do not (Mooney & Baum, 2024). This raises questions about the responsibility of corporations to ensure equitable access to opportunities and mitigate the social disparities exacerbated by business travel.

Overall, the examination of sustainable business travel practices within the context of multinational corporations engenders rich dialogues that encompass economic, environmental, technological, and social dimensions. These polarizing viewpoints underscore the complexities and trade-offs inherent in balancing corporate objectives with sustainability ambitions. All of which highlights the need for holistic and context-specific approaches to address the challenges of business travel in an increasingly interconnected world.

2.2.8. Marketing and Promotion: Catalysts for Sustainable Tourism Advocacy

In the dynamic realm of global tourism, effective marketing and promotion stand as powerful catalysts for steering the industry towards sustainability. These strategies play an important role in shaping tourists' choices and perceptions, emphasizing responsible travel practices, and fostering collaborations that transcend traditional boundaries. Here are key dimensions within marketing and promotion that contribute to the promotion of sustainable tourism (Dinkoksung et al., 2023). They are as follows:

First, Marketing campaigns that focus on educating tourists about responsible travel choices contribute to sustainability. Strategic campaigns that focus on imparting knowledge about responsible travel choices empower tourists to make informed decisions. These campaigns delve into the ecological, social, and cultural dimensions of sustainability, illuminating the consequences of choices and encouraging tourists to adopt practices that minimize negative impacts (Tkaczynski et al., 2020).

Second, the branding of destinations as bastions of sustainability is a strategic lever that marketing employs to shape tourists' perceptions. By emphasizing eco-friendly practices, conservation efforts, and cultural preservation, destination branding positions locales as stewards of responsible tourism. This not only attracts environmentally conscious tourists but also contributes to the overarching narrative of sustainability in the tourism industry (Ancheary et al., 2023).

Third, collaboration between governments, private sector stakeholders, and local communities in marketing efforts can amplify the message of sustainability. Collaborative campaigns can span diverse platforms, leveraging the strengths of each participant to create a unified front that promotes responsible tourism practices. This approach fosters a sense of shared responsibility and encourages collective action for the greater good (Rasoolimanesh et al., 2023).

2.3. Crisis Communication

In an era marked by unforeseen challenges and crises, effective marketing plays a crucial role in crisis communication. Transparent communication about the impact of crises on destinations, coupled with clear strategies for recovery and sustainable rebuilding, fosters trust among tourists. Sustainable tourism marketing during crises emphasizes resilience, community support, and the enduring commitment to responsible practices (Blasco-Arcas et al., 2022).

The environmental footprint of the tourism industry is substantial, with significant implications for climate change. The primary contributors to this impact include:

2.3.1. Transportation

Air travel, a fundamental component of international tourism, accounts for a substantial portion of carbon emissions. Additionally, cruise ships and other forms of transportation contribute significantly to the industry's carbon footprint. The aviation industry's reliance on fossil fuels releases a considerable volume of greenhouse gases into the atmosphere, exacerbating climate change. Beyond air travel, cruise ships and other transportation modes also contribute significantly to the industry's carbon footprint, highlighting the need for sustainable practices (Gössling et al., 2023).

2.3.2. Infrastructure Development

The construction of hotels, resorts, and tourist facilities often leads to deforestation, habitat destruction, and increased energy consumption, contributing to greenhouse gas emissions. The rapid expansion of hotels, resorts, and tourist facilities often comes at the expense of natural landscapes. Construction activities contribute to deforestation, habitat destruction, and increased energy consumption, leading to elevated greenhouse gas emissions (Raihan, 2023).

2.3.3. Waste Generation

Tourism results in heightened waste generation, impacting local ecosystems and contributing to environmental degradation. Tourism, while fostering enjoyment and leisure, often results in heightened waste generation. The influx of visitors to popular destinations leads to increased waste disposal, impacting local ecosystems and contributing to environmental degradation (Paul & Roy, 2023).

2.3.4. Resource Consumption

High water usage, overexploitation of natural resources, and increased energy demands in popular tourist destinations exacerbate the industry's environmental impact. The demand for water, natural resources, and energy in popular tourist destinations strains local ecosystems. Overexploitation of these resources, coupled with increased energy demands, intensifies the industry's environmental impact (Xu & Zhao, 2023).

3. Discussion

One of the primary challenges in the quest for sustainable tourism is the pervasive lack of awareness among both tourists and industry stakeholders. Many individuals remain blissfully ignorant of the environmental repercussions of their travel choices. The carbon footprint of air travel, the strain on local ecosystems, and the consequences of excessive waste generation often go unnoticed. Bridging this awareness gap is crucial for instigating a paradigm shift towards responsible and sustainable tourism. Inadequate and loosely enforced policies and regulations contribute to the environmental challenges posed by the tourism sector. Existing frameworks often fall short in addressing key issues such as waste management, carbon emissions, and the protection of natural habitats. Strengthening and effectively enforcing these policies is essential for curbing the negative impact of tourism on the environment. Resistance from certain sectors within the tourism industry itself poses a significant barrier to progress. Established practices, even if environmentally detrimental, may be deeply ingrained, and the prospect of change can be met with reluctance. Overcoming this resistance requires a concerted effort to demonstrate the long-term benefits of sustainable practices and the positive impact they can have on both the environment and the industry's bottom line.

A beacon of hope in the quest for environmentally responsible tourism is the increasing adoption of sustainable practices. This includes the establishment of eco-friendly accommodations that prioritize energy efficiency, waste reduction, and responsible water usage. Sustainable tourism practices extend beyond lodging to encompass activities, transportation, and overall resource management. The industry is witnessing a growing commitment to minimizing its ecological footprint and promoting responsible tourism experiences. Governments and international organizations recognize the urgency of addressing the environmental impact of global tourism. As a response, there is a concerted effort to implement more stringent policies and regulations. These initiatives aim to regulate tourism activities, incentivize sustainable practices, and penalize environmental transgressions. Through collaborative efforts, policymakers seek to create a framework that balances the economic benefits of tourism with the imperative to protect natural ecosystems. Technology plays a pivotal role in shaping the future of sustainable tourism. Innovations are driving the development of eco-friendly modes of transportation, energy-efficient infrastructure, and sustainable tourism solutions. From electric vehicles and renewable energy sources to smart tourism management systems, technology is enabling the industry to minimize its environmental impact while enhancing the overall travel experience. These innovations not only contribute to sustainability but also showcase the industry's potential for positive change. As the challenges and opportunities in addressing the environmental impact of global tourism continue to unfold, it is crucial to envision the road ahead. A sustainable future for the industry necessitates collective action, innovative solutions, and a fundamental shift in mindset.

Tackling the lack of awareness requires comprehensive education and awareness campaigns. Engaging both tourists and industry stakeholders through targeted initiatives can illuminate the environmental consequences of travel choices. By fostering a sense of responsibility and mindfulness, these campaigns can contribute to a more informed and conscientious approach to tourism. Overcoming policy and regulation gaps demands collaborative governance at the local, national, and international levels. Governments, industry associations, and environmental organizations must work together to create robust frameworks that prioritize sustainability. This collaborative approach can ensure that policies are not only comprehensive but also effectively enforced, promoting responsible tourism practices across the board. To overcome resistance to change within the industry, it is essential to create incentives for sustainable practices. Governments and organizations can offer financial incentives, certifications, and recognition for businesses that demonstrate a commitment to environmental responsibility. This can shift the narrative, showcasing sustainability as a key factor in business success rather than a hindrance. Continued investment in research and development is crucial for driving technological innovations that reduce the environmental impact of tourism. This includes advancements in renewable energy, waste management solutions, and sustainable infrastructure. By fostering a culture of innovation, the industry can continually evolve towards more eco-friendly practices. Sustainable tourism cannot exist in isolation; it must actively engage and benefit local communities. Empowering communities through tourism, involving them in decision-making processes, and ensuring that the economic benefits are shared locally can foster a sense of ownership and pride. This, in turn, contributes to the preservation of cultural heritage and the protection of natural environments.

The chapter's exploration of sustainable tourism practices presents an opportunity to delve into some rich discourse surrounding the adequacy of current initiatives in offsetting the environmental footprint of the tourism sector, particularly in light of the expansive scale of global tourism (Sard & Valle, 2024). One key debate centers on the effectiveness of existing sustainable tourism practices in addressing the sector's environmental impact (Higham et al., 2021). Critics argue that while initiatives such as eco-certifications, carbon offsetting, and sustainable tourism guidelines have proliferated, their overall impact remains limited. They contend that these measures often fail to address systemic issues such as over-tourism, habitat destruction, and carbon emissions associated with transportation (Nygaard, 2024). This raises questions about the efficacy of piecemeal approaches to sustainability and the need for more comprehensive and transformative solutions. Another contentious issue is the tension between economic growth and environmental conservation in the context of sustainable tourism. Proponents of economic growth argue that tourism is a vital engine of economic development, providing livelihoods for millions of people worldwide (Das & Chatterjee, 2015). They assert that strict environmental regulations or limitations on tourism growth could stifle economic opportunities, particularly in developing regions heavily reliant on tourism revenue (Dong et al., 2024). Conversely, environmental advocates stress the urgency of prioritizing environmental protection over economic interests, warning of the irreversible consequences of unchecked tourism expansion on fragile ecosystems and biodiversity (Wolters, 2023).

Moreover, there is ongoing debate about the role of technology in fostering sustainable tourism. Supporters highlight the potential of technological innovations such as smart destination management systems, renewable energy solutions, and digital platforms for promoting responsible travel behavior and reducing environmental impact (Firman et al., 2023). However, skeptics caution against over-reliance on technology, warning of unintended consequences such as digital pollution, privacy concerns, and the reinforcement of socio-economic inequalities in access to technology (Oo et al., 2023).

Additionally, there is discourse surrounding the social dimensions of sustainable tourism, including issues of equity, cultural preservation, and community empowerment. Critics argue that mainstream sustainable tourism practices often prioritize environmental concerns over social justice, perpetuating inequalities and marginalizing local communities. They advocate for approaches that prioritize community-led development, cultural authenticity, and equitable distribution of tourism benefits (Bazrafshan et al., 2023).

Overall, the paper's exploration of sustainable tourism practices presents an opportunity to engage in rich debates surrounding the efficacy, equity, and implications of current approaches to sustainability in the tourism sector. By critically examining these issues, researchers can contribute to the development of more robust and holistic strategies for mitigating the environmental footprint of global tourism while promoting social equity and economic development.

Incorporating a spectrum of critical perspectives into the discourse on sustainable tourism is essential to provide a more balanced critique of current practices and proposed solutions. By rigorously challenging the underlying assumptions and methodologies, researchers can foster a deeper understanding of the complexities and nuances inherent in sustainable tourism.

One key debate centers on the effectiveness of mainstream sustainable tourism practices in achieving their intended outcomes. Advocates argue that initiatives such as eco-certifications, community-based tourism, and sustainable development goals have made significant strides in promoting environmental conservation, cultural preservation, and economic development (Deb et al., 2023). However, critics contend that these initiatives often prioritize profit over sustainability, leading to greenwashing and tokenistic gestures that fail to address the root causes of unsustainable tourism practices (Peeters et al., 2024). They advocate for more radical approaches that challenge the status quo and prioritize systemic change over incremental improvements.

Another contentious issue is the tension between conservation and development goals in sustainable tourism. Proponents of conservation argue that strict regulations and limitations on tourism activities are necessary to protect fragile ecosystems and biodiversity hotspots (Adetola, 2023). They caution against the commodification of nature and cultural heritage, warning of the irreversible consequences of unchecked tourism growth on local communities and the environment. Conversely, proponents of development argue that tourism is a vital engine of economic growth and poverty alleviation, providing livelihoods for millions of people worldwide. They advocate for a balanced approach that maximizes the benefits of tourism while minimizing its negative impacts on the environment and society (Baloch et al., 2023).

Moreover, there is ongoing debate about the role of power dynamics and social justice in sustainable tourism. Critics argue that mainstream sustainable tourism practices often perpetuate inequalities and marginalize local communities, particularly in developing countries (Bazrafshan et al., 2023). They highlight issues such as land grabbing, displacement of indigenous peoples, and cultural appropriation as systemic injustices that need to be addressed. Conversely, proponents of empowerment argue that sustainable tourism has the potential to empower marginalized communities, promote cultural diversity, and foster intercultural understanding (Baiocco et al., 2023). They advocate for approaches that prioritize community-led development, equitable distribution of tourism benefits, and meaningful participation in decision-making processes.

Overall, incorporating a spectrum of critical perspectives into the discourse on sustainable tourism is essential to provide a more nuanced and comprehensive understanding of the field. By challenging existing paradigms and proposing alternative frameworks, researchers can contribute to the development of more inclusive, equitable, and sustainable tourism practices that benefit both people and the planet.

4. Conclusion

The exponential growth of global tourism has fundamentally transformed economies, facilitated cultural exchange, and presented several prospects. The expansion has posed difficulties, namely in terms of maintaining environmental sustainability and preserving cultural heritage. It is crucial for the tourism industry to handle these concerns in a sustainable manner during this critical period. Sustainable tourism is both morally upright and essential in order to alleviate the adverse impacts of climate change on various locations across the globe. The main focus is on minimizing ecological footprints, safeguarding vulnerable ecosystems, and conserving cultural heritage. Sustainable tourism safeguards the integrity of natural and cultural resources through the promotion of robust infrastructure, responsible waste management, and the empowerment of local communities. The advancement of technology and connectivity has greatly enhanced tourism and fostered a more profound cultural interaction among diverse communities. In order to have a stable and enduring future, the business must confront the issues of excessive tourism, ecological deterioration, and socio-cultural disruptions. Adopting sustainable tourism methods ensures a harmonious combination of economic progress and environmental stewardship in our future endeavors. In order to enhance the sustainability and resilience of the tourism sector, it is imperative for governments, local communities, businesses, and visitors to collaborate. This collaboration should prioritize environmental conservation, cultural preservation, and the overall welfare of all parties involved. Moreover, the environmental impact of global tourism presents both challenges and opportunities. The lack of awareness, policy gaps, and resistance to change are hurdles that demand thoughtful and concerted efforts. However, initiatives centered around sustainable tourism practices, policy reforms, and technological innovations showcase a positive trajectory toward a more responsible industry. As we navigate the intricate landscape of global tourism, it is imperative to envision a future where economic prosperity coexists harmoniously with environmental and social well-being. The road ahead requires commitment, collaboration, and a collective determination to transform the tourism industry into a force for positive change. By embracing the challenges and seizing the opportunities, the vision of sustainable tourism can become a reality, ensuring that the exploration of our world becomes a catalyst for its preservation and enhancement. Understanding and addressing the environmental footprint of the tourism industry demand a concerted effort from stakeholders across the globe. As the tourism sector continues to evolve, balancing the economic benefits with environmental stewardship becomes imperative for the preservation of the planet's ecological integrity and is in line with the Sustainable Development Goals, United Nations 2030 agenda (UN, 2015).

References

Abdelfattah, F., Al-Alawi, A., Abdullahi, M. S., & Salah, M. (2023). Embracing the industrial revolution: The impact of technological advancements and government policies on tourism development in Oman. *Journal of Policy Research in Tourism, Leisure and Events*, 1–25. https://doi.org/10.1080/19407963.2023.2294789

Adetola, B. O. (2023). Biodiversity conservation and tourism sustainability in Africa. In S. C. Izah & M. C. Ogwu (Eds.), *Sustainable utilization and conservation of Africa's biological resources and environment* (pp. 35–60). Springer Nature Singapore.

Ancheary, T., Rajauria, S., & Gupta, M. (2023). Regional destination branding in NCR: An integrative literature review. *International Journal for Multidimensional Research Perspectives*, *1*(3), 115–127.

Altinay, L., & Kozak, M. (2021). Revisiting destination competitiveness through chaos theory: The butterfly competitiveness model. *Journal of Hospitality and Tourism Management*, *49*, 331–340.

Baloch, Q. B., Shah, S. N., Iqbal, N., Sheeraz, M., Asadullah, M., Mahar, S., & Khan, A. U. (2023). Impact of tourism development upon environmental sustainability: A suggested framework for sustainable ecotourism. *Environmental Science and Pollution Research*, *30*(3), 5917–5930.

Baloch, Q. B., Shah, S. N., Maher, S., Irshad, M., Khan, A. U., Kiran, S., & Shah, S. S. (2022). Determinants of evolving responsible tourism behavior: Evidences from supply chain. *Cogent Social Sciences*, *8*(1), 2099565.

Baiocco, S., Leoni, L., & Paniccia, P. M. A. (2023). Entrepreneurship for sustainable development: Co-evolutionary evidence from the tourism sector. *Journal of Small Business and Enterprise Development*, *30*(7), 1521–1546.

Banso, A. A., Coker, J. O., Uzougbo, N. S., & Bakare, S. S. (2023). The nexus of law and sustainable development in south west Nigerian public policy: A review of multidisciplinary approaches in policy formation. *International Journal of Applied Research in Social Sciences*, *5*(8), 308–329.

Bazrafshan, S., McIntosh, A., & Cockburn-Wootten, C. (2023). Advancing a social justice-orientated agenda through research: A review of refugee-related research in tourism. *Journal of Sustainable Tourism*, *32*(3), 1–19.

Blasco-Arcas, L., Falkheimer, J., & Heide, M. (2022). Crisis communication in service ecosystems: Perspectives and future challenges. *Journal of Service Management*, *33*(4/5), 601–613.

Büchs, M., & Mattioli, G. (2024). How socially just are taxes on air travel and 'frequent flyer levies'?. *Journal of Sustainable Tourism*, *32*(1), 62–84.

Buhalis, D., Leung, X. Y., Fan, D., Darcy, S., Chen, G., Xu, F., Wei-Han Tan, G., Nunkoo, R., & Farmaki, A. (2023). Tourism 2030 and the contribution to the sustainable development goals: The tourism review viewpoint. *Tourism Review*, *78*(2), 293–313.

Chauka, L. J., & Nyangoko, B. P. (2023). Climate change impacts outweigh conservation efforts in coral reefs that are highly exposed to thermal stresses in Zanzibar, Tanzania. *Ocean & Coastal Management*, *238*, 106575.

Chewpreecha, U., Pollitt, H., & Mercure, J. F. (2023). Combining short-term economic recovery with long-term sustainability. In *Prospects and policies for global sustainable recovery: Promoting environmental and economic sustainability* (pp. 215–263). Springer International Publishing.

Das, M., & Chatterjee, B. (2015). Ecotourism: A panacea or a predicament?. *Tourism Management Perspectives*, *14*, 3–16.

de Mello, F. P. (2024). Voluntary carbon offset programs in aviation: A systematic literature review. *Transport Policy*, *147*(6), 158–168.

Deb, M., Sharma, V. K., & Panchapakesan, P. (2023). Sustainable practices, mindfulness, tranquility, and well-being: A mixed-method approach. *Journal of Destination Marketing & Management*, *30*, 100816.

Dinkoksung, S., Pitakaso, R., Khonjun, S., Srichok, T., & Nanthasamroeng, N. (2023). Modeling the medical and wellness tourism supply chain for enhanced profitability: An open innovation approach. *Journal of Open Innovation: Technology, Market, and Complexity*, *9*(3), 100137.

Dong, Z., Zhou, Z., Ananzeh, M., Hoang, K. N., Shamansurova, Z., & Luong, T. A. (2024). Exploring the asymmetric association between fintech, clean energy, climate policy, natural resource conservations and environmental quality. A post-COVID perspective from Asian countries. *Resources Policy*, *88*, 104489.

Du Cros, H. (2023). The partnership between tourism and cultural heritage management: Key actions at the international level. In P. L. Yu, T. Lertcharnrit, & G. S. Smith (Eds.), *Heritage and Cultural Heritage Tourism: International Perspectives* (pp. 193–205). Springer International Publishing.

Dwyer, L. (2023). Tourism development and sustainable well-being: A beyond GDP perspective. *Journal of Sustainable Tourism*, *31*(10), 2399–2416.

Elers, C. H., & Dutta, M. (2024). Local government engagement practices and Indigenous interventions: Learning to listen to Indigenous voices. *Human Communication Research*, *50*(1), 39–52.

Elkhwesky, Z., El Manzani, Y., & Elbayoumi Salem, I. (2024). Driving hospitality and tourism to foster sustainable innovation: A systematic review of COVID-19-related studies and practical implications in the digital era. *Tourism and Hospitality Research*, *24*(1), 115–133.

Firman, A., Moslehpour, M., Qiu, R., Lin, P. K., Ismail, T., & Rahman, F. F. (2023). The impact of eco-innovation, ecotourism policy and social media on sustainable tourism development: Evidence from the tourism sector of Indonesia. *Economic Research-Ekonomska Istraživanja*, *36*(2), 2143847.

Gautam, V., & Bhalla, S. (2024). Exploring the relationships among tourism involvement, residents' empowerment, quality of life and their support for sustainable tourism development. *Journal of Cleaner Production*, *434*, 139770.

Glyptou, K. (2024). Operationalising tourism sustainability at the destination level: A systems thinking approach along the SDGs. *Tourism Planning & Development*, *21*(1), 95–121.

Gordon, J. E. (2023). Climate change and geotourism: Impacts, challenges, and opportunities. *Tourism and Hospitality*, *4*(4), 514–538.

Gössling, S., Balas, M., Mayer, M., & Sun, Y. Y. (2023). A review of tourism and climate change mitigation: The scales, scopes, stakeholders and strategies of carbon management. *Tourism Management*, *95*, 104681.

Hertel, T., Elouafi, I., Tanticharoen, M. & Ewert, F. (2021). Diversification for enhanced food systems resilience. *Nature Food*, *2*(11), 832–834.

Higham, J., Font, X., & Wu, J. (2021). Code red for sustainable tourism. *Journal of Sustainable Tourism*, *30*(1), 1–13.

Hoang, S. D., Dey, S. K., Tučková, Z., & Pham, T. P. (2023). Harnessing the power of virtual reality: Enhancing telepresence and inspiring sustainable travel intentions in the tourism industry. *Technology in Society*, *75*, 102378.

Jahanger, A., Hossain, M. R., Awan, A., Adebayo, T. S., & Chishti, M. Z. (2023). Linking tourist's footprint and environmental tragedy through transportation, globalization and energy choice in BIMSTEC region: Directions for a sustainable solution using novel GMM-PVAR approach. *Journal of Environmental Management*, *345*, 118551.

Jatav, S. (2023). Current trends in sustainable tourism in the Indian context. In H. Sezerel & B. Christpiansen (Eds.), *Handbook of Research on Sustainable Tourism and Hotel Operations in Global Hypercompetition* (pp. 391–412). IGI Global.

Joshi, A., Kataria, A., Rastogi, M., Beutell, N. J., Ahmad, S., & Yusoff, Y. M. (2023). Green human resource management in the context of organizational sustainability: A systematic review and research agenda. *Journal of Cleaner Production*, *430*(3), 139713.

Kamb, A., Lundberg, E., Larsson, J., & Nilsson, J. (2021). Potentials for reducing climate impact from tourism transport behavior. *Journal of Sustainable Tourism*, *29*(8), 1365–1382.

Kapoor, D., & Jain, A. (2023). Sustainable tourism and its future research directions: A bibliometric analysis of twenty-five years of research. *Tourism Review*, *79*(3), 541–567.

Katelieva, M., Muhar, A., & Penker, M. (2020). Nature-related knowledge as intangible cultural heritage: Safeguarding and tourism utilisation in Austria. *Journal of Tourism and Cultural Change*, *18*(6), 673–689.

Kefi, H., Besson, E., Zhao, Y., & Farran, S. (2024). Toward museum transformation: From mediation to social mediation and fostering omnivisit experience. *Information & Management*, *61*(1), 103890.

Khizar, H. M. U., Younas, A., Kumar, S., Akbar, A., & Poulova, P. (2023). The progression of sustainable development goals in tourism: A systematic literature review of past achievements and future promises. *Journal of Innovation & Knowledge, 8*(4), 100442.

Kunst, J. R., Lefringhausen, K., & Zagefka, H. (2024). Delineating the boundaries between genuine cultural change and cultural appropriation in majority-group acculturation. *International Journal of Intercultural Relations, 98*, 101911.

Lasisi, T. T., Odei, S. A., & Eluwole, K. K. (2023). Smart destination competitiveness: Underscoring its impact on economic growth. *Journal of Tourism Futures*. https://doi.org/10.1108/JTF-09-2022-0243

Latinovic, Z., & Chatterjee, S. C. (2024). Value co-creation: Balancing B2B platform value and potential reverse-value effects. *Journal of Business Research, 175*, 114518.

Lee, Z., Gordon-Wilson, S., Davies, I., & Pring, C. (2024). Communicating about sustainability in fashion: A construal level theory approach. *European Journal of Marketing, 58*(1), 313–341.

Li, Y., Liu, Y., & Solangi, Y. A. (2023). Analysis of factors and strategies for the implementation of sustainable tourism in a green economic structure in China. *Journal of Cleaner Production, 434*, 140011.

Liburd, J., Menke, B., & Tomej, K. (2023). Activating socio-cultural values for sustainable tourism development in natural protected areas. *Journal of Sustainable Tourism, 32*(8), 1–19.

Lin, C. Y., Li, C., Mahmood, S., Guo, F., & Qian, Z. (2024). Integrating culture and tourism: A resilient scale for digital transformation innovation. *Journal of the Knowledge Economy*, 1–34. https://doi.org/10.1007/s13132-024-01753-y

Lopes, H. S., Remoaldo, P. C., Ribeiro, V., & Martin-Vide, J. (2022). Pathways for adapting tourism to climate change in an urban destination–Evidences based on thermal conditions for the Porto Metropolitan Area (Portugal). *Journal of Environmental Management, 315*, 115161.

Lupoae, O. D., Wilk, V., & Radu, R. I. (2024). Sustainable entrepreneurship in equine services. *Journal of Business Research, 170*, 114361.

Lv, Z., & Shang, W. (2023). Impacts of intelligent transportation systems on energy conservation and emission reduction of transport systems: A comprehensive review. *Green Technologies and Sustainability, 1*(1), 100002.

Ma, X., & Su, W. (2024). Local government intervention in tourism-driven rural gentrification: Types and interpretative framework. *Tourism Management, 100*, 104828.

Manzoor, F., Wei, L., Asif, M., Haq, M. Z. U., & Rehman, H. U. (2019). The contribution of sustainable tourism to economic growth and employment in Pakistan. *International Journal of Environmental Research and Public Health, 16*(19), 3785.

Marcouiller, D. W. (2023). *Advanced introduction to tourism economics*. Edward Elgar Publishing.

Mariam, K., Singh, M., Yaja, M., & Kumar, A. (2024). Negative perception of the local community towards tourism development. *Tourism and Hospitality Management, 30*(1), 15–25.

Martins, L. P., & Ribeiro, S. (2023). The enduring political viability of governance models in tourist destinations and their implications for people and planet. *Worldwide Hospitality and Tourism Themes, 15*(6), 669–676.

Mooney, S., & Baum, T. (2024). Intersectionality in hospitality. In A. Manoharan, J. M. Madera, & M. Singal (Eds.), *The Routledge handbook of diversity, equity, and inclusion management in the hospitality industry* (pp. 140–153). Routledge.

Moorman, T. G., Ditchkoff, S. S., Dunning, K., Mnguni, P., Morris, D., & Ward, C. (2024). Participatory study of policy process for wildlife conservation and ecotourism in Eswatini: A multiple streams analysis. *Environmental Science & Policy, 151*, 103637.

Morrison, A. M., & Buhalis, D. (Eds.). (2023). *Routledge handbook of trends and issues in global tourism supply and demand*. Taylor & Francis.

Nicola-Gavrilă, L. (2023). Empirical approaches regarding interdependency between technology and sustainable tourism. *Journal of Environmental Management and Tourism (JEMT)*, *14*(4 (68), 2140–2147.

Nguyen, D. T., Kuo, K. C., Lu, W. M., & Nhan, D. T. (2023). How sustainable are tourist destinations worldwide? An environmental, economic, and social analysis. *Journal of Hospitality & Tourism Research*, *48*(4), 698–711.

Nygaard, A. (2024). Business ethics, the stakeholder perspective and circular business models. In *Green Marketing and Entrepreneurship* (pp. 15–33). Springer International Publishing.

Ofélia de Queiroz, F. A., Morte, I. B. B., Borges, C. L., Morgado, C. R., & de Medeiros, J. L. (2024). Beyond clean and affordable transition pathways: A review of issues and strategies to sustainable energy supply. *International Journal of Electrical Power & Energy Systems*, *155*, 109544.

Oo, K. T., Jonah, K. A. Z. O. R. A., & Thin, M. M. Z. (2023). A systematic review of the pros and cons of digital pollution and its impact on the environment. *Journal of Sustainability and Environmental Management*, *2*(1), 61–73.

Pan, S. Y., Gao, M., Kim, H., Shah, K. J., Pei, S. L., & Chiang, P. C. (2018). Advances and challenges in sustainable tourism toward a green economy. *Science of the total environment*, *635*, 452–469.

Paraskevas, A., Pantelidis, I., & Ludlow, J. (2022). Duty of care for business travel: How do employers assess and manage business travel risk?. *International Journal of Contemporary Hospitality Management*, *34*(12), 4438–4457.

Paul, I., & Roy, G. (2023). Tourist's engagement in eco-tourism: A review and research agenda. *Journal of Hospitality and Tourism Management*, *54*, 316–328.

Pawalluri, T. (2023). The impact of tourism on social change in Indonesia. *International Journal of Education, Vocational and Social Science*, *2*(04), 104–117.

Pencarelli, T. (2020). The digital revolution in the travel and tourism industry. *Information Technology & Tourism*, *22*(3), 455–476.

Peeters, P., Çakmak, E., & Guiver, J. (2024). Current issues in tourism: Mitigating climate change in sustainable tourism research. *Tourism Management*, *100*, 104820.

Radovanović, D. (2023). Rethinking digital literacy. In *Digital literacy and inclusion: Stories, platforms, communities* (pp. 1–12). Springer International Publishing.

Raihan, A. (2023). A review of the global climate change impacts, adaptation strategies, and mitigation options in the socio-economic and environmental sectors. *Journal of Environmental Science and Economics*, *2*(3), 36–58.

Ranasinghe, J. P. R. C., Ranatunga, R. A. D. D. N., & Pattiyagedara, P. G. S. S. (2021). Operating carbon-neutral tourism: Systematic review & bibliometric analysis. *Journal of Sustainable Tourism and Entrepreneurship*, *3*(2), 139–158.

Rani, P., Sharma, P., & Gupta, I. (2024). Toward a greener future: A survey on sustainable blockchain applications and impact. *Journal of Environmental Management*, *354*, 120273.

Rasoolimanesh, S. M., Ramakrishna, S., Hall, C. M., Esfandiar, K., & Seyfi, S. (2023). A systematic scoping review of sustainable tourism indicators in relation to the sustainable development goals. *Journal of Sustainable Tourism*, *31*(7), 1497–1517.

Ray, A., Sachdeva, I., Rana, N. P., Nunkoo, R., & She, L. (2024). Is the information on green hotel websites aligned with the drivers affecting customers' intention to visit green hotels? A mixed-methods approach. *Journal of Hospitality Marketing & Management*, *33*(1), 1–32.

Rivera, J., Pastor, R., & Gomez Punzon, J. (2024). The impact of the Covid-19 on the perception of DMOs about the sustainability within destinations: A European empirical approach. *Tourism Planning & Development*, *21*(1), 41–69.

Roberto, R., Penna, M., Felici, B., & Rao, M. (2023). Smart working and flexible work arrangements: Opportunities and risks for sustainable communities. *In* P. Droege (Ed.), *Intelligent environments* (pp. 243–283). North-Holland.

Ruiz-Ballesteros, E. (2023). What community for community-based tourism?. *Current Issues in Tourism*, *26*(16), 2664–2677.

Ruiz-Castillo, M., Aragón-Correa, J. A., & Hurtado-Torres, N. E. (2024). Independent directors and environmental innovations: How the visibility of public and private shareholders' environmental activism moderates the influence of board independence. *Business Strategy and the Environment*. *33*(2), 424–440.

Sahni, H., Nsiah, C., & Fayissa, B. (2023). Tourism specialization, growth stage, and economic growth. *Tourism Economics*, *29*(6), 1533–1550.

Sard, M., & Valle, E. (2024). Tourism degrowth: Quantification of its economic impact. *Current Issues in Tourism*, 1–16. https://doi.org/10.1080/13683500.2024.2316201

Sathiyan, S. P., Pratap, C. B., Stonier, A. A., Peter, G., Sherine, A., Praghash, K., & Ganji, V. (2022). Comprehensive assessment of electric vehicle development, deployment, and policy initiatives to reduce GHG emissions: Opportunities and challenges. *IEEE Access*, *10*, 53614–53639.

Shafiee, S., Jahanyan, S., Ghatari, A. R., & Hasanzadeh, A. (2023). Developing sustainable tourism destinations through smart technologies: A system dynamics approach. *Journal of Simulation*, *17*(4), 477–498.

Sheller, M. (2021). Reconstructing tourism in the Caribbean: Connecting pandemic recovery, climate resilience and sustainable tourism through mobility justice. *Journal of Sustainable Tourism*, *29*(9), 1436–1449.

Shrestha, R. K., L'Espoir Decosta, J. P., & Whitford, M. (2024). Indigenous knowledge systems and socio-cultural values for sustainable tourism development: Insights from Indigenous Newars of Nepal. *Journal of Sustainable Tourism*, 1–25. https://doi.org/10.1080/09669582.2024.2316298

Siegel, L. A., Tussyadiah, I., & Scarles, C. (2023). Exploring behaviors of social media-induced tourists and the use of behavioral interventions as salient destination response strategy. *Journal of Destination Marketing & Management*, *27*, 100765.

Silva, L. F., Ribeiro, J. C., & Carballo-Cruz, F. (2023). Towards sustainable tourism development in a small protected area: Mapping stakeholders' perceptions in the Alvão natural park, Portugal. *Tourism Planning & Development*, 1–23. https://doi.org/10.1080/21568316.2023.2274581

Stoelhorst, J. W., & Vishwanathan, P. (2024). Beyond primacy: A stakeholder theory of corporate governance. *Academy of Management Review*, *49*(1), 107–134.

Susila, I., Dean, D., Harismah, K., Priyono, K. D., Setyawan, A. A., & Maulana, H. (2023). Does interconnectivity matter? An integration model of agro-tourism development. *Asia Pacific Management Review*, *29*(1), 104.

Tao, Y., You, S., Zhu, J., & You, F. (2024). Energy, climate, and environmental sustainability of trend toward occupational-dependent hybrid work: Overview, research challenges, and outlook. *Journal of Cleaner Production*, *442*, 141083.

Tkaczynski, A., Rundle-Thiele, S., & Truong, V. D. (2020). Influencing tourists' pro-environmental behaviours: A social marketing application. *Tourism Management Perspectives*, *36*, 100740.

Tomej, K., & Liburd, J. J. (2019). Sustainable accessibility in rural destinations: A public transport network approach. *Journal of Sustainable Tourism*, *28*(2), 222–239.

UN (2015). Transforming Our World: The 2030 Agenda for Sustainable Development. Resolution Adopted by the General Assembly on 25 September 2015, 42809, 1–13.

Varnajot, A., Makanse, Y., Huijbens, E. H., & Lamers, M. (2024). Toward antarctification? Tourism and place-making in Antarctica. *Polar Geography*, *47*(1), 49–70.

Vu, K., & Hartley, K. (2022). Drivers of growth and catch-up in the tourism sector of industrialized economies. *Journal of Travel Research*, *61*(5), 1156–1172.

Wolters, T. (2023). From green growth to post-growth approaches. In *Sustainable value creation: An inevitable challenge to business and society* (pp. 125–150). Springer Nature Switzerland.

Wang, J., & Brown, G. (2024). A critical analysis of volunteer tourism in post-socialist China. *Current Issues in Tourism*, 1–16. https://doi.org/10.1080/13683500.2024.2309143

Wang, H., & Guo, J. (2024). New way out of efficiency-equity dilemma: Digital technology empowerment for local government environmental governance. *Technological Forecasting and Social Change*, *200*, 123184.

Wani, M. D., Dada, Z. A., & Shah, S. A. (2024). The impact of community empowerment on sustainable tourism development and the mediation effect of local support: A structural equation Modeling approach. *Community Development*, *55*(1), 50–66.

Waris, I., Suki, N. M., Ahmed, A., & Barkat, W. (2024). Environmental corporate social responsibility initiatives and green customer citizenship behavior in tourism industry: The mediating roles of green trust, customer-company identification and green corporate image. *Social Responsibility Journal*. *20*(6), 1138–1156.

Wu, Z., & Ma, J. (2024). Chinese rural residents' identity construction with tourism intervention. *Tourism Management Perspectives*, *51*, 101218.

Xiao, X., Li, P., & Seekamp, E. (2024). Sustainable adaptation planning for cultural heritage in coastal tourism destinations under climate change: A mixed-paradigm of preservation and conservation optimization. *Journal of Travel Research*, *63*(1), 215–233.

Xu, Y., & Zhao, F. (2023). Impact of energy depletion, human development, and income distribution on natural resource sustainability. *Resources Policy*, *83*, 103531.

Xu, J., & Ho, P. S. (2024). Whether educational tourism can help destination marketing? An investigation of college students' study tour experiences, destination associations and revisit intentions. *Tourism Recreation Research*, *49*(1), 131–146.

Zhu, J., & Zhu, Z. (2023). The space-time evolution and driving mechanism of coordinated development of modern logistics industry and tourism industry. *Journal of Cleaner Production*, *423*, 138620.

Chapter 6

Carbon Footprint of Tourism Activities Including Transportation, Accommodation, and Infrastructure: A Critical Analysis

Noreen Kanwal, Chong Soon Meng, Motoki Watabe, Nasaruddin Zainal and Jolyne Khor Kuan Siew

Department of Management, Sunway Business School (SBS), Sunway University, Selangor Darul Ehsan, Malaysia

Abstract

The thriving global tourism industry is a pivotal player in the global economy. It significantly contributes to a substantial carbon footprint through emissions from transportation, lodging, and infrastructure. In this study we have conducted a critical analysis of carbon emissions within the tourism industry using previous academic literature and institutional databases. Addressing the importance of measuring carbon footprints, the study delves into the definition of the carbon footprint in this sector, emphasizing the urgent need to comprehend and address its effects. This investigation also centers on the aspect of transportation, which is contributing to carbon emissions in the tourism industry. It explores variations based on transportation modes and discusses mitigation strategies, such as green transit. It also concentrates on accommodation, scrutinizing its environmental impact by evaluating carbon emissions from energy consumption, water usage, and waste generation in hotels and resorts. Emphasis is placed on promoting sustainable and green lodging facilities. Furthermore, this study examines the environmental impact of the tourism industry. Specifically, it examines the carbon footprint associated with infrastructure development. The analysis includes the impact of construction and development, resource consumption, and the role of communities and stakeholders in sustainable tourism planning. Furthermore, an assessment is conducted to

The Need for Sustainable Tourism in an Era of Global Climate Change: Pathway to a Greener Future, 95–120

doi:10.1108/978-1-83608-668-020241015

examine the obstacles associated with adopting sustainable practices and assess the effectiveness of implemented mitigation strategies within the tourism industry. The concluding section outlines future perspectives and recommendations for research.

Keywords: Carbon footprint; green tourism; green infrastructure; net zero; sustainability

1. Introduction

In recent times, the concept of "carbon footprint" has garnered significant focus, primarily driven by concerns related to climate change (Wiedmann & Minx, 2008). According to Huang and Tang (2021), carbon footprint refers to "all relevant sources, sinks, and storage within the spatial and temporal boundaries of a population, system or activity." Carbon footprint is characterized as a quantification of the comprehensive volume of carbon dioxide emissions directly and indirectly attributed to an activity or accumulated across the life stages of a product (University of Michigan). According to Brainy Insights (2023), the projected trajectory of the global tourism market indicates an anticipated expansion to USD 16.9 Trillion by the year 2030. This growth is expected to occur at a Compound Annual Growth Rate (CAGR) of 5.4% within the forecast period spanning from 2022 to 2030. Notably, the Asia Pacific region is anticipated to exhibit the highest growth rate during this forecast period. Tourism has not only made a significant contribution to promoting social progress but has also been an important reagent for expanding socioeconomic and cultural interactions in human history. Numerous developing and developed countries have embraced tourism as a viable avenue for achieving sustainable national development. The global recognition of tourism's pivotal role in fostering economic growth is evident (Gwenhure & Odhiambo, 2017). This acknowledgement is emphasized by continuous and substantial global investments devoted to its development (Brida et al., 2020). This reinforces tourism's standing as a priority for numerous nations, reaffirmed annually.

Since 1960s, tourism has evolved from a modest undertaking into not only one of the world's largest industries but also the fastest-growing economic sector worldwide (Sofronov, 2018). This transformation highlights the substantial growth and impact of tourism over the years. It plays a pivotal role in the prosperity of numerous economies worldwide, offering various advantages to host destinations. Tourism plays a crucial role in boosting economic revenue. It creates many job opportunities, improves a country's infrastructure, and promotes cultural exchange between visitors and residents (Garsous et al., 2017; Seyfi et al., 2020). The employment impact of tourism extends beyond the tourism sector itself, reaching into diverse areas such as agriculture, communication, healthcare, and education. Many jobs are created across these sectors due to tourism's influence. Numerous travelers seek to immerse themselves in the culture, traditions, and culinary experiences of the host destination (Wondirad et al., 2021). It leads to profitable outcomes for local restaurants, shopping centers, and stores (Shah & Shende, 2017).

However, in 2020, when COVID-19 affected the world, the tourism industry was severely impacted (Duro et al., 2021). As per a report, commencing in March 2020, approximately 96% of global destinations curtailed travel in response to the pandemic, resulting in a virtual cessation of tourism (Gossling et al., 2020). Yet, 3 years post COVID-19, we can see a positive aspect for this industry. As the World Economic Forum (WEF) indicated on their website that by the end of 2023, international tourism is poised to regain 90% of its pre-pandemic levels.

The latest information from the World Tourism Organization (UNWTO) specifies that around 975 million international tourists set out on journeys from January to September 2023, marking a notable 38% surge compared to the same months in 2022. According to UNWTO Secretary-General Zurab Pololikashvili, the most recent data from UNWTO indicates that international tourism has fully recovered from the unprecedented challenges posed by COVID-19. Many destinations have not only reached but have even surpassed pre-pandemic levels in terms of arrivals and receipts. This recovery is of utmost importance for destinations, businesses, and communities that rely heavily on the tourism sector as a vital lifeline.

After three years of reduced greenhouse gas (GHG) emissions, the world will once again be confronted with more of these emissions. Tourism significantly contributes to global GHG emissions, comprising approximately 8% of the world's total emissions (Khatib, 2023). The surge in vacation activities is complicating global efforts to mitigate climate change, as highlighted by many scientists (Gössling et al., 2023).

A study was conducted in the University of Sydney, in which 160 nations were examined. This study unveiled tourism-related emissions, primarily propelled by domestic travelers. It finds that domestic tourism-related emissions were more prominent in the United States, China, Germany, and India (Lenzen et al., 2018). The findings further indicated that in 2013, the tourism sector, encompassing elements such as flights, hotels, food, and souvenir production, discharged the equivalent of 4.5 billion tons of carbon dioxide – constituting 8 percent of all anthropogenic GHGs. This represented a noteworthy increase from 3.9 billion tons recorded in 2009 (Lenzen et al., 2018).

Contrary to earlier approximations indicating that tourism contributed merely 2.5 to 3.0 percent of global emissions, this more comprehensive assessment reveals a more pronounced impact (Doyle, 2018). If current patterns persist, tourism is expected to release 6.5 billion tons of carbon emissions by 2025, which indicates the tourism sector's solidifying position as one of the rapidly growing sources of planet-warming gases. In this regard, governments are actively working to diminish these warming gases (Lenzen et al., 2018). This study, conducted by scientists from Australia, Taiwan, and Indonesia, identifies flights as the principal and most significant contributor to emissions linked with tourism.

Yang et al. (2021) delineated five primary sectors within the travel and tourism domain that play a role in carbon emissions: transportation, accommodation, and tourism activities. These sectors exert both direct and indirect influences on emissions. In line with findings by Tang et al. (2018), the transport sector stands out as the major contributor, accounting for 75% of CO_2 emissions in the industry, while lodgings and catering contribute 21%, and tourism activities contribute 4%. Additionally, the OECD (2010) assessed the overall contribution of the tourism industry to global GHG emissions at 5.3%, with transport making up a substantial 75% of this total.

Direct carbon emissions arise from specific sources such as transportation using fuel and energy, and the restaurant and accommodation sectors relying on electricity and gas. On the other side, indirect carbon emissions result from broader processes, including manufacturing, infrastructure development, and the operations of airline offices and travel agents indirectly associated with tourism (Ma et al., 2022). Thus, it is essential to consider and address both direct and indirect carbon emissions when evaluating the carbon footprint of the tourism industry, recognizing their collective impact on global warming.

Holden (2016) pointed out that the association between tourism and the environment is complex due to the potential harm that many tourism activities can cause to nature, posing a risk of destruction. The construction of infrastructure, including roads and airports, and the establishment of tourism-related facilities such as resorts, hotels, restaurants, shops, golf courses, and marinas, exemplify how tourism can have adverse effects on the environment.

The current landscape of sustainable tourism is characterized by numerous small entities and individual advisors competing for limited resources, managing modest, underfunded projects that result in only marginal achievements (Jenkins & Wearing, 2003). In both the public and private sectors, the emphasis is on planning and reporting rather than making substantial changes in organizational or operational practices (Kanwal et al., 2023; Streimikiene et al., 2021). Unfortunately, the hindrance to the implementation and continuous dedication to sustainability initiatives stems primarily from the perceived scarcity of resources.

This circumstance initiates heightened competition for constrained resources, resulting in a dearth of genuine collaboration (Streimikiene et al., 2021). Consequently, no singular organization attains success at a proximate level commensurate with the requisite and feasible impact. Despite authentic intentions, the aggregate initiatives fall short of the threshold necessary for substantial advancements in the realm of sustainable tourism.

This study evaluates the carbon footprint of the global tourism industry, focusing on transportation, accommodation, and infrastructure. It quantifies emissions, defines the carbon footprint, and explores mitigation strategies such as green transit and sustainable lodging. It identifies obstacles to sustainability, evaluates mitigation effectiveness, and concludes with future approaches and research recommendations. The study aims to promote understanding and address the environmental impacts of tourism-related carbon emissions.

Additionally, tourism holds the promise of positively impacting the environment by facilitating the implementation of protective and conservation measures.

The establishment of sustainable practices requires collaborative endeavors involving various stakeholders, such as the tourism industry, governments, and local communities (Popescu & Popescu, 2019). This collective effort is crucial for alleviating the adverse environmental consequences of tourism and fostering a more sustainable coexistence between industry and the natural environment.

A thriving and sustainable tourism industry can contribute to socio-economic development, benefiting communities (Pranita et al., 2022). It is essential to safeguard the environment and the constructed infrastructure for the sector's well-being and the overall health of communities (Kanwal & Isha, 2022). Hence, global initiatives to promote low-carbon alternatives are essential to ensure the enduring sustainability of tourism as a key industry.

2. Carbon Footprints in Transportation

2.1. Overview of Transportation-Related Emissions in Tourism

Transportation-related emissions in tourism contribute significantly to global GHG emissions. The tourism industry heavily relies on various modes of transportation, including air, road, rail, and maritime, all of which contribute to carbon dioxide (CO_2) emissions (Mishra et al., 2020). Addressing transportation-related emissions in tourism requires a multi-faceted approach. This includes promoting sustainable transport options, improving energy efficiency in vehicles, adopting alternative fuels, and encouraging behavioral changes among tourists and industry stakeholders (Streimikiene et al., 2021). Furthermore, investments in public transportation infrastructure and the development of low-carbon transport systems can contribute to reducing emissions in the tourism sector.

2.2. Statistical Data on the Contribution of Transportation to the Carbon Footprint

The influx of tourists to different tourist destinations has a significant impact on the environment. Notably, air travel constitutes 12% of tourism's global CO_2 footprint (Lenzen et al., 2018). The tourism sector has emerged as one of the major contributors to world pollution, approximately 8% of global CO_2-equivalent emissions in 2013 (Lenzen et al., 2018). Multiple researchers propose that the combined impact of transportation, accommodation, and activities constitutes approximately 5% of global direct energy use and emissions (Gössling, 2002). In 2016, transport-related CO_2 emissions, encompassing passenger and freight transport, reached 7,230 million tons globally, constituting 23% of all human-caused CO_2 emissions. Of this total, 64% (4,650 million tons) emanated from passenger transport. An analysis by Lenzen et al. (2018) suggests tourism sector contributed an alarming 8% of overall global warming. Here is a summary of the statistical data of the transport sector, which shows transports contribution to the carbon footprint based on region and transport types.

(i) In the United States, the transportation sector emerges as the foremost contributor to carbon dioxide (CO_2) emissions, even more than the electric power sector. Motor vehicles, encompassing both cars and trucks, constitute approximately 83% of the CO_2 emissions emanating from the transportation sector (CBO, 2023).

(ii) Three-quarters of global transportation emissions are attributed to road travel. Passenger vehicles, for instance, cars and buses, contribute around 45.1% of these emissions (WorldinData, 2023).

(iii) Aviation, while accounting for only 2.5% of global CO_2 emissions, has a significant impact on individual contributions to climate change due to air travel's prevalence. The overall contribution of aviation to climate change is higher when considering other gases and pollutants, resulting in a net warming effect (EPA, 2023; WorldinData, 2023).

Efforts to reduce emissions in the transportation sector, such as increasing vehicle efficiency, adopting electric vehicles, and implementing stricter fuel economy and emissions standards, are crucial in mitigating climate change and reducing the carbon footprint associated with transportation (CBO, 2023; EPA, 2023; Gössling, 2002).

2.3. Comparison of different modes of transportation based on carbon emission

(i) *Air Travel*: airplane found as a mode of traveling with highest carbon footprint because of significant amount of fuel is required for its functionality. On average air travel emits around 285 g (about 10.05 oz) of CO_2 per passenger kilometer (STO, 2023) as shown in Fig. 6.1. Various researchers suggested that most of emission is identified in developing countries because most of tourist's arrival in the developing countries through air traveling mode (Gössling, 2002). It is also projected that interregional air traveling will increase to 96.5% by 2030, and it will consequently increase carbon emission (Velthof et al., 2016). Aviation plays a pivotal role in the rise of emissions within the tourism sector, experiencing a notable surge at the rate of 6.8 from 1960 to 2018 (Lee et al., 2021). Consequently, it is imperative to tackle the carbon footprint linked to air travel to attain sustainable development in tourism. Various stakeholders such as policymakers, industry stakeholders, and travelers are encouraged to join in the effort to minimize the pollution from aviation industry. This joint effort will ensure that future growth in tourism is sustainable and ecologically responsible. However, advancements in aircraft technology and the use of sustainable aviation fuels (SAFs) are being explored to reduce emissions in the aviation industry.

(ii) *Road Travel*: road transport is essential to the functioning of the tourism sector. According to UNWTO (2019), in 2016, it is estimated that 41% of international tourists arrived at their destinations through road transport. Passenger cars emit around 120 to 150 g (about 5.29 oz) of CO_2 per kilometer traveled (STO, 2023), shown in Fig. 6.1. However, road transportation carbon emission varies. The carbon emissions from road travel depend on factors such as vehicle type, fuel efficiency, and driving conditions depending on the fuel type, whereas electric vehicles produce zero carbon. According to sustainable travel organization among road vehicles, SUV, and cars are ranked as first and second largest contributors to carbon emission, while most of the tourists (30%) prefer cars, 7% travel by buses, and 3% by rail (STO, 2023). According to Giannakis et al. (2020), the largest proportion of emissions are caused by land transport. However, among land transportation electric vehicles are ecofriendly, which have the potential to net zero carbon emission. Replacing traditional fossil combustion vehicles with electric vehicles is a positive change toward the environment. Such a move would align with global efforts to minimize the impact of road transport on the environment. The latest trend of adopting electric vehicles is changing the landscape of road transport. These advancements are critical to achieve zero carbon emission goal in the tourism sector.

(iii) *Rail Transport*: rail travel is considered to have lower carbon emissions than air and road transportation (Graver & Frey, 2016; WorldinData, 2023). However, diesel fueled trains emit 40 to 80 g of CO_2 per passenger per-kilometer, electric powered trains emit zero carbon (WorldinData, 2023), represented in Fig. 6.1. Overall, it also matters how electricity is being produced. If electricity is being produced by sustainable energy sources like thermal or solar, only then carbon emission can be controlled.

(iv) *Maritime Travel*: maritime travel, particularly cruise ships, can contribute significantly to carbon emissions. The emissions per passenger-kilometer can vary widely depending on the ship's size, speed, and fuel type. On average, cruise ships emit around 250 to 2200 g (about 4.85 lb.) of CO_2 per passenger-kilometer (Howitt et al., 2010), depicted in Fig. 6.1. In 2019, the cruise industry carried 30 million passengers (about the population of Texas) and the industry is projected to rebound to pre-COVID levels and could achieve

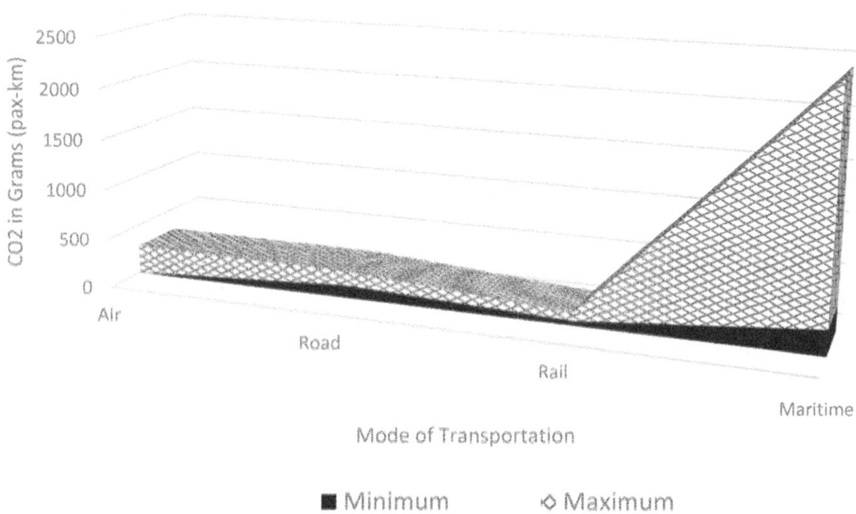

Fig. 6.1. Carbon Emission in Transportation Sector.

a significant growth to 12% above pre-COVID level by 2026 (Garzon et al., 2022; Palmer, 2022), which shows CO_2 emission will increase with the increase in the number of ship passengers. The cruise industry has faced criticism for their negative impact on the environment and residence around the ports (Simonsen et al., 2019). Many large cruise liners are powered by giant diesel engines and could emit copious amounts of pollutants into the environment. Despite being heavy carbon emitters, Lloret-Lloret et al. (2021) reveal alarming statistics about cruise ships' carbon footprint. A solitary large cruise ship has the potential to exceed the carbon emissions produced by more than 12,000 cars, and passengers embarking on Antarctic cruises generate a comparable amount of CO_2 during a seven-day voyage as the average European does throughout an entire year (Lloret-Lloret et al., 2021). The cruise industry appears as a substantial generator of waste, with estimates by Palmer (2022) proposing that a sizable cruise ship could generate as much as a ton of rubbish daily. Environmental organizations, including Friends of the Earth (FoE), have assumed a critical stance toward the cruise industry, releasing annual Cruise Ship Report Cards. These report cards assess cruise lines based on many factors such as sewage treatment, reduction in air pollution, compliance with water quality standards, and transparency. As highlighted in FoE's 2021 report, "Regrettably, everything that cruise ships come into contact with is likely to be adversely affected along their journey. The report underscores the detrimental impact on air quality, water resources, fragile habitats, coastal communities, and wildlife," signaling the urgent need for all stakeholders to address this issue promptly. This collective effort is imperative to mitigate pollution and safeguard the well-being of both ecosystems and communities affected by maritime tourism.

It is crucial to acknowledge that these figures are estimations and may fluctuate depending on many factors, including the efficiency of the vehicles, occupancy rates, and the distance covered. Also, transportation emissions' impact on climate change extends beyond CO_2, with other GHGs and pollutants also playing a role. Efforts to reduce transportation emissions include improving fuel efficiency, transitioning to electric and hybrid vehicles, promoting public transportation, and cycling, and exploring alternative fuels such as biofuels and hydrogen. These measures are crucial in mitigating the environmental impact of transportation.

2.4. Trends and innovations in sustainable transportation within the tourism industry

The tourism industry is witnessing several trends and innovations in sustainable transportation, aiming to control carbon emissions and foster environmentally friendly travel options. Some notable trends and innovations include:

(i) *Electric Vehicles (EVs)*: the adoption of electric vehicles is increasing, both in personal transportation and in the tourism industry. Electric cars, buses, and even electric bikes are being used as sustainable modes of transportation. The development of charging infrastructure and incentives for EV adoption further support this trend (Higueras-Castillo et al., 2021).

(ii) *Bike-Sharing and E-Scooters*: many cities and tourist destinations are implementing bike-sharing programs and e-scooter rentals. These options provide convenient and eco-friendly transportation for short distances (Flores & Jansson, 2021), reducing the reliance on fossil fuel-powered vehicles.

(iii) *Public Transportation Improvements*: investments in public transportation systems, such as buses, trams, and rails, are being made to improve accessibility and reduce congestion. Efficient public transportation networks encourage tourists to use sustainable modes of travel while exploring destinations.

(iv) *Sustainable Aviation*: the aviation industry is exploring initiatives to reduce carbon footprint. This includes the use of SAFs, advancements in aircraft technology for better fuel efficiency, and the adoption of carbon offset programs to compensate for emissions.

(v) *Green Infrastructure*: the development of green infrastructure, such as dedicated bike lanes and pedestrian-friendly spaces, promotes sustainable modes of transportation within tourist destinations. These initiatives encourage walking and cycling, reducing the need for motorized transportation.

(vi) *Multi-Modal Transportation System*: combined transportation systems that integrate different modes of traveling, such as buses, trains, and bicycles, are being promoted. This allows travelers to switch between modes seamlessly, minimizing the use of private vehicles and reducing emissions.

(vii) *Smart Mobility Solutions*: technology-driven solutions are emerging to optimize transportation efficiency and reduce congestion. These include real-time transportation apps, ride-sharing services, and smart traffic management systems, enabling travelers to make informed choices and reduce unnecessary travel.

(viii) *Sustainable Tourist Transport Operators*: tour companies and transportation providers are adopting sustainable practices, for instance, using fuel-efficient vehicles, implementing waste depletion measures, and supporting local communities. Certifications and eco-labels help identify and promote these sustainable operators.

These trends and innovations in sustainable transportation within the tourism industry contribute to reducing carbon emissions, preserving natural resources, and enhancing the overall travel experience for tourists. By embracing these practices, destinations can attract environmentally conscious travelers and promote a greener and more sustainable tourism sector.

Examples of eco-friendly transport initiatives are as follows:

(i) *Electric Vehicles (EVs)*: many destinations and transportation companies are promoting the use of electric vehicles, which produce zero tailpipe emissions. This includes electric cars, buses, and even electric bikes and scooters. EVs can help reduce carbon emissions and improve air quality in urban areas.

(ii) *Bike-sharing and Bike-friendly infrastructure*: Bike-sharing programs and the development of bike-friendly infrastructure, such as dedicated cycling lanes and bike parking facilities, encourage tourists and locals to use bicycles as a sustainable mode of transportation. This reduces reliance on motorized vehicles and promotes active and eco-friendly travel.

(iii) *Pedestrianization*: cities and tourist destinations are implementing pedestrian-friendly initiatives by creating car-free zones or pedestrian-only streets. This allows visitors to explore on foot, reducing vehicle emissions and improving the overall walking experience.

(iv) *Public Transportation Improvements*: investments in public transportation systems, such as buses, trams, and trains, can help reduce congestion and encourage more people to use public transport instead of private vehicles. Improvements may include increased frequency, expanded coverage, and the use of clean and energy-efficient vehicles.

(v) *Carpooling And Ridesharing*: carpooling and ride-sharing services promote vehicle sharing, reducing the number of cars on the road, and decreasing emissions. Platforms and apps connecting drivers and passengers are becoming increasingly popular, making it easier to coordinate shared rides.

(vi) *Sustainable Aviation Initiatives*: the aviation industry is exploring and implementing various eco-friendly initiatives. Such as SAFs, more efficient aircraft designs, and carbon offset programs. These measures aim to reduce the carbon footprint of air travel.

These eco-friendly transport initiatives contribute to reducing carbon emissions, promoting sustainable travel options, and improving the overall environmental impact of transportation in the tourism industry.

2.5. Challenges and Opportunities in Implementing Sustainable Transportation

The implementation of sustainable transportation in the tourism industry has numerous challenges and opportunities, as stated below.

Challenges:

(i) *Infrastructure and Investment*: it requires significant investment and coordination among different stakeholders to build the necessary infrastructure for sustainable transportation. Such as public transport systems, cycling lanes, and charging stations for electric vehicles,

(ii) *Behavioral Change*: encouraging travelers and local communities to adopt sustainable transportation practices can be challenging. Many people are accustomed to private vehicle use and may be resistant to change. Education and awareness campaigns are needed to promote the benefits of sustainable transportation.

(iii) *Limited Accessibility*: sustainable transportation options may not be readily available or accessible in all destinations. Rural areas and remote locations often lack public transport infrastructure, making it difficult for travelers to choose sustainable options.

Opportunities:

(i) *Economic Benefits*: implementing sustainable transportation can create new job opportunities, stimulate local economies, and attract environmentally conscious tourists who prefer sustainable travel options.
(ii) *Innovation and Technological Advancements*: advancements in technology, such as electric vehicles, smart mobility solutions, and renewable energy sources, provide opportunities for more sustainable transportation options. Embracing these innovations can lead to reduced emissions and improved efficiency.
(iii) *Collaboration and Partnerships*: collaboration among governments, transport operators, tourism industry stakeholders, and local communities is crucial for successful implementation. Public–private partnerships can drive investment and create synergies to overcome challenges and implement sustainable transportation solutions.

By addressing these problems and leveraging the opportunities, the tourism industry can make considerable progress in implementing sustainable transportation, reducing carbon emissions, and promoting environmentally friendly travel practices.

2.6. Case studies illustrating successful carbon footprint reduction initiatives in transportation

This exploration delves into two distinctive case studies, each illuminating innovative strategies employed by global cities to address carbon emissions within the realm of tourism transportation. In Case Study 1, we venture to Copenhagen, Denmark, renowned for its pioneering cycling infrastructure and bike-sharing programs (Taylor, 2018). This Nordic city has become a paradigm of sustainable urban mobility, exemplifying the positive impact of prioritizing cycling as a mode of transportation on carbon emission reduction. Transitioning to Case Study 2, we traverse to Bogotá, Colombia, where the implementation of the Bus Rapid Transit (BRT) system, TransMilenio, stands as a testament to effective public transportation initiatives (Urban Sustainable Exchange Metropolis, Organization). This case underscores the significant reduction in carbon emissions achieved through a strategic and efficient mass transit system, offering valuable insights for cities grappling with the dual challenges of tourism growth and environmental sustainability. Together, these case studies contribute valuable lessons to the ongoing discourse on mitigating carbon emissions in the realm of tourism transportation.

Case Study 1: Copenhagen, Denmark – Cycling Infrastructure and Bike-Sharing Programs.

Copenhagen is renowned for its cycling culture and has successfully implemented initiatives to reduce carbon emissions in transportation (Taylor, 2018).

Initiatives:

(i) *Extensive Cycling Infrastructure*: the city has developed an extensive network of dedicated cycling lanes, ensuring safe and convenient cycling routes throughout the city. This infrastructure encourages residents and tourists to choose cycling as a sustainable mode of transportation.

(ii) *Bike-Sharing Programs*: copenhagen introduced bike-sharing programs, such as the popular "Bycyklen" system, which provides easy access to bicycles for short trips. These programs promote sustainable transportation by making bikes readily available and reducing reliance on cars.

Results:

(i) *Increased Cycling Usage*: these initiatives have led to a significant increase in cycling as a mode of transportation. Around 62% of Copenhagen residents commute to work or education by bike, resulting in reduced carbon emissions and improved air quality.

(ii) *Improved Health and Well-Being*: the emphasis on cycling has also led to improved public health outcomes, as regular exercise through cycling has numerous health benefits. This comprehensive approach to sustainable transportation has had positive impacts on the well-being of the city's residents.

Case Study 2: Bogotá, Colombia – Bus Rapid Transit System: Bogotá implemented a successful Bus Rapid Transit (BRT) system known as TransMilenio, which has significantly reduced carbon emissions and improved transportation efficiency (Urban Sustainable Exchange Metropolis, Organization).

Initiatives:

(i) *Dedicated Bus Lanes*: transMilenio established dedicated bus lanes, separate from regular traffic, allowing buses to move quickly and efficiently. This helps reduce congestion and travel times.

(ii) *Integrated Fare System*: the system introduced an integrated fare system, allowing passengers to transfer seamlessly between buses without additional charges. This encourages the use of public transportation and reduces the reliance on private vehicles.

Results:

(i) *Reduced Emissions*: by providing an efficient alternative to private vehicles, TransMilenio has reduced traffic congestion and the associated carbon emissions. It is estimated that the BRT system saves around 3.6 million tons of CO_2 emissions annually.

(ii) *Improved Mobility*: the BRT system has improved mobility for residents, reducing travel times and increasing accessibility to various parts of the city. This has had positive social and economic impacts on the community.

In summary, the case studies of Copenhagen and Bogotá exemplify the efficacy of sustainable transportation solutions in reducing carbon emissions associated with tourism. These case studies highlight how sustainable transportation initiatives, such as cycling infrastructure and bus rapid transit systems, can successfully reduce carbon emissions, improve transportation efficiency, and enhance the quality of life for residents and visitors. These models provide valuable insights for cities worldwide seeking practical and impactful approaches to address environmental concerns within the realm of tourism transportation.

3. Carbon Footprint in Accommodation

Accommodation establishments play a crucial role in the tourism industry, and their numbers have consistently risen alongside the expansion of both international and domestic tourism. One estimate suggests that in 2018, the global count of beds in commercial accommodation establishments reached 119 million (Gössling & Peeters, 2015). Disruptive innovations within the platform economy, exemplified by services like Airbnb, alongside a sustained interest in second homes, constitute significant additional capacity (Zervas et al., 2017). Accommodation constitutes a sizable portion of GHG emissions in the tourism sector. According to a joint report by UNWTO, UNEP, and WMO in 2008, the sector is accountable for 21% of CO_2 emissions when considering transportation, accommodation, and activities collectively. On a global scale, it is estimated that each guest night results in 14 kg CO_2 emissions, primarily attributable to direct emissions related to energy consumption, excluding gastronomy (Gössling & Peeters, 2015). But the literature presents a considerable difference in the documented range of emissions, spanning from 0.1 kg CO_2 to 260 kg (about 573.2 lb.) CO_2 per guest night, depending upon the type of accommodation. According to Lenzen et al. (2018) calculation, global accommodation presently contributes approximately 282 million metric tons of CO_2 to annual GHG emissions. This calculation is derived from input–output analysis and excludes gastronomy.

The carbon dioxide (CO_2) emissions associated with hotel rooms are typically linked to various operational activities and energy usage within the hotel. For instance, energy consumption, water heating, water usage, waste management, building design and construction, operational practices, and guests' activities. Energy consumption is one of the main contributors to CO_2 in the hotel rooms in the form of Heating, Ventilation, and Air Conditioning (HVAC) Systems. Energy consumption is related to other activities such as water heating, lighting, temperature maintenance in swimming pools and spas, and laundry operations. Water usage is another critical factor influencing the carbon footprint of tourist accommodations. The hospitality industry, by nature, requires substantial water for various purposes, including guest amenities, laundry, and landscaping. The energy-intensive processes of water treatment and heating contribute to the overall carbon emissions associated with water consumption. Implementing water conservation measures, such as low-flow fixtures, rainwater harvesting, and efficient irrigation systems, not only reduces water usage but also contributes to a lower carbon footprint.

Other than that food and waste generation also causes CO_2 (Jeswani et al., 2021). Waste generation within accommodations poses a dual challenge in terms of environmental impact. The disposal of waste, particularly if sent to landfills, contributes to emissions through transportation and the decomposition of organic matter. Additionally, the production and disposal of single-use items, common in the hospitality industry, contribute to the overall carbon footprint. Adopting circular economy principles, such as recycling and upcycling, can help reduce the carbon impact of waste generated by accommodations (Sharma et al., 2021).

To effectively manage carbon emissions, it is essential to initially measure the emission rates associated with each factor contributing to carbon emissions in tourist accommodations. Currently, there are four methods employed for calculating carbon footprints. The first method, the life cycle assessment (LCA), employs a "top-down" calculation model. This model encompasses the entire lifecycle of a product or service, obtaining total carbon emissions by collecting input and output data throughout its existence. The resulting "carbon footprint" (CFP) serves as an indirect indicator of the environmental friendliness of the product. For instance, Salehi et al. (2021) employed a life cycle energy analysis (LCEA) methodology to evaluate the energy and carbon performance of six luxury five-star hotels. The findings of the energy and carbon assessment for luxury hotels in Iran were juxtaposed with the values reported in previous hotel research. The current investigation reveals that luxury hotels in Iran exhibit energy intensity up to three to four times higher and carbon intensity seven times greater than comparable hotels examined in earlier research (study in detail can be accessed through the link).

The second method involves calculating carbon emissions from energy and fossil fuels, following the guidelines set off by the Intergovernmental Panel on Climate Change (IPCC). The IPCC method, a GHG inventory guide established by the United Nations Climate Change Commission, comprehensively incorporates GHG emissions into the calculation process. It employs different carbon footprint calculation methods for various sectors, with the most used formula being carbon emissions = activity data × emission factor. For example, Abeydeera and Karunasena (2019) measured carbon emission of five Sri Lankan hotels through IPCC. The results revealed each hotel released more than 7000 tons of carbon annually. Electricity and heat were identified as the major contributors to carbon emission.

The third method is the input–output method (IO), characterized by a top-to-bottom calculation approach utilizing input and output for its computations. For instance, Xia et al. (2022) quantified carbon emission in Gansu China through IO method. The findings reveal that "(1) carbon emissions from tourism hotels, primarily attributed to the supply chain, are the most significant within the tourism sector, notably emanating indirectly from the production of food and tobacco; (2) the eco-efficiency of tourism sectors in Gansu follows a U-shaped curve, aligning with Kuznets' theory; and (3) enhancing the eco-efficiency of tourism sectors is contingent on advancements in energy technology." However, the IO method gets criticism because of its measurement inaccuracy. According to Towa et al. (2020), the results of this method may lack precision, it is commonly employed for macro-level calculations.

The fourth method is the Kaya carbon emission identity. Kaya's approach utilizes a straightforward mathematical formula to connect economic, policy, and population factors with the carbon dioxide produced by human activities. This method serves as a holistic framework linking essential elements influencing carbon emissions. Which shows that with the passage of time carbon emission is increasing. This further indicates the absence of adopting alternative technological measures to reduce GHG emission.

A case study by Caribbean hotel energy efficiency action program 2012, described electricity consumption as one of the major GHG emission factors and concluded that hotels with less rooms consume more electricity than hotels with more rooms. Further it was identified that air conditioning is the largest factor of GHG emission in hotel accommodations. As per Sustainable Travel Organization, accommodations with contemporary amenities, particularly resorts and hotels, typically exhibit higher emissions, whereas smaller lodgings like homestays and guest houses tend to have lower carbon footprints.

Numerous studies have explored possibilities for engaging guests in energy conservation. For instance, considerable attention has been directed toward towel and bed-linen exchanges, given their pivotal roles in accommodation, their inherent energy intensity, and their connections to guest expectations and their (un)willingness to forgo perceived benefits (Gössling et al., 2019). Similarly, there has been significant scrutiny of food consumption patterns as highly pertinent subsector of tourism-related GHG emissions, wherein restaurant offerings assume crucial significance (Juvan et al., 2018). Hence, different researchers recommended different methods to measure carbon emission. In this regard, further research is required to suggest an appropriate method to calculate carbon emission.

3.1. Sustainable Practices in the Hotel and Hospitality Industry

The hotel and hospitality industry can adopt several sustainable practices to minimize its environmental impact and promote responsible tourism. Some of these practices include:

(i) *Energy Efficiency*: implementing energy-efficient technologies such as LED lighting, motion sensors, and smart thermostats to reduce energy consumption. Installing renewable energy systems like solar panels can further reduce reliance on fossil fuels.
(ii) *Water Conservation*: installing low-flow fixtures, such as faucets and showerheads, and implementing water reuse systems to minimize water consumption. Encouraging guests to participate in towel and linen reuse programs can also save water and energy.
(iii) *Waste Management*: implementing comprehensive recycling and waste management programs that encourage waste reduction, recycling, and proper disposal. Minimizing single-use plastics and providing recycling bins in guest rooms and public areas can help reduce landfill waste.
(iv) *Local Sourcing*: supporting local farmers and suppliers by sourcing food and products locally. This reduces the carbon footprint associated with transportation and supports the local economy.

(v) *Sustainable Food Practices*: offering organic, locally sourced, and seasonal food options. Implementing practices such as composting food waste, reducing food waste through portion control, and donating excess food to local charities can also promote sustainability.
(vi) *Green Building Design*: constructing or renovating hotels using sustainable building materials, such as recycled or reclaimed materials, and incorporating energy-efficient design principles. Obtaining green building certifications, like LEED, can demonstrate a commitment to sustainability.
(vii) *Community Engagement*: engaging with local communities and supporting social initiatives, such as volunteering, promoting local cultural events, and collaborating with local organizations for sustainable tourism development.
(viii) *Employee Training*: providing sustainability training to employees to raise awareness and encourage sustainable practices in their daily operations. Engaging staff in sustainability initiatives fosters a culture of environmental responsibility.
(ix) *Guest Education*: educating guests about sustainable practices through in-room materials, signage, and interactive experiences. Providing information on local attractions, public transportation options, and eco-friendly activities can encourage responsible tourism.
(x) *Monitoring and Reporting*: regularly monitoring and measuring environmental performance, setting targets, and reporting progress. This helps identify areas for improvement and demonstrates transparency and accountability.

By implementing these sustainable practices, the hotel and hospitality industry can reduce its environmental footprint, enhance guest experiences, and contribute to the overall well-being of local communities and ecosystems.

3.2. Case Studies of Eco-Friendly Accommodation Models and Their Impact on Carbon Footprint

Case Study 1: Soneva Fushi, Maldives (Soneva.com)
Soneva Fushi, a luxury resort in the Maldives, is renowned for its commitment to sustainability and low-carbon operations. The resort has been carbon neutral since 2012 (Soneva Foundation). The resort has implemented numerous measures to minimize its environmental impact. It utilizes solar power extensively, with a large solar photovoltaic system providing a sizable portion of its energy needs. Soneva Fushi also employs innovative waste management practices, including composting and recycling, to reduce waste generation. Moreover, the resort promotes sustainable transportation by utilizing electric vehicles and bicycles for guest transportation. Through its eco-friendly initiatives, Soneva Fushi highlights that luxury accommodations can be operated in a low-carbon manner without compromising guest comfort and experience.

Case Study 2: Whitepod, Switzerland (Frameweb)
"Whitepod, located in the Swiss Alps, is an exceptional example of low-carbon accommodations. The resort offers guests a unique experience by providing

eco-friendly and energy-efficient geodesic dome-shaped pods. These pods are equipped with energy-efficient heating systems and use locally sourced wood for fuel, reducing reliance on fossil fuels. Whitepod encourages guests to minimize their environmental impact by promoting sustainable practices such as waste reduction and recycling. The resort also implements renewable energy sources, including solar panels and hydroelectric power, to power its facilities. Through its innovative design and sustainable practices, Whitepod demonstrates that nature-based tourism can be enjoyed with minimal carbon emissions."

Case Study 3: Proximity Hotel, United States
"The Proximity Hotel, located in Greensboro, North Carolina, is a LEED Platinum-certified hotel recognized for its sustainable design and low-carbon operations. The hotel incorporates energy-efficient features such as geothermal heating and cooling systems, solar panels, and LED lighting. It utilizes a rainwater harvesting system to reduce water consumption and employs a comprehensive recycling and waste management program. The Proximity Hotel actively promotes sustainable practices to its guests, including providing electric vehicle charging stations and offering bicycles for guest use. The hotel's commitment to sustainability has resulted in significant energy and water savings while providing a comfortable and eco-friendly experience for its guests."

These case studies highlight successful examples of low-carbon accommodations in tourism destinations. They demonstrate that through sustainable design, energy-efficient technologies, renewable energy sources, and eco-friendly practices, accommodations can significantly reduce their carbon footprint while offering unique and enjoyable experiences for guests. These initiatives also inspire and educate guests about sustainable living, encouraging them to adopt similar practices in their own lives. Overall, these accommodations exemplify that sustainability and luxury can go hand in hand, setting a positive example for the wider tourism industry.

4. Carbon Footprint in Infrastructure

Carbon footprint in infrastructure refers to the total amount of GHG emissions. It is typically measured in carbon dioxide equivalents (CO_2e), which is associated with the construction, operation, and decommissioning of physical structures and facilities. Infrastructure projects play a significant role in contributing to carbon emissions, and assessing their carbon footprint is crucial for implementing sustainable practices and mitigating environmental impact.

4.1. Analysis of the Role of Tourism-Related Infrastructure in Carbon Emissions

Examining the role of tourism-related infrastructure in carbon emissions involves analyzing the different facets of infrastructure development and operation within the tourism industry that contribute to the release of GHGs. Tourism-related infrastructure includes facilities and structures designed to support and enhance the tourism experience. Such as hotels, resorts, transportation systems, airports, and recreational facilities. Here are key factors to consider in this analysis:

1. Airports
Airport infrastructure significantly contributes to GHG emissions throughout its life cycle, encompassing construction, operation, and maintenance. Several factors within airport operations contribute to the carbon footprint. Thus, it is essential to analyze and address these sources to achieve more sustainable aviation practices. The key aspects of airport infrastructure and their impact on GHG emissions are given below:

1. Construction and Expansion:

   (i) *Materials And Construction*: the building and enlargement of airport infrastructure encompasses energy-intensive procedures, which include the extraction, fabrication, and transportation of materials like concrete and steel. The carbon embodied in construction materials contributes to the initial carbon footprint of the infrastructure.
   (ii) *Land Use Change*: airport expansion often requires changes in land use, potentially leading to the release of stored carbon in vegetation and soil, contributing to GHG emissions.

2. Energy Consumption:

   (i) *Terminal Operations*: airport terminals consume significant amounts of energy for lighting, heating, cooling, and numerous services. The source of this energy, whether from fossil fuels or renewable sources, affects the overall GHG emissions.
   (ii) *Runway Lighting And Navigation Systems*: the energy-intensive nature of runway lighting, airfield ground lighting, and navigation systems contributes to the overall carbon footprint.

3. Ground Transportation:

   (i) *Vehicle Emissions*: airport ground transportation, including shuttle buses, service vehicles, and taxis, contributes to GHG emissions. The type of fuel used in these vehicles influences the level of emissions.
   (ii) *Aircraft Ground Operations*: the infrastructure and equipment used for aircraft ground operations, such as tugs, baggage handling, and ground power units (GPUs), contribute to emissions.

4. Aviation Fuel Use:

   (i) *Aircraft Operations*: while the primary occurrence of aircraft emissions takes place during flight, the emissions associated with take-off, landing, and taxiing also play a part in shaping the overall carbon footprint of the airport. The infrastructure that facilitates these operations, including runways and taxiways, contributes to emissions at ground level.

(ii) *Fueling Facilities*: the infrastructure for storing and fueling aircraft also contributes to emissions, particularly if energy-intensive processes are involved.

5. Waste Management:

   (i) *Terminal Waste*: the disposal of waste generated within airport terminals, including food waste and packaging materials, contributes to GHG emissions. Incineration and landfill disposal methods release carbon dioxide and other pollutants.
   (ii) *Aircraft Waste*: the disposal of waste generated during flight, including catering waste and toilet waste, adds to the overall carbon footprint.

6. Maintenance and Renovation:

   (i) *Facility Upkeep*: the ongoing maintenance, renovation, and retrofitting of airport infrastructure contribute to emissions. Construction activities associated with these processes can release carbon if not managed sustainably.
   (ii) *Climate Control And Ventilation*: climate control and ventilation systems in airports contribute to carbon emissions primarily through their heavy reliance on energy, often derived from non-renewable sources. The energy-intensive nature of HVAC systems, along with the operation of ground support equipment, requires substantial fossil fuel consumption. Additionally, the need for auxiliary power units (APUs) or GPUs to provide climate control for parked aircraft can further contribute to emissions. While essential for passenger comfort and safety, the challenge lies in transitioning these systems to cleaner and more sustainable energy sources to reduce their environmental impact and align with global efforts to combat climate change.

Mitigating GHG emissions from airport infrastructure requires a multifaceted approach that incorporates sustainable practices, energy efficiency measures, and the adoption of cleaner technologies. Airports worldwide are increasingly focusing on initiatives such as energy-efficient designs, the use of renewable energy sources, and the implementation of green building standards to reduce their environmental impact. Additionally, innovations in aircraft technology, air traffic management, and alternative fuels contribute to more sustainable aviation practices. Certification initiatives, exemplified by the Airport Carbon Accreditation program, offer airports a structured framework to assess, oversee, and diminish their carbon emissions.

2. Roads and transportation networks

Roads contribute to CO_2 emissions in several ways. First, the construction of roads requires the extraction and processing of raw materials, such as asphalt and concrete, which release CO_2 during manufacturing. Second, vehicle emissions from cars, trucks, and motorcycles contribute significantly to CO_2 emissions. The burning of fossil fuels, such as gasoline and diesel, leads to the emission of carbon dioxide (CO_2) into the atmosphere. Additionally, traffic congestion and idling vehicles on roads lead to increased fuel consumption and emissions. Lastly, the maintenance and repair of roads also require energy and resources, which indirectly contribute to CO_2 emissions. Overall, the transportation sector, including roads, is a significant source of CO_2 emissions.

4.2. Sustainable Design and Development of Tourism Infrastructure

The sustainable design and development of tourism infrastructure encompasses the thoughtful planning, construction, and operation of tourism facilities with the goal of minimizing adverse environmental and social impacts while maximizing economic benefits. This approach entails integrating sustainability principles at every phase of infrastructure development, including site selection, design, construction, operation, and maintenance. The primary objective of sustainable design and development in tourism infrastructure is to achieve a harmonious equilibrium between the requirements of tourism and the conservation of natural resources, cultural heritage, and the overall well-being of local communities. The following outlines essential aspects of sustainable tourism infrastructure:

- (i) *Site Selection*: choosing locations that have minimal ecological and cultural impacts, such as brownfield sites or previously developed areas, rather than pristine environments.
- (ii) *Energy Efficiency*: integrating design elements that enhance energy efficiency in buildings, including strategies for passive heating and cooling. The implementation of efficient lighting systems, and the utilization of renewable energy sources such as solar or wind power.
- (iii) *Water Conservation*: implementing water-saving measures like rainwater harvesting, efficient irrigation systems, and water recycling to reduce water consumption and minimize impacts on local water resources.
- (iv) *Waste Management*: implementing effective waste management systems, including recycling and composting facilities, to minimize the generation of waste and promote responsible waste disposal practices.
- (v) *Local Materials and Labor*: using locally sourced materials and labor to support the local economy, reduce transportation-related emissions, and preserve the local cultural identity.
- (vi) *Biodiversity Conservation*: designing infrastructure in a way that minimizes impacts on local ecosystems, protects wildlife habitats, and promotes biodiversity conservation.
- (vii) *Community Engagement*: engaging local communities in the planning and decision-making processes, ensuring their active participation, and creating economic opportunities through the development of tourism.
- (viii) *Cultural Heritage Preservation*: honoring and safeguarding the cultural heritage of the local community through the incorporation of traditional architectural styles, encouragement of local arts and crafts, and active support for cultural events and traditions.

(ix) *Accessibility and Inclusivity*: crafting infrastructure to be accessible and inclusive for everyone, including individuals with disabilities, to guarantee equal opportunities for engaging in tourism experiences.

(x) *Long-Term Viability*: strategizing for the enduring sustainability of tourism infrastructure through the incorporation of elements such as climate change resilience, adaptability to evolving visitor demands, and the future-proofing of designs.

By adopting sustainable design and development practices, tourism infrastructure can contribute to the conservation of natural resources, protection of cultural heritage, and development of local communities' well-being, while still providing pleasant and memorable experiences for tourists.

4.3. Case Studies Highlighting Successful Examples of Low-Carbon Infrastructure in Tourism Destinations

To reduce carbon emission of tourism infrastructure successful case studies can be followed. Many case studies have been conducted to reduce carbon emission from tourism infrastructure. For instance, the case study of Samso, Denmark (Nielsen & Jørgensen, 2015) and Cost Rica (UNO Environment Programme) as elucidated below.

Case Study 1: Samso, Denmark

"Samso, a small island in Denmark, serves as an excellent example of low-carbon infrastructure in a tourism destination. The island has successfully transitioned to renewable energy sources and has become a carbon-neutral community. Samso's sustainable infrastructure includes wind turbines, solar panels, and biomass plants that provide clean energy for the island's electricity, heating, and transportation needs. The community has also implemented energy-efficient buildings, electric vehicle charging stations, and cycling infrastructure to promote sustainable transportation options for tourists. Samso's commitment to low-carbon infrastructure has not only reduced CO_2 emissions but has also attracted eco-conscious tourists seeking sustainable travel destinations."

Case Study 2: Costa Rica

"Costa Rica is renowned for its commitment to sustainability and has implemented various low-carbon infrastructure initiatives in its tourism sector. The country generates over 99% of its electricity from renewable sources, primarily hydropower, wind, and geothermal energy. This clean energy infrastructure has helped Costa Rica to reduce carbon footprint significantly. The country has also invested in sustainable transportation, with a vigorous network of electric buses and the promotion of electric vehicles. Additionally, Costa Rica has established numerous protected areas and national parks to conserve its rich biodiversity, attracting tourists interested in eco-tourism and wildlife conservation."

These case studies highlight successful examples of low-carbon infrastructure in tourism destinations. It shows that through a combination of renewable energy sources, energy-efficient buildings, sustainable transportation options, and conservation efforts can significantly reduce carbon footprint while attracting environmentally conscious tourists. These initiatives not only contribute to mitigating climate change but also promote sustainable economic development and preserve the natural and cultural heritage of the tourist destinations.

5. Conclusion

The tourism industry is recognized as a significant sector that positively impacts health and well-being. If electricity is being produced by sustainable energy sources like thermal or solar, it also presents critical challenges. On the one hand, the travel and tourism sector play a pivotal role in boosting global GDP and fortifying the employment market. According to statistics from the World Travel and Tourism Council, the travel and tourism sector contributed 7.6% to the global GDP in 2022, marking a substantial 22% increase compared to that in 2021. Additionally, domestic visitor spending witnessed a notable rise of 20.4% in 2022. On the other hand, the travel and tourism industry is a contributor to GHG emissions. As reported by the Sustainable Travel Organization, approximately 8% of the world's carbon emissions stem from travel and tourism activities. Despite the positive economic impact, the environmental consequences of increased travel and tourism activity highlight the need for sustainable practices within the industry. It is essential for the tourism sector to find a balance between economic growth and environmental responsibility. Sustainable tourism practices, such as promoting eco-friendly transportation, reducing carbon footprints, and encouraging responsible travel behavior, are integral for mitigating the adverse environmental effects associated with the industry. Striking this equilibrium is crucial for ensuring the long-term sustainability of the tourism sector while minimizing its negative impact on the environment.

Considering the social and economic benefits of travel and tourism there is a pressing need to prioritize zero carbon emissions in the tourism industry. It has been identified that the primary source of carbon emissions in industry is energy consumption. Therefore, there is an overarching imperative to transition to sustainable energy sources, such as wind and solar energy, and concurrently reduce energy consumption in all business practices associated with tourism.

Various facets of travel and tourism, including accommodation, transportation, infrastructure, and tourists' behavior, necessitate distinct measures to curtail respective carbon emissions. In the realm of accommodation, effective strategies should be employed to regulate heat loss and gain in hotel buildings. This includes the implementation of energy management systems (utilizing sensor technology), adoption of energy-efficient insulation (such as high-performance insulation material, insulated glazing, and thermal bridging), utilization of smart HVAC systems, and incorporation of efficient building envelope designs (sealing air leaks, application of energy-efficient window coatings like low emissivity glass, and the integration of green roofs).

Concerning transportation, it has been identified that airplanes are the largest contributors to CO_2 emissions at a rate of 0.82 per passenger per mile, followed by SUVs (CO_2=0.68), small cars (CO_2=0.55), motorcycles (CO_2=0.40), transit buses (CO_2=0.37), and subsequently, trains, tour buses, and ferries (CO_2=0.13,

0.10, 0.07) (STO, 2023). In response, aviation authorities should prioritize the adoption of SAF to reduce CO_2 emissions. According to the World Economic Forum, SAF has the potential to decrease carbon emissions from flights by 80% (WEF, 2023). Boeing, in alignment with this approach, aims to achieve a 100% reduction in flight carbon emissions by 2030. Furthermore, the carbon emissions from road transport can be significantly mitigated through the widespread adoption of electric vehicles. Statistics clearly indicate that the tourism industry contributes to carbon emissions. Achieving carbon emission reduction in the tourism sector necessitates substantial investment, a responsibility shared by all stakeholders involved – government entities, the hotel industry, transporters, tourists, and researchers. Future studies in this domain are imperative. Firstly, identifying factors that attract tourists to zero-carbon accommodations, transportation, and destinations is essential. This identification can lead to increased demand for zero-carbon tourist packages, subsequently driving up the supply of such packages. Secondly, zero-carbon transportation and accommodation prove to be cost-efficient. Motivating hotel owners and transporters to update their businesses with carbon-neutral products and operations in the tourism industry is crucial. Governments can play a pivotal role by providing subsidies to proprietors to establish and update tourism businesses based on zero-carbon strategies.

National and international authorities should measure carbon emissions associated with tourism transportation, accommodations, and infrastructure. Highlighting emission rates based on tourist destinations will raise awareness among tourists. This information assists tourists in making informed decisions about their travel plans, emphasizing the significance of zero-carbon tourist destinations. It is paramount for governments and international bodies to take proactive measures in this regard to guide the industry toward sustainable practices and enhance environmental consciousness among tourists.

References

Abeydeera, L. H. U. W., & Karunasena, G. (2019). Carbon emissions of hotels: The case of the Sri Lankan hotel industry. *Buildings*, *9*(11), 227.

Brainy Insights. (2023). *Tourism market. Market Research Report*. https://www.thebrainyinsights.com/report/tourism-market-13245

Brida, J. G., Gómez, D. M., & Segarra, V. (2020). On the empirical relationship between tourism and economic growth. *Tourism Management*, *81*, 104131.

CBO. (2023). https://www.cbo.gov/publication/58861

Doyle, A. (2018). *Booming tourism emits 8 percent of greenhouse gases, study shows*. Reuters. https://www.reuters.com/article/idUSKBN1I90YQ/

Duro, J. A., Perez-Laborda, A., Turrion-Prats, J., & Fernández-Fernández, M. (2021). Covid-19 and tourism vulnerability. *Tourism Management Perspectives*, *38*, 100819.

EPA. (2023). *Transportation air pollution and climate change*. https://www.epa.gov/transportation-air-pollution-and-climate-change/carbon-pollution-transportation

Flores, P. J., & Jansson, J. (2021). The role of consumer innovativeness and green perceptions on green innovation use: The case of shared e-bikes and e-scooters. *Journal of Consumer Behaviour*, *20*(6), 1466–1479.

Garsous, G., Corderi, D., Velasco, M., & Colombo, A. (2017). Tax incentives and job creation in the tourism sector of Brazil's SUDENE area. *World Development*, *96*, 87–101.

Garzon, A. R. G., Lee, T., Merry, K., Bektas, V., Cruise-Palmer, J., & Bettinger, P. (2022). Urban land use cover changes in three developed cities of the United States: San Diego, Denver, and Buffalo. *Heliyon*, *8*(3), e09093.

Giannakis, E., Serghides, D., Dimitriou, S., & Zittis, G. (2020). Land transport CO_2 emissions and climate change: Evidence from Cyprus. *International Journal of Sustainable Energy*, *39*(7), 634–647.

Gössling, S. (2002). Global environmental consequences of tourism. *Global Environmental Change*, *12*(4), 283–302.

Gössling, S., Araña, J. E., & Aguiar-Quintana, J. T. (2019). Towel reuse in hotels: Importance of normative appeal designs. *Tourism Management*, *70*, 273–283.

Gössling, S., Balas, M., Mayer, M., & Sun, Y.-Y. (2023). A review of tourism and climate change mitigation: The scales, scopes, stakeholders and strategies of carbon management. *Tourism Management*, *95*, 104681. https://doi.org/10.1016/j.tourman.2022.104681

Gössling, S., Scott, D., & Hall, C. M. (2020). Pandemics, tourism and global change: A rapid assessment of COVID-19. *Journal of Sustainable Tourism*, *29*(1), 1–20. https://doi.org/10.1080/09669582.2020.1758708

Gössling, S., & Peeters, P. (2015). Assessing tourism's global environmental impact 1900–2050. *Journal of Sustainable Tourism*, *23*(5), 639–659.

Graver, B. M., & Frey, H. C. (2016). Highway vehicle emissions avoided by diesel passenger rail service based on real-world data. *Urban Rail Transit*, *2*, 153–171.

Gwenhure, Y., & Odhiambo, N. M. (2017). Tourism and economic growth: A review of international literature. *Tourism: An International Interdisciplinary Journal*, *65*(1), 33–44.

Higueras-Castillo, E., Guillén, A., Herrera, L.-J., & Liébana-Cabanillas, F. (2021). Adoption of electric vehicles: Which factors are really important? *International Journal of Sustainable Transportation*, *15*(10), 799–813. https://doi.org/10.1080/15568318.2020.1818330

Holden, A. (2016). *Environment and tourism*. Routledge.

Howitt, O. J. A., Revol, V. G. N., Smith, I. J., & Rodger, C. J. (2010). Carbon emissions from international cruise ship passengers' travel to and from New Zealand. *Energy Policy*, *38*(5), 2552–2560.

Huang, T., & Tang, Z. (2021). Estimation of tourism carbon footprint and carbon capacity. *International Journal of Low-Carbon Technologies*, *16*(3), 1040–1046.

Jenkins, J., & Wearing, S. (2003). Ecotourism and protected areas in Australia. In R. K. D. David & A. Fennell (Eds.), *Ecotourism policy and planning* (p. 205). CABI. https://books.google.com.my/books?id=ywH2jUwOCfkC

Jeswani, H. K., Figueroa-Torres, G., & Azapagic, A. (2021). The extent of food waste generation in the UK and its environmental impacts. *Sustainable Production and Consumption*, *26*, 532–547.

Juvan, E., Grün, B., & Dolnicar, S. (2018). Biting off more than they can chew: Food waste at hotel breakfast buffets. *Journal of Travel Research*, *57*(2), 232–242.

Kanwal, N., & Isha, A. S. N. b. (2022). Identifying the moderating effect of hyperconnectivity on the relationship between job demand control imbalance, work-to-family conflict, and health and well-being of office employees working in the oil and gas industry, Malaysia. *Human Behavior and Emerging Technologies*, *2022*, 5260866. https://doi.org/10.1155/2022/5260866

Kanwal, N., Nizam Bin Isha, A. S., Salleh, R. B., Kanwal, N., & Al-Mekhlafi, A.-B. A. (2023). Paradoxical effects of social media use on workplace interpersonal conflicts. *Cogent Business & Management*, *10*(1), 2200892.

Khatib, A. N. (2023). Climate change and travel: Harmonizing to abate impact. *Current Infectious Disease Reports*, *25*(4), 77–85.

Lee, D. S., Fahey, D. W., Skowron, A., Allen, M. R., Burkhardt, U., Chen, Q., Doherty, S. J., Freeman, S., Forster, P. M., Fuglestvedt, J., Gettelman, A., De León, R. R., Lim, L. L., Lund, M. T., Millar, R. J., Owen, B., Penner, J. E., Pitari, G., Prather, M. J., … Wilcox, L. J. (2021). The contribution of global aviation to anthropogenic climate forcing for 2000 to 2018. *Atmospheric Environment*, *244*, 117834.

Lenzen, M., Sun, Y.-Y., Faturay, F., Ting, Y.-P., Geschke, A., & Malik, A. (2018). The carbon footprint of global tourism. *Nature Climate Change*, *8*(6), 522–528.

Lloret-Lloret, E., Pennino, M. G., Vilas, D., Bellido, J. M., Navarro, J., & Coll, M. (2021). Main drivers of spatial change in the biomass of commercial species between summer and winter in the NW Mediterranean Sea. *Marine Environmental Research*, *164*, 105227.

Ma, X.-W., Wang, M., Lan, J.-K., Li, C.-D., & Zou, L.-L. (2022). Influencing factors and paths of direct carbon emissions from the energy consumption of rural residents in central China determined using a questionnaire survey. *Advances in Climate Change Research*, *13*(5), 759–767.

Mishra, S., Sinha, A., Sharif, A., & Suki, N. M. (2020). Dynamic linkages between tourism, transportation, growth and carbon emission in the USA: Evidence from partial and multiple wavelet coherence. *Current Issues in Tourism*, *23*(21), 2733–2755.

Nielsen, S. N., & Jørgensen, S. E. (2015). Sustainability analysis of a society based on exergy studies – A case study of the island of Samsø (Denmark). *Journal of Cleaner Production*, *96*, 12–29.

OECD. (2010). *OECD tourism trends and policies 2010*. https://doi.org/10.1787/tour-2010-en

Palmer, C. (2022). *Cruise industry faces choppy seas as it tries to clean up its act on climate*. Reuters. https://www.reuters.com/business/sustainable-business/cruise-industry-faces-choppy-seas-it-tries-clean-up-its-act-climate-2022-07-27/

Popescu, C. R. G., & Popescu, G. N. (2019). An exploratory study based on a questionnaire concerning green and sustainable finance, corporate social responsibility, and performance: Evidence from the Romanian business environment. *Journal of Risk and Financial Management*, *12*(4), 162.

Pranita, D., Sarjana, S., & Musthofa, B. M. (2022). Mediating role of sustainable tourism and creative economy to improve community wellbeing. *African Journal of Hospitality, Tourism and Leisure*, *11*(2), 781–794.

Salehi, M., Filimonau, V., Asadzadeh, M., & Ghaderi, E. (2021). Strategies to improve energy and carbon efficiency of luxury hotels in Iran. *Sustainable Production and Consumption*, *26*, 1–15.

Seyfi, S., Hall, C. M., & Rasoolimanesh, S. M. (2020). Exploring memorable cultural tourism experiences. *Journal of Heritage Tourism*, *15*(3), 341–357.

Shah, G., & Shende, K. (2017). A study on the importance of food tourism and its impact on creating career opportunities amongst the residents of Pune city. *International Journal of Research in IT and Management*, *7*(3), 192–208.

Sharma, H. B., Vanapalli, K. R., Samal, B., Cheela, V. R. S., Dubey, B. K., & Bhattacharya, J. (2021). Circular economy approach in solid waste management system to achieve UN-SDGs: Solutions for post-COVID recovery. *Science of the Total Environment*, *800*, 149605.

Simonsen, M., Gössling, S., & Walnum, H. J. (2019). Cruise ship emissions in Norwegian waters: A geographical analysis. *Journal of Transport Geography*, *78*, 87–97.

Sofronov, B. (2018). The development of the travel and tourism industry in the world. *Annals of Spiru Haret University. Economic Series*, *18*(4), 123–137.

STO. (2023). *carbon footprint of tourism*. Sustainable Travel Organization. https://sustainabletravel.org/issues/carbon-footprint-tourism/#:~:text=Tourism%20is%20responsible%20for%20roughly,the%20top%20of%20the%20list.

Streimikiene, D., Svagzdiene, B., Jasinskas, E., & Simanavicius, A. (2021). Sustainable tourism development and competitiveness: The systematic literature review. *Sustainable Development*, *29*(1), 259–271.

Tang, S., Liu, Y., & Lei, A. (2018). Electrochemical oxidative cross-coupling with hydrogen evolution: A green and sustainable way for bond formation. *Chem*, *4*(1), 27-45.

Taylor, L. (2018). *Cycling city Copenhagen sprints to become first carbon-neutral capital*. Reuters. https://www.reuters.com/article/idUSKCN1MF060/

Towa, E., Zeller, V., & Achten, W. M. J. (2020). Input–output models and waste management analysis: A critical review. *Journal of Cleaner Production*, *249*, 119359.

Urban Sustainable Exchange. (2000). *TransMilenio bus rapid transit system*. Urban Sustainable Exchange, Metropolis Organization. https://use.metropolis.org/case-studies/transmilenio-bus-rapid-transit-system

Velthof, G. L., Bruggen, C. v., Groenestein, C. M., Huijsmans, J. F. M., Luesink, H. H., van der Sluis, S. M., van der Kolk, J. W. H., Voshaar, S. V. O., Vonk, J., & van Schijndel, M. W. (2016). Reference estimate from emission to the air originating from agriculture until 2030. Background document of the National Energy survey 2015, including emissions of ammonium, methane, laughing gas, nitrogen oxide and fine dust from agriculture until 2030. *Alterra-rapport-Wageningen University and Research Centre* (2746).

Wiedmann, T., & Minx, J. (2008). A definition of 'carbon footprint'. *Ecological Economics Research Trends*, *1*(2008), 1–11.

Wondirad, A., Kebete, Y., & Li, Y. (2021). Culinary tourism as a driver of regional economic development and socio-cultural revitalization: Evidence from Amhara National Regional State, Ethiopia. *Journal of Destination Marketing & Management*, *19*, 100482.

WorldinData. (2023). *Which form of transport has the smallest carbon footprint?* https://ourworldindata.org/transport

Xia, B., Dong, S., Li, Z., Zhao, M., Sun, D., Zhang, W., & Li, Y. (2022). Eco-efficiency and its drivers in tourism sectors with respect to carbon emissions from the supply chain: An integrated EEIO and DEA approach. *International Journal of Environmental Research and Public Health*, *19*(11), 6951.

Yang, X., Zhao, C., Xu, H., Liu, K., & Zha, J. (2021). Changing the industrial structure of tourism to achieve a low-carbon economy in China: An industrial linkage perspective. *Journal of Hospitality and Tourism Management*, *48*, 374–389.

Zervas, G., Proserpio, D., & Byers, J. W. (2017). The rise of the sharing economy: Estimating the impact of Airbnb on the hotel industry. *Journal of Marketing Research*, *54*(5), 687–705.

Chapter 7

Technological Advancements and Innovations in the Tourism Industry: Driving Sustainable Tourism

Kaniz Fatema[a], *Punitha Sinnappan*[b], *Chong Soon Meng*[a] and *Motoki Watabe*[a]

[a]*Department of Management, Sunway Business School (SBS), Sunway University, Malaysia*
[b]*Department of Marketing Strategy & Innovation, Sunway Business School (SBS), Sunway University, Malaysia*

Abstract

Sustainability entails promoting social development and meeting current needs without jeopardizing future life on Earth. It encompasses responsible resource stewardship and a consideration of environmental, social, and economic impacts. Sustainable tourism, as defined by the World Tourism Organization (WTO), addresses these impacts and balances economic prosperity, social well-being, and environmental preservation. Despite the tourism industry's growth, its carbon-intensive nature significantly contributes to climate change, necessitating the adoption of sustainable practices. The global carbon footprint from tourism has risen, projecting to reach 6.5 $GtCO_2e$ by 2025, further stressing the environment. Sustainable tourism aligns with the United Nations' Sustainable Development Goals, particularly SDG 8 and SDG 12, which emphasize inclusive and sustainable economic growth, and responsible consumption and production patterns. Technological advancements, such as AI and the Internet of Things, offer innovative data-driven solutions that not only revolutionize the tourism ecosystem but also mitigate its environmental impact. These technologies can help monitor and manage resource use in real time, promote renewable energy, and facilitate eco-friendly tourism practices.

The Need for Sustainable Tourism in an Era of Global Climate Change: Pathway to a Greener Future, 121–149

Published under exclusive licence by Emerald Publishing Limited
doi:10.1108/978-1-83608-668-020241019

Embracing Industry 5.0, which integrates human capabilities with intelligent machines, the tourism sector can achieve human-centric, resilient, and sustainable growth. Future research should explore technology's role in sustainable tourism, focusing on its environmental benefits and alignment with sustainability goals.

Keywords: Sustainable tourism; environmental impact; technological advancements; smart tourism; responsible travel; sustainable development goals (SDGs), sharing economy

1. Introduction

Sustainability revolves around the notion of promoting social development and meeting present needs without endangering life on Earth. It emphasizes responsible stewardship of resources and consideration for environmental, social, and economic impacts. This concept has gained significant traction across various industries as practitioners, policy makers, and individuals recognize the importance of conserving the environment as a whole.

1.1. Sustainable Tourism

The tourism industry has not been exempted from the influence of sustainability. As the global community is recognizing the impact of tourism industry growth in a more profound way, the concept of 'sustainable tourism' is getting increasing importance in both academia and industry. The WTO defines sustainable tourism as 'tourism that takes full account of its current and future economic, social and environmental impacts, addressing the needs of visitors, the industry, the environment, and host communities' (UNWTO, 2005). While it is rooted in the principles of responsible travel and conscientious sector management, it highlights the importance of balancing economic prosperity, social well-being, and environmental preservation. In the same vein, it has become imperative to embrace sustainability principles throughout the tourism industry, given that the scope of the industry is vast, encompassing numerous sectors such as transportation, accommodation, recreation, entertainment, food and beverages, as well as other service-related activities.

1.2. Sustainable Tourism Growth and Environmental Sustainability

Continued and sustainable growth in the tourism industry is essential for economic, social, and cultural reasons. However, it is necessary to acknowledge its significant negative impact on the environment. Tourism is one of the biggest sectors and the fastest growing sector of the world economy, but it remains as one of the least progressed sectors in respect to environmentally sustainable practices (Loureiro, 2019). Due to its carbon-intensive nature, the global tourism industry is considered as one of the major contributors to climate change. The emissions of pollutants such as carbon and other greenhouse gases (GHGs) from the tourism industry have increased over the years, exacerbating the devastating impacts of climate change that includes extreme heat waves, widespread hunger and drought, and rising sea levels. This is supported by a study conducted by Adedoyin and Bekun (2020) who have confirmed the unidirectional causal relation from tourism (i.e., volume of tourists' arrival) to carbon emissions and energy consumption.

In addition, the global carbon footprint (CF) due to tourism growth has increased from 3.9 to 4.5 $GtCO_2e$ in between 2009 and 2013, this accounts for around 8% of the global GHG emissions (Lenzen et al., 2018). The global CF continues to rise due to growth in tourism and is forecasted to reach a value of 6.5 $GtCO_2e$ by 2025 which would account for around 12% of current GHG emissions (Adetunji, 2018). The adverse effect of GHG emissions produced from the tourism industry (75% due to transportation, 21% due to accommodation, and 4% due to other services) poses a huge burden to the local environment. So, to combat climate change, the development strategy of the tourism sector should focus on innovative approaches from all its stakeholders to make the carbon cost of trips as low as possible. Furthermore, the irresponsible use of energy and natural resources associated with tourism activities has endangered essential ecological processes, natural heritage, and biodiversity of our planet. An increasing mass of travellers in a tourist destination also often contributes to substantial wastage (e.g., food waste and single-use plastics) and eventually places a significant burden on the local waste management systems.

The myriad environmental challenges we face today have eventually impacted the tourism industry, prompting a shift towards prioritizing environmental sustainability as a fundamental pillar within sustainable tourism practices. In reference to this, the United Nations has embedded sustainable tourism within its overarching framework, notably through the Global Goals or Sustainable Development Goals (SDGs) with a great emphasis on SDG Goal 8 and SDG Goal 12 (*THE SDGS IN ACTION*, n.d.). By integrating SDG 8 and SDG 12 into tourism practices, the industry can actively contribute to global efforts for sustainable development. SDG 8 is mentioned as '*Decent Work and Economic Growth*' which focuses on promoting continuing, comprehensive, and sustainable economic development. SDG 8 highlights the fact that economic development must create decent and fulfilling jobs without damaging our planet's environment. In the context of tourism, this goal aims to provide fair and equitable socio-economic advantages to all the stakeholders in the tourism ecosystem through making substantial contributions to alleviating poverty and improving their living standards. On the other hand, SDG 12 is mentioned as '*Responsible Consumption and Production*' which essentially serves as a critical guidepost for ensuring sustainability within the tourism sector. Given that our planet is already affected by climate change, it is imperative that we adopt responsible practices in both production and consumption of resources to offset the negative impact on the environment. Sustainable tourism initiatives, in harmony with SDG 12, emphasize mindful consumption of energy, water and other resources associated with travel activities. Basically, eco-friendly accommodations, green transportations, renewable energy utilization, efficient resource management, effective waste management, and finally community involvement initiatives collectively form the foundation of sustainable tourism growth.

1.3. Technological Advancements Driving Environmental Sustainability in the Tourism Industry

The significant boom in the tourism industry has led to a burden on tourist destinations, exerting pressure on local environments, host communities, and the authenticity of these destinations. According to Loureiro (2019), the potential growth in the tourism industry can be turned into an opportunity for the welfare of local population and economy rather a threat for the local environment if only supported by technological advancement and continuous innovation processes. In recent years, technological advancements have improved the efficiency, productivity, and competitiveness of a business. Advanced IR 4.0 technologies such as Artificial Intelligence (AI), and the Internet of Things (IoT) have boosted the production flow and business processes in various industries, thus achieving an overall improvement in operational efficacy (Zhang et al., 2019).

The impact of technological advancements now extends beyond economic gains to achieving sustainability objectives. For instance, the use of natural resources can be monitored and managed in real time through AI or block-chain based innovative solutions integrated in the production line or supply chain management. As a greater mass of tourists consumes more energy (Adedoyin & Bekun, 2020), ensuring higher production of sustainable energy is imperative for sustainable tourism growth. Due to advancement in power generation technology, ecologically sustainable and renewable energy alternatives like solar, wind, or biomass power are now available. These energy generation plants have minimal environmental impact as they do not burn fossil fuels. This greatly resonates with the UN global goal of 'Responsible Consumption and Production'. Tourism ecosystem, if managed through smart technology, can interconnect and engage all its stakeholders in value co-creation and bring sustainable gains for all (Buhalis, 2020).

> As such, technologies should be seen as an enabler rather than the ultimate solution to create tangible impacts. This indicates prioritizing solutions that not only support technology but also consider the social and environmental aspects, taking into account the diversified demand and aspirations of all stakeholders involved. Given this, technological advancement and digitalization in tourism have been studied for decades focusing mostly on its potential use for business or consumer outcomes. However, these studies have given insufficient attention to the implications of these technologies on environmental dimensions and combating the effects of climate change. Therefore, this chapter discusses how the knowledge, and applications of advanced technologies and innovations can promote sustainable growth in the tourism industry and mitigate its environmental implications in both the short and long term.

2. Technological Evolution in Tourism and its Impact on Sustainable Practices

The convergence of technology and tourism has redefined how individuals explore and experience the world around them. Over the years, advancements in technology have consistently reshaped the landscape of the tourism industry, altering how people plan and embark on their journeys. The following sections outline the historical development of technological innovations in three key areas related to tourism and their associated environmental implications.

2.1. Transportation Technologies in Tourism

Since the invention of the internal combustion engine in the 19th century, the trajectory of transport technologies has been on a relentless path of evolution. Concurrently, the tourism industry is undergoing a profound transformation with the advent of revolutionary transportation technologies. These advancements are reshaping how people travel, explore, and engage with destinations worldwide. Today, trains, automobiles, and airplanes are not merely modes of transportation; they represent the gateways to new horizons and unparalleled experiences. The increased accessibility and efficiency of travel allowed people to venture beyond their immediate surroundings, marking the beginning of tourism as a mass phenomenon.

Notably, half a century ago, air travel was prohibitively expensive for most people; for instance, the fare for a flight between London and New York in the 1960s was ten times higher than today (Houston, 2019). With the advancement in aircraft technologies, today's planes are more energy efficient than before, they are more aerodynamic with reduced flight time due to the evolution of state-of-the-art wing and computerized cockpit technologies. Air travel has also become more affordable and convenient for people due to the Budget-Airline model like Air Asia. Websites like Expedia, Priceline, and Orbitz offer competitive air travel prices and reward programs like purchase price back, early bird price, etc.

As transportation infrastructure continued to develop, tourism flourished, and a global community of travellers began to emerge. However, the massive growth in transport infrastructure largely contributes to the global rise of Carbon emissions. In less than two decades, transportation in the tourism sector has made the tourism industry a leading environmental polluter, surpassing even the construction industry. The invention of electric vehicles (EVs) is an important milestone achieved in the innovation of cleaner transport solutions, The EV has a great potential to significantly and positively impact environmental sustainability because of its minimal lifetime CF compared with gasoline vehicles. But EV's short range and expensive charging stations are limiting its prospect as a sustainable transport solution.

The environmental impact of transportation extends beyond just the emissions of pollutants. The preservation of biodiversity and the safeguarding of the natural and cultural heritage of tourist destinations are also at risk due to the mass gathering of people. In response to these challenges, distant travel solutions are emerging as potential remedies. Drones equipped with video cameras can be used for virtual tourism, and the adoption of Augmented/Virtual/Mixed reality in the tourism sector offers an immersive digital travel experience to address these concerns.

A recent groundbreaking development in tourism is space tourism. Due to the tremendous advancement in space technologies (rocket propulsion systems, crewed spacecraft design and engineering, etc.), space tourism has become a reality, enabling individuals to venture beyond Earth's atmosphere and experience space travel firsthand. The startup company 'Axiom Space' or Jeff Bezos' Blue Origin' has already organized commercial space tourism flights to the ISS (International Space Station) (Wattles, 2022). Elon Mask's SpaceX is playing a leading role in carrying out business activities in space, the investors for futuristic space tourism (for building satellite hotels, cheaper rockets, etc.) should greatly consider the environmental sustainability of the planet Earth and space beyond.

2.2. Communication Technologies in Tourism

Humans are constantly evolving innovative ways to better communicate with each other. They witnessed the greatest advancement in communication in 1831 with the invention of the electric Telegraph. The introduction of teletype systems and telephone reservations streamlined the booking process, making it more convenient for individuals to plan their trips. The 1960s saw the rise of computer reservation systems, marking a significant leap forward in the industry. These systems not only facilitated more efficient management of bookings but also laid the groundwork for the computerization of travel agencies, a trend that would reshape the industry's operational landscape.

The late 20th century brought about the most transformative shift in communication technology with the introduction of the internet. The World Wide Web revolutionized the way people gathered information, planned trips, and made reservations. Later, mobile technology started leading communication technology; in the 2000s, the number of mobile subscriptions was 740 million worldwide, and within two decades it exceeded 8 billion (Agenda, 2020, November 18). From 2010 onwards, the widespread adoption of smartphones, mobile, and digital technology has become an integral part of the travel experience, offering real-time navigation, instant communication, and on-the-go booking capabilities. Mobile apps from airlines, hotels, and travel services enable smartphone users to avail themselves of seamless travel experiences.

Till today, social media platforms stand out as the most popular addition to the communication landscape. In the context of tourism, social media serves not only as a source of inspiration for potential travellers but also as a platform for value co-creation within the tourism ecosystem (Buhalis, 2020). The advent of social media platforms such as Facebook, Instagram, and Twitter has redefined the way individuals share their travel experiences. User-generated content became a powerful influencer in shaping travellers' perceptions and decisions in the context of sustainable practices. Social media not only functions as a ubiquitous platform for social communication but also helps in raising travellers' awareness of environmental protection (Haque et al., 2021).

However, the pervasive use of smartphones, the internet, and social media is explosively raising our appetite and enthusiasm for travel. While this surge benefits the tourism industry, it also places a considerable burden on many tourist destinations, especially when considering their deteriorating environmental conditions. In response, the integration of disruptive technologies such as AI, Big Data Analytics, Machine Learning, Augmented Reality (AR), and Virtual reality (VR) in the tourism industry is emerging as a transformative force, especially in the realm of communication, paving the way for more sustainable tourism practices.

2.3. Technology-driven Accommodation Solutions in Tourism

The environment has to suffer badly from huge burdens related to the construction, use, and demolition of buildings, hence technological progress and innovation in the building industry is mandatory. One of the milestones in the advancement of building technology was made from the 1930s through the 1960s when revolutionary cooling methods were adopted in the buildings. Although the new air conditioning system together with reflective glass, and structural steel made the enclosed glass and steel buildings very popular in modern cities, those skyscrapers had a large negative impact on the environment due to the use of massive heating, ventilation, and air conditioning (HVAC) systems that ran on fossil fuels. As energy use by air conditioning continued to rise, in the 1970s, the idea of modern green building was initiated to improve energy efficiency and reduce its impacts on the environment and human health. The new air conditioning system that we use at present consumes about 50% less energy compared with their energy consumption in 1990 (Lester, 2015).

According to the World Economic Forum (2021), decarbonization, electrification, efficiency, and digitalization are the four critical movements for ensuring zero-carbon buildings. Within the context of the tourism industry, accommodation sector development is a fundamental component of the overall planning process of a tourism destination. Modern hotel structures if designed with sustainable bio-architecture; constructed with bio-based, recycled, reused, and carbon-neutral materials; and equipped with renewable energy sources like solar panels and energy-efficient appliances, can lower their CF and improve the structural energy efficiency, cost-efficiency, and operational efficiency of the hotels. The adoption of these practices conveys the impression that the hotels are demonstrating their social responsibilities (Sinnappan et al., 2023). The concept of eco-lodge and eco-amenities also emerge to ensure a green image of tourism accommodation.

The adoption of AI for energy management in hospitality establishments represents a paradigm shift towards sustainability. The conceptual framework incorporates IoT sensors, actuators, and AI algorithms to monitor and optimize energy consumption in real time. In-room technology, controlling temperature, lighting, and curtains is generally accepted by customers with minimal resistance (Nam et al., 2021). The implications are far-reaching, as AI-driven energy management systems contribute not only to cost savings for businesses but also to a reduced environmental impact. The hospitality industry, particularly in the food and beverage sector, generates substantial waste. AI emerges as a potent tool for identifying and mitigating patterns of wastage within hospitality services. These AI Data Analytics involve an in-depth analysis of consumption patterns and enable the hotel industry to optimize their inventory, reduce over-ordering, and simultaneously minimize food waste. In the subsequent sections, we will delve deeper into the emerging technological trends in the tourism sector, examining them from different perspectives, and highlighting pathways to sustainability.

3. Current Technological Landscape of the Tourism industry and a route towards greener environment

Technology has become an indispensable tool for industrial development, and the tourism industry is no exception. The continuous innovations in technologies and advancement in ICT have great influence on transforming the tourism industry into a novel network of '*interconnected and phygital*' elements, known as tourism 4.0. This section will discuss the current technological trends in four main areas that can significantly contribute to the sustainable growth in tourism by mitigating the environmental impact of travel-related activities. These areas include:

- *Communication and Marketing.*
- *Transportation.*
- *Accommodation.*
- *Visitor Experience and Destination Management.*

3.1. Technological Innovations in Communication and Marketing for Sustainable Tourism

The Communication and Marketing sectors are among the major sectors undergoing radical transformation due to the incorporation of state-of-the-art technologies. In the context of the tourism industry, technological advancement is capable of bringing all its stakeholders into the tourism service ecosystems and interconnecting them for value co-creation (Buhalis, 2020). Big data, accumulated in the tourism ecosystems through social media, smartphones, IoT/5G devices or sensors, etc., acts as the key source of value creation in the tourism industry. This huge volume of data is monitored, extracted, analysed and visualized through sophisticated analytical tools and data-mining techniques to reveal patterns of the travellers' attitudes, views, demands, preferences, etc. and offer contextualized products and services more suitable to the tourists' needs. These technology-based advancements can help counterbalance the environmental impact of traditional travel-related activities by promoting and facilitating the choice of sustainable travel options through diverse communication and marketing avenues, as discussed below.

3.1.1. Smartphones and Digital Platforms in Communication and Marketing for Sustainable Tourism

Smartphones – the smart computers in hand and the unstoppable access to the internet have nearly replaced conventional communication with online communication. In 2020, over 4.5 billion people used the internet, among them 3.8 billion were social media users (Lin et al., 2020). As of the fourth quarter of 2023, 94% of global users access the internet via their smartphones (Petrosyan, 2024). Given the extensive user base of the internet and social media, smartphones are serving as essential tools for maintaining seamless connectivity and facilitating information flow within the tourism ecosystem.

These days, smartphones and their seamless access to wide-ranging data networks have established a 'digital infostructure' shifting the marketing paradigm in the tourism industry (Buhalis, 2020). Today's tourism operators and travel service providers can directly promote their tourist destinations, hotels, airlines, and other services to both local and international tourists through various mobile applications and websites. This direct engagement reduces tourists' dependency on travel agencies and enhances competitiveness among the suppliers. Additionally, marketing campaigns on digital platforms like social media prove to be effective, as regular users of social media often receive unexpected recommendations based on their prior interactions that influence their travel-related behaviour.

Travellers now use online platforms in every step of their travel – starting from planning their trips to consumption, from searching for information to posting reviews of their travel experience. These digital platforms and tools influence people on how they plan and embark on their journeys. Different Travel apps and Travel aggregator apps for Android and iPhones and web apps such as Trip.com, Hotels.com, Booking.com, Tripadvisor, Expedia, Trivago, flights.google.com, etc. assist travellers to have smoother and less stressful travel experiences by alerting them about the best flight deals or even finding for them a last-minute accommodation. Nonetheless, these travel apps enhance the overall travel experience by offering innovative ways to discover destinations, interact with local cultures, and access real-time travel information and recommendations. Since today's travellers are environmentally conscious, they are now more mindful than ever in choosing sustainable travel options through these travel apps.

3.1.2. Value Creation from 'Social Big Data'

Within the smart tourism ecosystem, social media platforms, as the biggest source of user-generated data, present a significant marketing opportunity for tourism operators. Social Big data, referring to the vast amount of information accumulated every day on social media platforms, can effectively be leveraged as competitive assets for tourist destinations. In particular, Electronic or Online word-of-mouth (e-WOM/OWOM) in social big data can serve as a major driving force for reshaping marketing strategies.

E-WOM/OWOM is personal communication posted online and circulated electronically through online social networking apps, online community forums, blogs, or review sites. According to Feng and Liu (2018), OWOM referral is a new communication strategy in marketing, and it considerably impacts new customer acquisition. Among all industrial sectors, the tourism and hospitality sector is the most impacted by eWOM. In this modern age, travellers no longer depend on traditional marketing activities like advertisements or promotions for planning a trip, rather the observations and opinions of other users in online platforms act as extensively reliable sources of information for them. Therefore, not only the direct star ratings but also the indirect impression of user sentiments from enormous data of user reviews and comments need to be scrutinized through complex algorithms.

'Data analytics' is a powerful means that enables tourism stakeholders to discover new avenues of progress by revealing consumer satisfaction through active listening on the internet or examining the online customer reviews representative of OWOM or any other form of unstructured data. For instance, a research work by Saura et al. (2018) deployed data analytics to learn the environmental sustainability-related key factors identified by hotel guests during their stay in some top-ranked Swiss hotels. To carry out the study, the user reviews on the travel website TripAdvisor were collected for a specific period, and then *'Sentiment Analysis'* and *'Textual Analysis'* were performed. The *'Sentiment analysis'* algorithm used the 'machine learning technique' trained with the 'data-mining process' and the *'Textual Analysis'* was performed through qualitative software to identify the most recurring texts in the user reviews. The study revealed that hotel guests perceive garbage, smoke, noise pollution, and huge gatherings of people as environmental concerns in hotels. There is also a perceived lack of respect for local food traditions, homegrown products, recycling, and sustainable energy practices in the hotels. The hotel managers can utilize these pinpointed factors to enhance and quality control their environmental sustainability practices.

3.1.3. Smart Technologies for Personalized Experiences

Personalization can be described as a procedure of collecting personal data about individual choices and preferences and utilizing those to offer contextualized services, i.e., to provide the right propositions at the right time that perfectly fit the user demand. Through advanced location-based technology, real-time marketing messages can be delivered to the customers right in a destination where they are most likely to avail themselves of the unique deal. From the customers' point of view, personalization of information supports them to decide efficiently as it displays only relevant information. Conversely, the decision-making process cannot be effective if it is based on a huge amount of information that becomes obsolete in no time.

Within the context of tourism, customers leave electronic traces of their travel-related activities starting from trip planning to service consumption and, finally, after service evaluation, this produces an enormous volume of data in the tourism ecosystem, and this 'big data' acts as the driving force in today's tourism marketing. For offering travellers a personalized experience, necessary user information can be obtained through both implicit and explicit methods. The implicit method contains traces from previous online activities like, such as uploaded trip-related photos, videos, posted reviews, comments, or ratings by tourists on various social networks or other online platforms. An explicit method can be an online survey or other user input to the system on demand.

The next big thing in digital marketing is the concept of 'relevance'. Instead of targeting unique marketing campaigns for individual customers or customer segments based on the search data available, marketers will focus on 'relevance' and anticipate through predictive analysis what customers will prefer and expect next. According to the latest report by Grominsky (2021) in Forbes, 'relevance' can deliver more compelling value propositions to individuals without requiring huge investment and resources like 'personalization'.

3.1.4. Prediction and Forecasting through Machine Language and AI-based models

In the context of tourism, the term 'Sustainability' itself has turned into a big value proposition. According to a study by Nowacki et al. (2023) conducted on Gen Z individuals finds that, the perceived green (eco-friendly) image of destinations which shows their commitment towards sustainability is the strongest predictor of their intention to travel to green holiday destinations. The most recent Sustainability Travel Report for Booking.com (GSTC, 2023) shows that 76% of global travellers want to avail more sustainable travel options which is an increase of 16% and 5% over the company's 2021 and 2022 data, respectively. The report further states that while nearly half (49%) of travellers perceive more sustainable travel options as being significantly more expensive, 43% are still willing to pay a premium for sustainable certification. Although low-cost transport and accommodation providers are offering tourists more budget-friendly options to bring low-cost evolution in the tourism industry, the tourists being the changemakers, are leading positive change to adopt responsible travel practices that benefit both local communities and the environment.

In recent years, AI-based models are growing in popularity for demand forecasting in the tourism sector. Machine learning (ML), which is a subset of AI is frequently used to accurately predict an outcome from real data based on previous learning (pattern recognition). A study by Sun et al. (2019) deployed the ML technique to precisely forecast the arrival of tourists in Beijing, this type of forecasting of tourist volume can provide important information for policy-making for future economic development. Another study by Perles-Ribes et al. (2020) also utilized ML to forecast potential issues of over-tourism in tourist destinations within Spanish Cities. This prediction recommends decentralizing tourism activities towards new destinations that do not experience high tourism demand.

3.2. Technological Innovations in the Transportation Sector for Sustainable Tourism

Transportation stands as the largest contributor to GHG emissions in the tourism industry, making it essential for all stakeholders to prioritize mitigating the environmental impact of travel-related activities. In response to this, policymakers may implement specific regulations to limit travel activities to certain tourism destinations, such as imposing a carbon tax on destination-based travel. Additionally, governments should offer incentives to companies to adopt sustainable transport solutions and encourage tourists to utilize them. Numerous providers of high-tech and green transport solutions operate in the market, with a primary focus on sharing economy approaches like car sharing and ride sharing. We will delve into these green transport solutions in detail in the following sections, focusing on their relevance to sustainable tourism.

3.2.1. Car Sharing (CS) Schemes

Environmental sustainability of sharing economies is globally recognized for increasing resource efficiency. 'Car sharing (CS)' is a smart transport solution based on this sharing economy concept and is similar to a car rental service but operated through a digital platform. The CS scheme facilitates shared access of a vehicle for individual use either from a pool of automobiles distributed within a region by organizations like Car2Go or Zipcar (business-to-consumer sharing) or peer-to-peer sharing through online platforms provided by organizations like Drivejoy or Snappcar. The CS users, through the application from their smartphones, can locate and reserve the vehicle that is closest to their location, can make online payments, and eventually receive a smart code to physically access the vehicles.

The sharing economy in transport services promotes efficient use of existing resources which essentially reduces carbon emissions. A study by Nijland and van Meerkerk (2017), aimed to quantify the effect of CS schemes, found that car sharers reduce their emissions by 240 to 390 kg of CO_2 per person per year. This reduction represents 13%–18% of the CO_2 emissions related to traditional car ownership and their use. A further study by Amatuni et al. (2020), conducted in three geographical locations (Netherlands, San Francisco, and Calgary), also confirms that the sharing habit of an average CS user significantly reduces mobility-related GHG emissions in business-to-consumer CS. The CS service promotes alternative mobility behaviours like cycling and walking, thus further contributing to the reduction of GHG emissions. Overall, CS schemes transform traditional vehicle ownership models and influence mobility behaviour, this can significantly contribute to the reduction of GHG emissions related to tourism transportation as well.

3.2.2. Car Pooling/Ride Sharing

Car pooling or ride sharing is another form of smart vehicle-sharing scheme where several people simultaneously share rides throughout their journey in the same direction like Bla-Bla-Car or on-demand eHailing transport services like Grab, Uber, or Lyft. The sharing of motorless vehicles through the bicycle-sharing scheme is a cheap and popular solution for tourism activities and is a more sustainable substitute for individual car trips (Szymańska et al., 2021). These bicycles have a GPS tracking system installed inside and are stationed at the Base station, users need to register and pay online to avail of the service.

Technological innovations like QR codes and electronic security locks have enabled the evolution of station-based bicycle-sharing services from free-floating services. There are other environment-friendly ride-sharing schemes for tourists, containing electrically propelled vehicles like e-bikes, e-scooters, and Segways which must be brought back to the charging station after use.

A study by Szymańska et al. (2021) conducted on the tourists in 'Białowieza Forest' region, a UNESCO World Heritage site, reveals that besides the first and foremost need for a flexible and reorganized public transport system, 40% of the tourists show soaring demand for hi-tech transport solutions comprising of e-mobility (E-car and E-bike sharing services) in the peripheral (remote) tourist areas that would reduce the negative effect of transport on the environment but increase the competitive capability of the regions.

A disruptive transport solution 'Self-driving or Autonomous Vehicle' is a combination of state-of-the-art technologies like AI, geolocation, vehicle-to-everything communication (V2X), etc. Autonomous vehicle is expected to empower vehicle-sharing services in support of environmental sustainability. Modern electric cars and hybrid cars are novel transport solutions deploying cleaner transportation technology, they greatly reduced GHG emissions and use of fossil fuel. But then its short range, lengthy charging time and expensive infrastructure of charging stations are limiting its prospect as a sustainable solution of greener transport in tourism.

3.3. Technological Innovations in Accommodation Sector for Sustainable Tourism

Over the decades, the tourism and hospitality industry has been aiding climate change by dramatically impacting energy consumption, water consumption, waste production, biodiversity destruction, and damage to the ecosystem for the construction and operation of accommodation venues for tourists. Now a days, eco-friendly hotels or other green accommodation solutions are gaining immense popularity among green travellers. Hence, a large number of hotels around the world are now deploying sustainability practices to reduce the negative impact of tourists' accommodation facilities on the environment. Globally, over 500,000 accommodations are labelled with a 'Travel Sustainable badge' on Booking.com as a recognition of their sustainability practices (GSTC, 2023). A green image for tourist accommodation can be achieved through a combined effort from both the public and private sectors. Various green initiatives, as outlined below, can be introduced to minimize both the short and long-term environmental impacts of accommodations in the context of sustainable tourism.

3.3.1. Sustainable Innovations in the Construction of Accommodation Facilities

As we enter a new era, the high-tech GenY and GenZ individuals, due to their climate-conscious habits and sensitivity towards biodiversity, look for the green image of a tourist destination. Eco-friendly accommodations can follow several sustainable innovative designs and alternative materials to ensure their green image.

- Buildings for hotels, guest houses, or resorts need to be designed with bio-architecture concepts including limited height, harmonious and plain rooms, etc.
- Create both healthier indoor and outdoor environments in sustainable or green buildings, and make green roofs for a low environmental impact.
- Use of alternative materials like lightweight concrete or alternative insulation materials will reduce the amount of energy and resources needed for transportation of materials associated with building construction. Alternative materials are mostly renewable, bio-based, and recycled, they divert waste materials from landfills. For example, insulation material 'rock wool' is made from waste during steel production, 'brown cellulose' is recycled from used newspaper pulp, and 'fiberglass' is recycled from used glass.
- Economically volatile and scarce materials should be replaced with innovative alternatives in construction.

3.3.2. Sustainable Innovations in Operational Activities of Hotels

Now a days, the concept of green hotels and other eco-centric accommodations is gaining more popularity among green travellers all around the world (Olorunsola et al., 2022). In their pursuit of sustainable tourism, green hotels are adopting state-of-the-art technologies to ensure their operations closely align with the Global Goals, particularly SDG 12. Their efforts encompass:

- Use renewable energy sources to serve some amount of hotels' energy demands.
- Provide energy-efficient electric lamps and lighting systems, air conditioning systems, water heating systems, etc.
- Use electric cars for maintenance and shuttle services within the hotel facility.
- Use smart technology like robotics to track the reusing of towels and linens, and clean those with green-certified standards.
- Serve locally sourced and organic foods to the guests. Use biodegradable and locally sourced products for decoration, packaging, etc. Advanced technologies like blockchain can ensure the sustainability of the supply chains.
- Use high-tech systems, enabled with 5G sensors, IoT devices, and AI technology, to ensure energy efficiency and environmental safety by monitoring, regulating, and optimizing the energy and water conservation systems, the waste management systems, as well as the recycling and reuse systems within hotels. The integration of AI and digital technologies has already been proven to be immensely valuable for the monitoring and management of natural resources (Shivaprakash et al., 2022).
- Travelling-related mobile applications can be used to provide necessary information and suggestions to the guests for lowering energy and water consumption, efficient food and waste management of the hotels, and raising customer awareness regarding sustainable tourism.

3.3.3. Innovative Solutions for Peer-to-Peer Accommodation Sharing

In the last decade, the accommodation sector, like the transport sector, has experienced rapid growth in sharing economy-based services. Within the context of tourist accommodation, sharing economy solutions are viewed as more environmentally sustainable. They allow travellers to utilize resources that would otherwise remain idle, promoting resource efficiency and reducing the environmental impact of travel. Peer-to-peer accommodation solutions not only maximize resource utilization but also result in lower energy consumption and reduced waste production compared to traditional hotel buildings.

The sharing economy has brought revolutionary changes to the accommodation sector by incorporating the high-tech solution of a paid online P2P platform, where 'Airbnb' plays a major role. Airbnb's potential to offset the environmental impact of tourism comes mostly from energy efficiency since small private rooms are more likely to use less energy than larger commercial structures. A study by Cheng et al. (2020), regarding the CF of Airbnb hosts in the greater Sydney region, finds that the Airbnb platform and its hosts produce CF ranging from 7.27 to 9.39 kg CO_2e/room-night, while the traditional hotels produce a significantly higher amount of CF ranging from 23.17 to 34.32 kg CO_2e/room-night. P2P accommodation sharing helps travellers to consume less resources and energy and facilitates them to use underutilized resources, thus this innovative solution ensures less pollution and less waste in the environment.

Now a days, there is a growing concern among travellers, especially the young ones about the reduction of carbon emissions related to their travel activities. Sustainability needs to be understood as a motivating force first to get the fullest benefit of the sustainability potential of the sharing economy in the accommodation sector. A study conducted among Airbnb users in the United Kingdom shows that the users' perceived sustainability of the Airbnb service significantly influences their loyalty towards that service (Garrod et al., 2023). Furthermore, the eco-friendly approach of peer-to-peer accommodation sharing can provide sustainable livelihoods to local communities, enhancing their quality of life through inclusive social integration. This aligns with Global Goal SDG 8 on promoting sustainable and inclusive economic growth.

3.4. Technological Innovations for Visitor Experience and Destination Management

In a smart tourism ecosystem, all stakeholders—including travellers, service providers, suppliers, destination management organizations, intermediaries, the public sector, and others—need to collaborate for value co-creation, leading to sustainable progress in tourism across economic, social, and environmental dimensions (Garanti, 2023). The integration of emerging technologies like AI, robotics, and even the metaverse within a smart tourism destination can revolutionize the traditional visitor experience. In the subsequent sections, we will explore various perspectives of integrating smart technologies to achieve sustainability in tourism destinations worldwide, with a focus on enhancing the visitor experience and ensuring effective destination management.

3.4.1. Smart AI for Smart Travel Experience

With the tremendous development in computing power and advanced analytics, 'Artificial Intelligence (AI)' is now being involved in almost every aspect of the travel and tourism industry to bring breakthrough excellencies for the service providers, and delightful and smart experiences for travellers. AI is mostly used in the tourism industry to swap humans with *robots* programmed for delivering customized services like robotic concierges, robotic chefs, robotic bartenders, etc., and also with *conversational systems* like chatbox, virtual guides, virtual hosts, digital personal assistants, or smart travel agents (Tussyadiah, 2020).

In the context of hotel stays, AI-enabled robots can accumulate data from the guests' previous experiences and comprehend the customer's needs and preferences without asking, so the hotels can offer seamless and personalized services to the guests. Besides, AI can handle multiple requests with higher accuracy and greater efficiency even if a large number of guests stay in a hotel and ensure the maximum utilization of hotel resources (Gupta et al., 2022). Anyway, there is a debate on the overall benefit of replacing humans in the tourism sector with AI-enabled solutions as it diminishes an important element of travel experience which is the social exchange between the hosts (service providers) and guests (customers).

Smart AI can greatly influence the effective decision-making process of an individual. The revolutionary generative AI-ChatGPT deploys a novel idea in an AI-driven decision-making system by empowering the individuals to dynamically probe the system for related information in a conversation mode. In the context of tourism, ChatGPT has the ability to improve the decision-making process of the tourists in throughout their trips including- pre-trip, en-route, and post-trip phases (Wong et al., 2023). Therefore, AI can proficiently support tourists to decide over mindful and responsible travel choices.

3.4.2. Immersive Digital Tourism: 'The Metaverse'

The Internet is now evolving into the next generation, known as Web 3.0. The emergence of the 'Metaverse' as a part of this new generation of the internet offers immersive digital experiences through technologies like AR, VR, and Mixed Reality (MR) (Buhalis et al., 2023). Metaverse is the conjunction of the physical and virtual universes through immersive technologies and allows its users to seamlessly navigate and interact within the physical and simulated environments and experience a hyper-realistic sense of presence through avatars and holograms. Although the metaverse is still a theoretical concept for most of people, the immersive games and the virtual ambiance offered by Fortnite, Second Life, Roblox or VRChat reveal the potential socio-impact of the fully operational metaverse platforms (Dwivedi et al., 2023).

In the context of tourism, the incorporation of high-tech Metaverse can offer an immersive digital experience to tourists. Furthermore, it is envisioned to revamp the travel experiences and has immense potential to disrupt tourism management and marketing ideas. The World Economic Forum has recognized 'tourism' as one of the vital sectors that can be benefitted tremendously from the adoption of the metaverse (Agenda, 2022, May 26). Metaverse tours can enable travellers to interact physiologically and psychologically with the agents and objects of the digital environment of a tourist destination from a distant place, and also allow individuals to travel in time (Chen & Yao, 2022). Therefore, travellers can safely experience hazardous natural phenomena like volcano eruptions or space explorations and can even encounter ancient events within immersive tourism.

Immersive digital tourism can be facilitated for tourist locations like Arctic regions or famous tourist destinations that are already endangered by climate change or over-tourism. Arctic tourism growth in recent years resulted in increased human activities in the Arctic regions, the increased number of cruise ships passing through this region increases the chance of catastrophes like oil spills or sewage leaks and severely damage the ecosystem. Again, the terrible transformation of Venice from inclusion in the List of 'UNESCO World Heritage' in 1987 to the List of 'World Heritage in Danger' in 2016 is caused by over-tourism. In particular, tourism in Venice is predominantly affected by the arrival of cruise ships, as large groups of cruise passengers crowd in the same location at the same time posing a negative impact on the environment and ecosystem of the city (Capocchi et al., 2019).

Additionally, the metaverse can also play a major role in preserving cultural heritage, it can limit the demand for physical visits to historical sites and artefacts, and offer the tourists to experience the same through digital museums. Digital museum characterizes an exceptional tourism venue that provides visitors with an immersive experience of eloquent learning and discovery. Famous digital museums like the *Mori Digital Art Museum in Tokyo* or the *Atelier des Lumières in Paris* offer visitors an interactive and multidimensional experience by leveraging multisensory cues, such as visual, auditory, olfactory, taste, and haptic cues, this unique form of experience potentially redefines the traditional museum experience that is dominated by static exhibitions. A study by Guo et al. (2023) conducted on the visitors of 'the *Mori Digital Art Museum in Tokyo*', found that the combination of visual and auditory stimuli elevates the visitors' emotional state and sense of presence in such a way that brings the best digital museum experience for the visitors.

According to Buhalis et al. (2023), immersive experience is not necessarily expected to always replace real travel activities, but rather motivate people for physical visits. Additionally, a pre-visit of a tourist destination through digital immersion allows travellers to be better prepared for physical trips. Due to the broader access to virtual resources than real travel, the metaverse offers an enhanced level of tourism experience and destination interpretation. As a result, metaverse would eventually alter the pattern of consumer behaviour creating demand for new marketing and operational strategies in tourism management (Gursoy et al., 2022).

3.4.3. Smart Tourism Destinations and Smart Destination Management

To start with, 'Smart Tourism' is a high-tech and data-driven approach to tourism management where a large volume of information is collected and processed.

This enables the stakeholders to gain valuable understandings about the travellers' attitudes, opinions, demands, and preferences, which can be turned into 'value co-creation' opportunities in the context of sustainable tourism growth. Correspondingly, the term 'Smart tourism destinations' is used to describe the travel sites that have widely adopted emerging technologies to exploit the information-intense nature of the tourism industry to promote destinations' competitiveness, innovativeness, and sustainability (Vecchio et al., 2018).

The concept of 'Smart tourism destinations' is fundamentally built on advanced digital platforms that dynamically interconnect different tourism stakeholders; thus, it can guarantee instant information exchange and better collaboration within a tourism ecosystem. Smart tourism destinations enable the destination makers to make data-driven decisions through real-time monitoring and management of the tourist destinations, as well as empowering them to responsibly plan and better manage tourism resources.

Anticipating tourist demands and understanding their potential environmental impact are essential aspects of value creation in smart tourism destinations. In light of this, destination managers can adapt their sustainable policies and practices, focusing on areas like resource management, waste reduction, pollution control, and the protection of local ecosystems and biodiversity within the destinations. A study by Vecchio et al. (2018) was performed regarding the regional tourist experiences in Apulia, Southern Italian region, to demonstrate how smart tourism destinations can discover value-creation opportunities through the management of social big data. The study derived several prospects for value creation in several aspects, such as refinement of the decision-making process with synchronized and real-time information flow, upgradation of the travel experience by offering more personalized products and services suitable to tourist needs, redefining marketing strategies and adopt data driven-business models, etc.

Additionally, the Real-Time-Response (RTR) is another competitiveness of smart tourism destinations that enables several actors in a smart tourism ecosystem to make more informed and efficient decisions in the face of crisis or disaster and enhance destination resilience and safety of the visitors (Bethune et al., 2022). Moreover, smart destination management can effectively address the issue of overtourism. By utilizing smart technologies, it can educate both tourists and host communities about responsible and sustainable tourism practices and encourage community involvement to mitigate the social and environmental impacts associated with tourist activities.

3.4.4. Leveraging AI for Decision-Making in Sustainable Tourism

Disruptive technologies like 'Big data' can portray the big picture of the tourists and the destination makers' commitment towards environmental sustainability by addressing and assessing its potential impact on 'tourist satisfaction', which is the ultimate driving force in the growth of a tourist destination. Artificial intelligence (AI) technology, based on advanced data analytics and intelligent computer programs, can act as an effective tool to empower travellers and service providers to make synchronized and real-time decision-making in the context of sustainable tourism.

Currently, AI-based models are widely used for prediction and forecasting purposes in the tourism sector through a variety of applications. For instance, a recent study by Mengual Torres et al. (2022) has applied AI to identify a range of possibilities to partially replace the most energy-consuming devices in some Mexican hotels of the intermediate category which are located in the tropical-climate region. Hotels in Mexico's tropical-climate regions use up to 52% more electricity than those in temperate areas, largely due to extensive air conditioning for guest comfort. Notably, 80% of these hotels rely on fossil fuels for their energy needs leading to increased GHG emissions. The study employed an Artificial Neuron Network (ANN) expert model to analyse energy consumption data from each hotel. The model could identify configurations for the hotels to reduce energy consumption by 9% to 12% per room annually without requiring a complete overhaul of lighting or air-conditioning systems. Thus, AI can provide attractive solutions to facilitate quicker and easier-to-implement decision-making for hotel management or other destination makers to promote environmental sustainability through their operational activities.

4. Challenges and Opportunities

4.1. Ethical Concerns and Risks of Overreliance on Technology

Technological advancements and innovations have undoubtedly revolutionized the tourism industry, making it more sustainable, efficient, and convenient for tourists to travel around the world and experience all it has to offer. As discussed earlier, they have reshaped the industry in a myriad of ways, including digital platforms and tools, accommodation, transportation, and destination management. However, as these technologies continue to shape the future of tourism and mold it into a more sustainable industry, they also present several challenges that necessitate careful thought. Whether a destination actively champions sustainable practices or not, the impact of technological advancements on the tourism landscape remains a pervasive and consequential aspect.

Genuine human interactions, which are important for meaningful travel experiences, may be overshadowed by digital experiences. While technological innovations work better for the benefit of tourists, they often come with a serious lack of human-centred design (Stankov & Gretzel, 2020). Even though chatbots, digital concierge services, and automated check-ins can be effective in some situations, they may not always provide the human touch or local knowledge that tourists usually look for when visiting new locations. The human touch encompassing personal stories, local perspectives, and shared experiences, which contributes significantly to the richness and authenticity of the entire journey, may not be present in the digital space. For instance, the CEO of the Global Hotel Alliance (GHA), Christopher Hartley, highlighted the significance of human interaction that increases proportionally with the scale of accommodation, with leisure travellers consistently placing a higher value on the personalized and human-centric aspects of their experiences (Carruthers, 2022).

Next, collecting extensive traveller data through surveillance and tracking technologies raises ethical questions about privacy and data security, leading to concerns about data misuse (Samala et al., 2020). This challenge remains significant and applicable across the entire spectrum of tourism locales, regardless of their active promotion of sustainability. According to privacy and data security lawyer, Mark McCreary, Big Data is big business in the travel industry (Read, 2020). Chatbots for reservations and databases for hotels, airlines, and vehicle rentals can appear very useful. However, there are ways in which the data could be mined and used by others through this technology. Travellers also frequently share information on social media and other travel applications, potentially exposing their data to third-party entities that may seek to monetize it. More significant threats can include an external party using travel technologies to hack into bank accounts and credit cards or can result in less-apparent invasions like sniffing, malware, and phishing. This assertion is substantiated by the findings of Sustacha et al. (2023), who mentioned that travellers articulate concerns regarding the perceived lack of security or loss of privacy when using smart technologies. These challenges hurt the overall tourism experience, signifying the crucial importance of security and privacy in influencing travellers' perceptions and interactions with smart technologies.

The digital divide is another challenge facing the travellers and also tourism organizations or destinations, affecting their ability to communicate effectively in a global tourism environment (Reverte & Luque, 2021). The digital divide becomes particularly pertinent as it highlights the disparities in accessing and leveraging cutting-edge digital tools and innovations. The divide is not merely about having internet access but also extends to the ability to utilize advanced technologies such as Big Data, AI, and analytics. Shin and Baek (2023) have presented empirical evidence that supports the digital divide between younger and older travellers in using smartphones for travel, highlighting the disparities in travel interest, planning engagement, independence of travel experiences, and social sharing. Moreover, constraints such as limited access to digital infrastructure and technology resources or insufficient digital literacy and skills among tourism organizations and destinations may also create disparities in the travel experience. This challenge transcends boundaries and pertains to destinations irrespective of their commitment to promoting sustainability.

4.2. Opportunities for Collaboration and Improvement

Besides the challenges, there are also opportunities for collaboration. Collaboration emerges as the linchpin for realizing the full potential of advanced technologies in fostering environmental sustainability within the tourism industry. A key component of this collaboration is bringing various stakeholders together to set shared objectives and develop a framework for joint action. Hence, collaboration between policymakers, tourism organizations, technology providers, and local communities is crucial for effective destination management. Fundamentally, the development of public policies for sustainable tourism and the support provided by policymakers to tourism organizations in the form of financial incentives may facilitate the implementation of smart infrastructure that enhances sustainability in their operations (UNWTO, 2023). This involves leveraging costly technologies like IoT sensors for intelligent waste management, smart water systems, automated climate control systems, AI-powered analytics for resource optimization, blockchain for transparent supply chain management, and AR in tourism establishments. Technology providers are responsible for bringing innovation to the forefront and contributing to the development of essential digital tools and infrastructure. As destinations increasingly rely on technology for various functions, ensuring the security and privacy of data has also become paramount. As such, technology providers can also play a crucial role in implementing robust cybersecurity measures and privacy protocols to protect both travellers and destination stakeholders (Paraskevas, 2022). On the other hand, community engagement contributes to the preservation of the social cohesion of the destination. Their participation ensures that the use of technology in tourism does not jeopardize or compromise the authenticity of the local ecosystems.

4.3. Balancing Innovation with Cultural Preservation

In addition, effectively balancing innovation with cultural preservation presents a dual challenge and opportunity for the industry to remain competitive. This is because there is a growing number of travellers who would like to discover the destinations in an authentic and concrete way (Lenoir, 2020). In relation to this, the travellers are mostly interested in the culture and history of the destination and enjoy meeting the local population. Given this scenario, cultural preservation and digital innovation are required to provide both travellers and locals with an authentic experience. Preserving the local identity and its uniqueness can include the central and historical structures as well as the cultural infrastructure. Jia et al. (2023) asserted that it is extremely important to maintain a balance between the preservation of cultural heritage and the advancement of technology. As a result of this preservation, it is ensured that technology will not be dominating the local traditions or deteriorating their authenticity.

In support of this claim, numerous studies have demonstrated the positive impact of modern technologies on the desirable qualities of cultural sites. There are numerous technologies that have been employed to promote cultural heritage tourism [e.g., 3D virtual, VR, AR, MR, and mobile augmented reality (MAR)]. Their use holds the promise of preserving cultural heritages and enhancing the management of heritage sites (Bec et al., 2019; Caciora et al., 2021). Liu (2020) has noted that digital display technologies are widely accepted by heritage visitors and have a favourable influence on certain aspects such as encouraging travellers to explore and learn more about the site, enhancing their visualization of the heritage's richness, and providing novel experiences in perceiving, engaging, and communicating about history. Guo and Zhuang (2017) have also stated that the use of digital technology is not a replacement for cultural heritage itself, but it serves as a medium for heritage communication and complements travellers' experience of heritage.

In the realm of environmental sustainability, Manglis et al. (2021) revealed that AR, VR, and other technologies can encourage travellers to engage in underwater cultural tourism and enhance their comprehension of the imperative to safeguard natural heritage. This underscores the significant potential of advanced technology to simultaneously preserve cultural heritage and bolster environmental sustainability.

5. Future Trends and Predictions

5.1. Emerging Technologies and Their Potential Impact

While innovations in technology have undoubtedly changed the way we travel and experience destinations, exploring future trends beyond just current technological evolution is vital for a better understanding of the trajectory of the tourism industry. Apparently, rapid innovation and technological advancements have caused paradigm shifts, and the most recent is 'Industry 4.0'. Tourism 4.0 is one of many sub-concepts under the broader Industry 4.0 umbrella and uses key supporting technologies such as IoT, big data, blockchain, AI, VR and AR to unlock the innovative potential of the industry (Peceny et al., 2019). Although the whole industry and organizations are still in the midst of this fourth revolution, the next revolution, Industry 5.0, is already well on its way. Industry 5.0 affirms the essential role and contribution of every industry to society, presenting an alternative perspective that goes beyond productivity and efficiency while respecting the production limits of the planet.

According to the European Commission (2023), Industry 5.0 has three key pillars: human-centric, resilient and sustainable. A human-centric approach prioritizes the needs, talents, and empowerment of individuals, shifting the focus from considering people as resources to recognizing them as the central purpose and beneficiaries of organizational strategies. Resilience implies being agile and adaptable, enabling organizations to anticipate, react, and learn from crises progressively for stable performance. Sustainability, signifies a commitment towards the environment, respecting planetary boundaries. In contrast to previous industrial revolutions, Industry 5.0 places a high priority on environmental protection. As a result, sustainability and resilience have emerged as new trends, specifically to also meet the demands of international organizations like the United Nations, most governments, and consumers (Aheleroff et al., 2022).

Fundamentally, Industry 5.0 represents a move away from economic value to societal value and towards wellbeing rather than welfare (Kraaijenbrink, 2022). This means that its applicability is far broader than Industry 4.0. Basically, Industry 5.0 is a new concept that has begun to permeate various industries but has yet to fully integrate into the tourism industry. However, the industry should savour the evolving era of Industry 4.0 and the technological innovations that continually emerge, while also being mindful to anticipate the potential values that Industry 5.0 might bring as it advances. This will be further elaborated in the subsequent section.

5.2. Forecasting the Evolution of Sustainable Tourism Technology

Industry 5.0 signifies a paradigmatic departure from Industry 4.0, highlighting the harmonious integration of human capabilities with intelligent machines.

This transformative shift can be expected to evolve within the tourism industry in the future. Collaborative robots, or cobots, have significantly impacted the manufacturing industry by working alongside humans to enhance efficiency and productivity. There is now the possibility of extending its flexible application to the tourism industry, with the key focus being on achieving sustainability goals. They can be programmed to carry out tasks such as waste sorting, cleaning, and maintenance. This helps to minimize the harmful effects of tourism on the environment and promotes the sharing of information among travellers.

One of the technologies includes Blockchain Middleware for decentralized Industrial Internet of Things (IIOT) from Industry 5.0. This technology could transform data management and security (Khan et al., 2023) and the benefits are further amplified when it is integrated with cloud for collaboration and data analytic capability. This method maximizes smart technologies for tourism businesses by combining data from several sources to create efficient value chains. It also protects against many issues (e.g., interoperability, heterogeneity and data piracy) and offers an encrypted platform for traveller data. Blockchain smart contracts in tourism can also improve procedures, remove outside intervention, and increase sustainability by eliminating paperwork.

Next, by implementing predictive maintenance, a commonly employed technique in manufacturing, there is the potential to significantly enhance operational efficiency and improve resource utilization (Yan et al., 2017). Predictive maintenance in tourism infrastructure, such as hotels, transportation, and destinations, enables proactive decision-making by leveraging asset performance data to efficiently control operating expenses, including maintenance costs. Tourism facilities can use sensor-collected data to forecast the lifespan of their equipment, which helps them plan and carry out maintenance tasks on a regular basis. This strategy reduces resource consumption, prolongs the lifespan of equipment, and promotes sustainable resource utilization in the tourism industry.

Hyper customization, rooted in Industry 4.0, offers a sustainable approach by leveraging real-time data through AI and Big Data (Khan et al., 2023). This technology allows precise tailoring of experiences, minimizing resource waste, reducing overconsumption, and promoting flexibility through modular and customizable frameworks. The open architecture models can support the integration of eco-friendly practices, and a decentralized, crowdsourced approach can foster community engagement and align with sustainable tourism goals.

5.3. Recommendations for Industry Stakeholders

As discussed before, the tourism businesses have been adopting the technologies well, but the changing landscape requires a proactive strategy. To effectively address global concerns, a human-centric, resilient, and sustainable approach is needed. Industry 5.0 has the potential to bolster tourist organizations, which will ultimately result in long-term benefits. As a result, it is absolutely necessary for businesses in the tourism industry to offer training to its staff members in order to cultivate a culture that is both technologically knowledgeable and flexible. In order to advance sustainability, it is essential to participate in collaborative projects that make it easier for various stakeholders to share their thoughts and best practices with one another. These businesses need to be proactive in anticipating upcoming market trends and technological breakthroughs in order to guarantee the continuous provision of travel experiences that are both environmentally sustainable and culturally satisfying. Through the implementation of this forward-looking approach, business operations will be protected against future issues, while at the same time, attempts to advance sustainable tourism on a global scale will be enhanced.

6. Conclusion

6.1. Recap of Key Points

The tourism industry is a dynamic landscape that is primed for considerable positive change, as revealed by the examination of sustainable innovations in the industry. It is possible to evaluate the deployment of technology in the tourism industry across a variety of domains. A summary of those can be seen below:

(i) *Digital Platforms and Tools*: smartphones have allowed tourism organizations to directly communicate with a wide spectrum of travellers, making them indispensable for marketing efforts. Traditional marketing methods have also given place to user-generated content on online platforms for travel decisions. This pattern shows how crucial it is for advanced algorithms to assess massive user reviews.
(ii) *Accommodation*: AI is exerting a substantial influence on accommodation. Due to its multifaceted function, encompassing the provision of tailored services via robotics and predictive modelling for energy-efficient solutions in hotels, this technology has the capacity to transform the tourism industry and contribute to environmental conservation through informed decision-making.
(iii) *Transportation*: A shift towards mobility that is both more efficient and less destructive to the environment is shown by the deployment of environmentally friendly technology in transportation (e.g., CS and carpooling).
(iv) *Visitor Experience and Destination Management*: the emerging concept of the Metaverse has the potential to significantly alter the travel experience by enabling users to navigate both the physical and virtual realms in a seamless manner through the utilization of immersive technology. This has the potential to significantly alter the travel experience. When current digital platforms, ICT infrastructure, and business analytics are utilized together, stakeholders are provided with the opportunity to make decisions that are both well-informed and driven by data. This new development contributes to the growth of environmentally responsible tourism.

While the technology brings unprecedented efficiencies, it introduces ethical concerns such as the potential erosion of genuine human interactions, privacy and data security issues associated with extensive traveller data collection, and the digital divide, which poses challenges in global communication and access to cutting-edge technologies. On the other hand, collaboration is identified as important for harnessing the full potential of advanced technologies, requiring joint efforts from policymakers, tourism organizations, technology providers, and local communities. Similarly, striking a delicate balance between technological innovation and cultural preservation is found crucial, ensuring that advancements contribute positively to cultural tourism experiences while preserving local authenticity. The transition from Industry 4.0 to the upcoming Industry 5.0, with a focus on human-centric, resilient, and sustainable practices, is identified as significant for the tourism industry. Cobots, Blockchain Middleware for Decentralized IIOT, predictive maintenance, and hyper customization are identified as examples of key technological trends in the future of sustainable tourism.

6.2. Closing Thoughts on the Future Direction

In a nutshell, technological advancement in sustainable tourism holds an enormous opportunity to create positive environmental effects and encourage responsible travel. By taking advantage of the technology, the tourism industry can efficiently provide and facilitate environmentally friendly options for travellers. This can considerably reduce the industry's environmental impact and contribute to developing a more sustainable future for tourism. The future trajectory of the tourism sector involves adopting Industry 5.0, which entails a significant transition towards harmoniously bringing together human capabilities with intelligent machines. This transformative approach promotes the three pillars of Industry 5.0 (i.e., human-centric, resilient and sustainable), directing the industry towards societal value and well-being. Future research can explore how technology in the tourism industry aligns with these three pillars. Further, examining the impact and integration of these emerging technologies in various areas (i.e., digital platforms and tools, accommodation, transportation, and destination management) can contribute valuable insights into the industry's sustainable growth.

References

Adedoyin, F. F., & Bekun, F. V. (2020). Modelling the interaction between tourism, energy consumption, pollutant emissions and urbanization: Renewed evidence from panel VAR. Environmental Science and Pollution Research International, *27*(31), 38881.

Adetunji, J. (2018, May 7). *The carbon footprint of tourism revealed (it's bigger than we thought)*. The Conversation UK. https://theconversation.com/the-carbon-footprint-of-tourism-revealed-its-bigger-than-we-thought-96200

Aheleroff, S., Huang, H., Xu, X., & Zhong, R. Y. (2022). Toward sustainability and resilience with Industry 4.0 and Industry 5.0. *Frontiers in Manufacturing Technology*, *2*, 951643.

Agenda, D. (2020, November 18). *Here's how technology has changed the world since 2000*. World Economic Forum. https://www.weforum.org/agenda/2020/11/heres-how-technology-has-changed-and-changed-us-over-the-past-20-years/

Agenda, D. (2022, May 26). *How the metaverse can be a force for good in an uncertain world.* World Economic Forum. https://www.weforum.org/agenda/2022/05/how-metaverse-can-be-a-force-for-good-in-an-uncertain-world/

Amatuni, L., Ottelin, J., Steubing, B., & Mogollón, J. M. (2020). Does car sharing reduce greenhouse gas emissions? Assessing the modal shift and lifetime shift rebound effects from a life cycle perspective. *Journal of Cleaner Production*, *266*, 121869.

Bec, A., Moyle, B., Timms, K., Schaffer, V., Skavronskaya, L., & Little, C. (2019). Management of immersive heritage tourism experiences: A conceptual model. *Tourism Management*, *72*, 117–120

Bethune, E., Buhalis, D., & Miles, L. (2022). Real time response (RTR): Conceptualizing a smart systems approach to destination resilience. *Journal of Destination Marketing and Management*, *23*(January), 100687.

Buhalis, D. (2020). Technology in tourism-from information communication technologies to eTourism and smart tourism towards ambient intelligence tourism: A perspective article. *Tourism Review*, *75*(1), 267–272.

Buhalis, D., Leung, D., & Lin, M. (2023). Metaverse as a disruptive technology revolutionising tourism management and marketing. *Tourism Management*, *97*(January), 104724.

Caciora, T., Herman, G. V., Ilieş, A., Baias, ş., Ilieş, D. C., Josan, I., & Hodor, N. (2021). The use of virtual reality to promote sustainable tourism: A case study of wooden churches historical monuments from Romania. *Remote Sensing*, *13*(9), 1758.

Capocchi, A., Vallone, C., Pierotti, M., & Amaduzzi, A. (2019). Overtourism: A literature review to assess implications and future perspectives. *Sustainability*, *11*(12), 3303.

Carruthers, M. (2022). *Human touch still crucial despite advances in travel technology*. TTG Asia. https://www.ttgasia.com/2022/05/18/human-touch-still-crucial-despite-advances-in-travel-technology/

Chen, C., & Yao, M. Z. (2022). Strategic use of immersive media and narrative message in virtual marketing: Understanding the roles of telepresence and transportation. *Psychology and Marketing*, *39*(3), 524–542.

Cheng, M., Chen, G., Wiedmann, T., Hadjikakou, M., Xu, L., & Wang, Y. (2020). The sharing economy and sustainability–assessing Airbnb's direct, indirect and induced carbon footprint in Sydney. *Journal of Sustainable Tourism*, *28*(8), 1083–1099.

Dwivedi, Y. K., Hughes, L., Wang, Y., Alalwan, A. A., Ahn, S. J., Balakrishnan, J., Barta, S., Belk, R., Buhalis, D., Dutot, V., Felix, R., Filieri, R., Flavián, C., Gustafsson, A., Hinsch, C., Hollensen, S., Jain, V., Kim, J., Krishen, A. S., … Wirtz, J. (2023). Metaverse marketing: How the metaverse will shape the future of consumer research and practice. *Psychology and Marketing*, *40*(4), 750–776.

European Commission. (2023). *Industry 5.0 community of practice*. https://research-and-innovation.ec.europa.eu/system/files/2023-06/ec_rtd_industry5-cop-terms-of-reference.pdf

Feng, J., & Liu, B. (2018). Dynamic impact of online word-of-mouth and advertising on supply chain performance. *International Journal of Environmental Research and Public Health*, *15*(1), 1–16.

Garanti, Z. (2023). Value co-creation in smart tourism destinations. *Worldwide Hospitality and Tourism Themes*, *15*(5), 468–475.

Garrod, B., Zhao, A. L., & Koenig-Lewis, N. (2023). A greener way to stay: The role of perceived sustainability in generating loyalty to Airbnb. *International Journal of Hospitality Management*, *110*(April 2022), 103432.

Grominsky, T. (2021, December 10). *The era of personalization in marketing gives way to relevance*. Forbes. Retrieved August 30, 2024, from https://www.forbes.com/sites/forbesbusinessdevelopmentcouncil/2021/12/10/the-era-of-personalization-in-marketing-gives-way-to-relevance/

GSTC. (2023, April 19). *Booking.com 2023 sustainable travel report:cost vs conscience: Booking.com delves into the dilemma dividing sustainable travel in 2023*. GSTC. Retrieved August 30, 2024, from https://www.gstcouncil.org/booking-com-2023-sustainable-travel-report/

Guo, H., & Zhuang, D. H. (2017). Research on user experience of digital communication of intangible cultural heritage in Huizhou Province. *Hundred Schools in Arts*, *33*(06), 239–240.

Guo, K., Fan, A., Lehto, X., & Day, J. (2023). Immersive digital tourism: The role of multisensory cues in digital museum experiences. *Journal of Hospitality and Tourism Research*, *47*(6), 1017–1039.

Gupta, S., Modgil, S., Lee, C. K., Cho, M., & Park, Y. (2022). Artificial intelligence enabled robots for stay experience in the hospitality industry in a smart city. *Industrial Management and Data Systems*, *122*(10), 2331–2350.

Gursoy, D., Malodia, S., & Dhir, A. (2022). The metaverse in the hospitality and tourism industry: An overview of current trends and future research directions. *Journal of Hospitality Marketing and Management*, *31*(5), 527–534.

Haque, E., Sungsuwan, T., & Sanglimsuwan, S. (2021). Can social media be a tool for increasing tourists' environmentally responsible behavior? *Geo Journal of Tourism and Geosites*, *38*(4), 1211–1222.

Houston, J. (2019, November 8). *How flights have become so cheap over the years*. Business Insider. https://www.businessinsider.com/why-air-travel-is-so-cheap-2019-11

Jia, S., Chi, O. H., Martinez, S. D., & Lu, L. (2023). When "Old" Meets "New": Unlocking the future of innovative technology implementation in heritage tourism. *Journal of Hospitality & Tourism Research*. (Published online).

Khan, M., Haleem, A., & Javaid, M. (2023). Changes and improvements in Industry 5.0: A strategic approach to overcome the challenges of Industry 4.0. *Green Technologies and Sustainability*, *1*(2), 100020.

Kraaijenbrink, J. (2022). *What Is Industry 5.0 And How It Will Radically Change Your Business Strategy?* Forbes. https://www.forbes.com/sites/jeroenkraaijenbrink/2022/05/24/what-is-industry-50-and-how-it-will-radically-change-your-business-strategy/?sh=72a1dd1320bd

Lenoir, A. (2020). *The rise of authentic experiences*. TouriScope. https://www.touriscope.ca/articles/authentic-experiences

Lenzen, M., Sun, Y. Y., Faturay, F., Ting, Y. P., Geschke, A., & Malik, A. (2018). The carbon footprint of global tourism. *Nature Climate Change*, *8*(6), 522–528.

Lester, P. (2015, July 20). *History of Air Conditioning*. ENERGY.GOV. https://www.energy.gov/articles/history-air-conditioning

Lin, M., Li, F. Y., & Zhou, H. (2020). A Research on the combination of oblique photography and mobile applications based on the sustainable development of tourism. *Sustainability (Switzerland)*, *12*(9).

Liu, Y. (2020). Evaluating visitor experience of digital interpretation and presentation technologies at cultural heritage sites: A case study of the old town, Zuoying. *Built Heritage*, *4*(1), 1–15.

Loureiro, A. (2019). Innovation and technology – the only answer for sustainable tourism growth. *Worldwide Hospitality and Tourism Themes*, *11*(6), 743–747.

Manglis, A., Fourkiotou, A., & Papadopoulou, D. (2021). A Roadmap for the sustainable valorization of accessible underwater cultural heritage sites. *Heritage*, *4*(4), 4700–4715.

Mengual Torres, S. G., May Tzuc, O., Aguilar-Castro, K. M., Castillo Téllez, M., Ovando Sierra, J., Cruz-y Cruz, A. del R., & Barrera-Lao, F. J. (2022). Analysis of energy and environmental indicators for sustainable operation of Mexican hotels in tropical climate aided by artificial intelligence. *Buildings*, *12*(8).

Nam, K., Dutt, C. S., Chathoth, P., Daghfous, A., & Khan, M. S. (2021). The adoption of artificial intelligence and robotics in the hotel industry: prospects and challenges. *Electronic Markets*, *31*(3), 553–574.

Sinnappan, P., Salim, F. A. A., Maideen, M. B. H., & Kunjiapu, S. (2023). The effect of environmental concern and global environmental connectedness on green purchase intention of hotel customers in Malaysia. *International Journal of Sustainable Society*, *15*(1), 1–19.

Nijland, H., & van Meerkerk, J. (2017). Mobility and environmental impacts of car sharing in the Netherlands. *Environmental Innovation and Societal Transitions*, *23*, 84–91.

Nowacki, M., Kowalczyk-Anioł, J., & Chawla, Y. (2023). Gen Z's Attitude towards green image destinations, green tourism and behavioural intention regarding green holiday destination choice: A study in Poland and India. *Sustainability (Switzerland)*, *15*(10), 7860.

Olorunsola, V. O., Saydam, M. B., Arasli, H., & Sulu, D. (2022). Guest service experience in eco-centric hotels: a content analysis. *International Hospitality Review*, *38*(1), 81–100.

Paraskevas, A. (2022). Cybersecurity in travel and tourism: a risk-based approach. In Z. Xiang, M. Fuchs, U. Gretzel, & W. Höpken (Eds.), *Handbook of e-tourism* (pp. 1605–1628). Springer International Publishing

Peceny, U. S., Urbančič, J., Mokorel, S., Kuralt, V., & Ilijaš, T. (2019). Tourism 4.0: Challenges in marketing a paradigm shift. In M. Reyes (Ed.), *Consumer behavior and marketing* (pp. 39–58). IntechOpen.

Perles-Ribes, J. F., Ramón-Rodríguez, A. B., Moreno-Izquierdo, L., & Such-Devesa, M. J. (2020). Machine learning techniques as a tool for predicting overtourism: The case of Spain. *International Journal of Tourism Research*, *22*(6), 825–838. https://doi.org/10.1002/jtr.2383

Petrosyan, A. (2024, April 15). *Share of users worldwide accessing the internet in 4th quarter 2023, by device*. Statista. Retrieved August 30, 2024, from https://www.statista.com/statistics/1289755/internet-access-by-device-worldwide/

Read, J. (2020). *Will new travel technology invade your privacy?* National Geographic. https://www.nationalgeographic.com/travel/article/will-new-travel-technology-invade-your-privacy-coronavirus

Reverte, F. G., & Luque, P. D. (2021). Digital divide in e-tourism. In Z. Xiang, M. Fuchs, U. Gretzel, & W. Höpken (Eds.), *Handbook of e-tourism* (pp. 1–21). Springer International Publishing

Samala, N., Katkam, B. S., Bellamkonda, R. S., & Rodriguez, R. V. (2020). Impact of AI and robotics in the tourism sector: A critical insight. *Journal of Tourism Futures*, *8*(1), 73–87.

Saura, J. R., Reyes-Menendez, A., & Alvarez-Alonso, C. (2018). Do online comments affect environmental management? Identifying factors related to environmental management and sustainability of hotels. *Sustainability (Switzerland)*, *10*(9), 3016. https://doi.org/10.3390/su10093016

Shin, H., & Baek, S. (2023). Unequal diffusion of innovation: Focusing on the digital divide in using smartphones for travel. *Journal of Hospitality and Tourism Management*, *55*, 277–281.

Shivaprakash, K. N., Swami, N., Mysorekar, S., Arora, R., Gangadharan, A., Vohra, K., Jadeyegowda, M., & Kiesecker, J. M. (2022). Potential for artificial intelligence (AI) and machine learning (ML) applications in biodiversity conservation, managing forests, and related services in India. *Sustainability (Switzerland)*, *14*(12), 1–20. https://doi.org/10.3390/su14127154

Stankov, U., & Gretzel, U. (2020). Tourism 4.0 technologies and tourist experiences: a human-centered design perspective. *Information Technology & Tourism*, *22*(3), 477–488.

Sun, S., Wei, Y., Tsui, K. L., & Wang, S. (2019). Forecasting tourist arrivals with machine learning and internet search index. *Tourism Management*, *70*(July 2018), 1–10.

Sustacha, I., Baños-Pino, J. F., & Del Valle, E. (2023). The role of technology in enhancing the tourism experience in smart destinations: A meta-analysis. *Journal of Destination Marketing & Management*, *30*, 100817.

Szymańska, E., Panfiluk, E., & Kiryluk, H. (2021). Innovative solutions for the development of sustainable transport and improvement of the tourist accessibility of peripheral areas: The case of the białowieża forest region. *Sustainability (Switzerland)*, *13*(4), 1–23.

THE SDGS IN ACTION. (n.d). UNDP. https://www.undp.org/sustainable-development-goals

Tussyadiah, I. (2020). A review of research into automation in tourism: Launching the annals of tourism research curated collection on artificial intelligence and robotics in tourism. *Annals of Tourism Research*, *81*(February), 102883. https://doi.org/10.1016/j.annals.2020.102883

UNWTO. (2005). *Making tourism more sustainable – a guide for policy makers*. https://www.unwto.org/archive/global/publication/making-tourism-more-sustainable-a-guide-for-policy-makers-0

UNWTO. (2023). *Tourism and the sustainable development goals – journey to 2030*. https://www.e-unwto.org/doi/pdf/10.18111/9789284419401

Vecchio, P. Del, Mele, G., Ndou, V., & Secundo, G. (2018). Creating value from social big data: Implications for smart tourism destinations. *Information Processing and Management*, *54*(5), 847–860. https://doi.org/10.1016/j.ipm.2017.10.006

Wattles, J. (2022). *SpaceX just launched paying customers to the ISS. Here's everything you need to know*. CNN Business. https://edition.cnn.com/2022/04/08/tech/spacex-iss-launch-friday-scn/index.html

Wong, I. K. A., Lian, Q. L., & Sun, D. (2023). Autonomous travel decision-making: An early glimpse into ChatGPT and generative AI. *Journal of Hospitality and Tourism Management*, *56*(May), 253–263. https://doi.org/10.1016/j.jhtm.2023.06.022

World Economic Forum. (2021). *How to build smart, zero carbon buildings - and why it matters*. World Economic Forum. (2021, September 8). https://www.weforum.org/agenda/2021/09/how-to-build-zero-carbon-buildings/

Yan, J., Meng, Y., Lu, L., & Li, L. (2017). Industrial big data in an industry 4.0 environment: Challenges, schemes, and applications for predictive maintenance. *IEEE access*, *5*, 23484–23491.

Zhang, Y., Khan, U., Lee, S., & Salik, M. (2019). The influence of management innovation and technological innovation on organization performance. a mediating role of sustainability. *Sustainability (Switzerland)*, *11*(2), 495. https://doi.org/10.3390/su11020495

Chapter 8

Assessing Tourism Sustainability: A Case Study of Ladakh Region

Ishika Jaiswal

Faculty of Economics, Indian Institute of Management (IIM), Sambalpur (Odisha), India

Abstract

The Ladakh region stands at a crossroads when it comes to sustainability, grappling with the repercussions of mass tourism and an oversight of carrying capacity in its development. This has triggered apprehensions among policymakers, residents, and visitors alike. While Ladakh boasts tourism potential, there is a growing acknowledgment that the trajectory of tourism development must shift towards sustainability. Encouragingly, certain Indian Himalayan Region (IHR) states offer promising examples in waste management, visitor flow control, limitations on pilgrim numbers, and the promotion of community-based tourism. However, these positive instances often remain confined to specific contexts and lack broader scalability. Addressing the concerns and risks associated with tourism development in Ladakh necessitates a multifaceted approach. Priorities include minimising the adverse impacts on communities and the environment, maximising tourism's positive contributions to local economies, and emphasising the conservation of natural and cultural heritage. Furthermore, there is a call to harness opportunities and adopt innovative measures to steer tourism onto a sustainable path. The government has taken various policy measures to promote sustainable tourism in Ladakh along with launching awareness campaigns, organising international events and rebranding of Ladakh as an eco-tourism destination. In essence, Ladakh's challenge lies in transforming its tourism landscape into a model that not only reaps economic benefits but also safeguards its ecological and cultural treasures, ensuring a harmonious coexistence between hosts and visitors.

The Need for Sustainable Tourism in an Era of Global Climate Change: Pathway to a Greener Future, 151–172

doi:10.1108/978-1-83608-668-020241023

Keywords: Tourism landscape; Ladakh eco-tourism; community participation; tourism sustainability; sustainable development

Introduction

Mountains cover around 27% of the Earth's land surface and contribute to the sustenance and wellbeing of 720 million people living in the mountains and billions more living downstream. The challenge is to identify new and sustainable opportunities that can bring benefits to both highland and lowland communities and help to eradicate poverty without contributing to the degradation of fragile mountain ecosystems (Food and Agriculture Organisation (FAO), 2018).

Tourism Sustainability

The tourism industry being one of the rapidly growing industries globally, is known for its potential to generate employment and alleviating poverty in various parts of the world. In fact, before the COVID-19 pandemic resulted in travel restrictions, the tourism industry contributed for 10.6% of employment and 10.4% of GDP at global level in 2019. As far as the tourism sector in the Indian Himalayan Region (IHR) is concerned, it has undergone significant expansion and diversification in recent decades, making it a leading contributor to India's economic growth as the sector is anticipated to have an average annual growth rate of 7.9% from NITI Aayog (2018). Although the tourism industry is being promoted both at community level and government level, it is important to adopt and integrate the sustainable practices in and by to reap the potential of the positive externalities and mitigate the negative externalities. The introduction of sustainable practices in the tourism industry will help to assess the socio-economic effects on the environment. According to the UN Environment Program and UN World Tourism Organisation, 'tourism that takes full account of its current and future economic, social and environmental impacts, addressing the needs of visitors, the industry, the environment and host communities'. Therefore, the sustainability brings a balance among various dimensions of environment, economic, and socio-cultural connections of tourism development. UNWTO has defined a framework through which sustainable practices can be introduced and integrated in travel and tourism sector. This framework contains 12 different components to work upon including:

- Ensuring economic viability.
- Promoting prosperity at local levels.
- Enhancing the quality of employment.
- Achieving social equity.
- Providing satisfying and fulfilling experience for visitors.
- Engaging and empowering local community.
- Ensuring wellbeing of local community.

- Maintaining cultural richness.
- Maintaining and enhancing the quality of landscapes.
- Preserving biological diversity.
- Minimising wastage of resources.
- Protecting environmental goods and services.

Sustainable Tourism is not only concerned with the interaction of socio-economic and cultural integrity with the environmental determinants but also covers development of destinations and markets, promoting innovation and competitiveness, managing interests of stakeholders, assessing consumer behaviours and perceptions apart from maintaining ecological contributions to region's GDP. It is importance to emphasise the utilisation of local resources, promotion of entrepreneurial activities and stakeholder collaboration for development of sustainable tourism, especially in rural regions. In addition, a study in Botswana by Stone et al. (2011) considers the challenges in devising, conniving, and putting into practice the community-based tourism in line with the matters pertaining to community resolution, business awareness and income distribution. Subsequently, socio-cultural sustainability concentrating upon conservation and preservation of the cultural heritage along with promotion of community development comes into picture. In this regard, Ghirardello et al. (2022) examine the linkage between tourism activities and intangible cultural heritage thereby proposing means to achieve sustainable governance which aims to ensure socio-cultural sustainability. Further, the need of the hour lies in diverting the attention of policymakers and experts the social effects of tourism development from exhaustive commentaries to realistic determinations. Although various measures have been taken indigenously, there is much left to be done in arena of public tourism. Brandão et al. (2014) have undertaken a study regarding public tourism regulatory framework for indigenous tourism in Brazil. One classic example is Airbnb houses which are micro-level entities to promote social sustainability and can be used for increasing community participation in tourism industry thereby enlarging the positives of tourism development via collaborating with other stakeholders. Apart from this, local festivals are organised to generate social capital, specifically at those tourist destinations which have popularised in past few years. These festivals not only generate revenues to the government but also promote community integration in the respective regions. Moreover, there are various socio-economics determinants considered by Alcay et al. (2020) when it comes to providing weightage to ecological forces in sustainable tourism suggests that socio-economic factors such as per capita income, environmental spending, and education level contribute to this variation. Further, tourism development creates value through cross-cultural interactions between local communities and tourists which emphasises on promoting local participation while adopting sustainable practices.

Tourism Sustainability in India

The 2023 Union Budget has significantly emphasised tourism as a key sector in the front burner. The Ministry of Tourism is comprehensively promoting the tourism sector through actively involving all States, converging the Government Programs, and setting up poignant Public-Private Partnerships. This will help panoramic development in at least 50 chosen destinations deploying a challenge mode process. Considering the significance of heterogeneity and women empowerment, the tourism industry is pioneering in providing job opportunities to women. The Ministry of Tourism has also regarded the Year 2023 as the 'Visit India' year, so that people all around the world can experience what a nation like India has to offer. The tourism sector has endless possibilities to explore including country's rich cultural heritage, architecture, enticing landscapes and mesmerising natural beauty. Since tourism sector is one of the major contributors in India's economy as it leads to economic growth and job creation. In line with the data released by the Ministry of Tourism, the Travel and Tourism sector has generated nearly 80 million jobs in 2022, both directly and indirectly. Additionally, around 15% of the employment opportunities generated in 2019-2020 were from this sector only and the sector has contributed 5.19% of GDP. The country's is placed at 54th spot in the World Economic Forum's Travel and Tourism Development Index (TTDI) in 2021. Moreover, it is predicted that the sector will grow at about 8% CAGR in coming decade and generate nearly 140 million jobs by 2030.

India has been experiencing a new dimension of tourism – Adventure tourism which is enlarging the landscapes of exploration by the tourists and specifically encompassing the development in the interiors and downtown. Adventure activities are now being promoted through different tourist packages and contributing largely to the local economy of respective regions. In view of integrating the same with inclusivity and sustainability, the government recognised the significance of spreading awareness among the youth for preserving and conserving natural beauty, cultural ethos, and spiritual heritage. In addition, government has launched initiatives like State Ranking Criteria for Adventure Tourism Sites, Mega Trails, Model Adventure Tourism Law, and Adventure Activity Guidelines with an aim to make the adventure tourism responsible and make it internationally competitive at premium levels.

The government has also introduced the *National Strategy for Sustainable Tourism* as framed by the Ministry of Tourism (2022) to concentrate on sustainability in country's tourism sector so that resilience, inclusiveness, carbon neutrality and resource efficiency can be ensured with an aim to conserve and preserve the natural and cultural resources. The Union Minister for Culture, Tourism and Development of Northeastern Region, during High-Level Political Forum 2023 – UN World Tourism Organisation, mentions

> India is actively advancing Sustainable Development Goals through tourism, implementing various activities and strategies. India is creating a conducive environment for the tourism sector to contribute to economic growth, environmental protection and social wellbeing in a sustainable manner. India is committed to sustainable tourism and green tourism.

The National Strategy for Sustainable Tourism, with an aim to make India a world-class destination of sustainable and responsible tourism, has comprehensively taken under its gamut seven strategic pillars including promotion and protection of environmental sustainability, economic sustainability, socio-cultural sustainability, scheme for certification of sustainable tourism, capacity building, preservation of biodiversity and responsible governance along with emphasising on creating the synergistic effects involving stakeholders from 16 ministries of the Central Government, State Governments, Local Governments, industry networks, non-government institutions as well as localised communities. Furthermore, the government has introduced several other initiatives to serve the purpose like National Strategy for Ecotourism, Travel for LiFE, Swadesh Darshan Scheme, Sustainable Tourism for India Criteria and Indicators (STCI) among others. Therefore, Sustainable Tourism in India is eco-friendly, youth-friendly, and responsible by its very nature based on reduce, reuse and recycle approach.

Ladakh – The Land of Passes

Ladakh is a region characterised by breathtaking natural attributes, situated in a vast and impressive landscape. Encircled by two of the planet's most formidable mountain ranges – the Karakoram to the north and the Great Himalaya to the south – it is intersected by two additional parallel mountain chains, namely the Ladakh Range and the Zanskar Range. In geological terms, Ladakh is a relatively young region, having taken shape a few million years ago. Its fundamental contours, shaped by tectonic forces, have undergone transformations through the ages, sculpted by erosion from wind and water into the present-day landscape. Ladakh, being situated in a high-altitude desert, remains shielded from the rainfall brought by the Indian monsoon due to the imposing barrier of the Great Himalaya. In the past, Ladakh was covered by an extensive lake system, remnants of which can still be found on the southeastern plateaux of Rupshu and Chushul, manifested in the drainage basins or lakes of Tso-moriri, Tso-kar, and Pangong-tso. However, the primary water source now is the winter snowfall. Regions like Dras, Zanskar, and the Suru Valley, located on the northern slopes of the Himalaya, receive substantial winter snow, contributing to glaciers that, when melted, irrigate fields during the summer. For the rest of the area, the snow on the mountain peaks is virtually the sole water source. While the crops grow, the villagers hope not for rain but for sunlight to thaw the glaciers and release their water. Ladakh spans altitudes from approximately 9,000 ft (2,750 m) in Kargil to 25,170 ft (7,672 m) at Saser Kangri in the Karakoram Range. Summer temperatures seldom exceed 27°C in the shade, and in winter, they can occasionally drop to minus 20°C, even in Leh. Surprisingly, the thin air amplifies the sun's heat, making it more intense than at lower elevations. The region has spectacular topography across its divisions encompassing Monasteries (Thiksey, Alchi, Shey, Hemis, Likir, Diskit, Spituk, Phugtal, Hundur, etc.) Valleys (Sham, Shyok, Nubra, Zanskar, etc.), Palaces (Leh and Shey among others), Tourist Villages (Turtuk thang, Hundur, Lamayuru, etc.), Gompas and Stupas, Hot Water Springs (Chumathang, Puga, Demchok, etc.), Cold Desert Changthang region, Lakes (Pangong, Tsomoriri, etc.), various Treks and Passes (ChangLa, KhardungLa, UmlingLa, etc.), War Memorials and Hall of Fame, Bactarian Camel Ride, National Park and Wildlife Sanctuary apart from Hanle, Astronomical Observatory and India's First Dark Sky Reserve. For imagery of this rich landscape see Figs. 8.1–8.3.

Recent Initiatives of Ladakh Administration and Ladakh Autonomous Hill Development Council (LAHDC) for Tourism Sustainability

Since the tourism is opened in Ladakh back in 1974, the number of tourists visiting the region every year has grown various folds which alarmingly calls for replacing the practices adopted for managing the tourism industry. To exemplify the exponential growth of tourists, when Ladakh initially welcomed tourists in 1974, government records indicate that there were 527 visitors, comprising 500 foreign and 27 domestics but in 2022, the tourist influx has surged to 4.5 lakhs within an 8-month period, spanning from January to August. The tourists, both domestic and foreign, visiting Ladakh are surpassing the entire population of the region and to accommodate them, construction has increased to build different facilities, the quantum of waste generation got quite high, and there is undue pressure on the mountains and landscapes. This situation asks for adoption of sustainable practices in tourism industry of Ladakh at large, quickly and effectively. In this regard, the Ladakh government has taken following initiatives for ensuring tourism sustainability in the territory:

Green Fee and Inner Line Permit

LAHDC Leh and District Administration Leh have passed the order directing that the tourists visiting the region, both domestic and foreign must pay a fixed

Fig. 8.1. The Snowy Mountains. *Source*: Author.

Fig. 8.2. The Shyok River. *Source*: Author.

amount as Green Fee which comprises of the environmental fee, contribution to Red Cross Fund and wildlife fee. This will help in promoting quality tourism and protecting environment and ecology in various divisions. The authorities have scrapped ILP requirement for domestic tourists, however foreign tourists still

Fig. 8.3. The Pangong Tso Lake. *Source*: Author.

need to obtain the Protected Area Permit. Recognising the concession of ILP system, the administration has decided to specifically notify areas where tourists will not strictly be allowed to enter. In recent years, the government authorities have taken several measures to introduce and integrate such practices which ensure tourism sustainability in Ladakh.

Ladakh Eco-tourism Policy 2017

The LAHDC, Leh has implemented the tourism policy for the Leh district based on eco-tourism considerations. The policy emphasises the crucial need to guide Ladakh's tourism industry comprehensively. Aligned with the vision for 2025, the administration embraces an eco-tourism policy to position Ladakh as a globally recognised eco-destination at pan level. The initiative focuses on allocating the revenue generated, before COVID-19 outbreak, from environmental fees (Rs. 5.30 Crore) to be used as seed money for policy implementation as it is important to overtake the individual interests through general welfare and demands. The said revenue is used for maintaining public convenience facilities and sanitation (1 crore), spreading awareness/publicity (50 lakhs), preserving and promoting cultural heritage (1 crore), exploring new destinations and trekking routes (50 lakhs), developing theme-based economy (50 lakhs), developing and supporting homestays (50 lakhs), managing different types of waste generated (50 lakhs), protecting water resources (30 lakhs), and promoting tourism in rural areas (50 lakhs). The policy is implemented strategically based on four aspects ensuring tourism sustainability in the region including securing a steady influx of quality tourists, equitable spending of revenue generated from tourism among local communities, executing environmental protection mechanisms effectively, preserving and promoting culture heritage and value system of Ladakh.

Tourism Incentive Policy 2020

The newly formed Union Territory of Ladakh has implemented the first Tourism Incentive Policy to ensure equitable and sustainable development of all regions. The policy focuses on preserving and promoting tourism assets in remote and untouched tourist destinations in the territory. This policy has helped people associated with tourism industry to develop skills by providing hands-on training. It has created a fund as Ladakh Tangible Cultural Heritage Conservation Fund with an aim to conserve the rich historic and heritage sites of the region. In addition, it provides subsidy through Capital Outright Investment Subsidy Scheme in lieu of the investments with low gestation period made for assets creation for new tourism units based on local traditions and architecture. Moreover, the policy has promoted adventure tourism via Support for Adventure or Recreation Tourism Scheme where subsidy is provided subject to a maximum of 20% of the cost of different types of equipment of standard specifications installed for adventure tourism activities. Further, the initiative has been taken to replace the dry toilets wherein the homestays will be given a pre-determined amount to install the flush

Fig. 8.4. The Leh City Night View. *Source*: Author.

toilets as well as for their repairs and maintenance. See Fig. 8.4 for a night view of Leh City.

Ladakh Homestay Policy 2023

The administration of the Union Territory has rolled out Ladakh Homestay Policy 2023 with a revised framework which has been made comprehensive based on the suggestions and recommendations collected from the concerned stakeholders on the Homestay Policy of 2020 with an aim to ensure tourism sustainability and economic development of rural and border areas in the region. In line with this, the administration has incentivised and expanded the development of homestays outside the protected areas via four dimensions focused on strengthening local communities – Economic Empowerment, Environmental Responsibility, Cultural Exchange, and Community Development. The policy will remain effective till five years where eligibility of the homestays is subject to assessment by the Selection Committees which include Field Level Physical Verification Committee, District Level Committee, UT Level Committee, and Technical Monitoring Committee. The administration includes both new and existing beneficiaries for fiscal and non-fiscal incentives under the policy.

District Environment Management Plan

In response to the order of National Green Tribunal, the administration of the union Territory of Ladakh has implemented the District Environment Management Plan (DEMP) recently. The DEMP comprises of elements including assessment of environmental effects, complying legal and regulatory requirements, deciding environmental objectives and targets, conducting environment management programs, promoting responsible and structured tourism, conducting training and awareness programs, controlling documentation, preparing for emergency situations, managing reporting, auditing, and reviewing formalities. This initiative will promote sustainable practices in the region by focusing on different types of waste management including solid waste management, plastic waste management, construction and demolition waste management, biomedical waste management, hazardous and other waste management along with e-waste management. Additionally, the government will take care of water and air quality as well as light and noise pollution in several areas.

Campaign for Sustainable Eco Tourism in Ladakh

Indian Institute of Development Studies has decided to run a campaign designed by the IISD Research and Knowledge Support Centre for Sustainable Development in Ladakh based on the research done by the organisation in the Hindukush Himalayas with an aim to identifying and reaping the benefits of promoting and preserving Sustainable Tourism in the region in collaborative manner with the respective stakeholders. The campaign ensures that the local communities should benefit from the tourism activities of all kinds, expansion of livelihood opportunities in the regions, publicity for eco-tourism and integrating sustainable tourism practices for Natural and Cultural Heritage Conservation. In line with this campaign, all those who are associated with the tourism industry along with tourists need to abide by the guidelines and behavioural practices to support eco-tourism.

Community-Based Eco-Tourism for Preserving Snow Leopards

Spotting the elusive snow leopard during a trek in the rugged and secluded mountains of Ladakh can be a dream realised for wildlife enthusiasts. Nevertheless, these predators have gained an intimating reputation among locals, who often perceive them as nuisances due to their inclination to prey on livestock. In some instances, these majestic cats have been responsible for the loss of multiple animals in a single incident. Consequently, certain local livestock herders have resorted to revengeful killings of these snow leopards. To mitigate losses stemming from livestock depredation and foster greater tolerance for snow leopards, the Himalayan Homestays Program (HHP) was initiated in various valleys of Leh in 2002 by the Snow Leopard Conservancy – India Trust. This program has given local residents an opportunity to supplement their income by hosting tourists and is currently operated in collaboration with the global wild cat conservation organisation, Panthera. Residents underwent training in housekeeping, while young individuals acquired skills as guides to lead nature tours, offering the chance for visitors to potentially encounter the captivating snow leopards. For assessing the impact of the HHP on the perceptions of local communities towards snow leopards, some interviews have been conducted across seven Ladakhi villages, with four situated in the western Sham valley and three in the eastern Rong valley. Employing qualitative semi-structured interviews lasting between 15 and 60 minutes, villagers were queried about their participation in the HHP and the importance they attributed to wildlife. They emphasised the significance of intrinsic, instrumental, and economic values associated with the program which they have realised during their involvement. Intrinsic values encompass an appreciation for nature even when it does not directly benefit humans, while aesthetic, spiritual, and ecological values fall under instrumental values. Economic values, on the other hand, are rooted in monetary and market transactions.

Rebranding of Ladakh as a Sustainable Eco-destination

The Ministry of Tourism revealed Ladakh's undiscovered potential and rebranded the region as a sustainable eco-destination with an aim to transform it into a high-value low-impact destination for tourists. The Ministry is in the view that tourism in the region is restricted to some divisional areas only while remote areas are untrodden resulting in skewed migration. Therefore, development and resources allocation must be holistic and equitable so that ample of employment opportunities would be created, local participation can be increased across the region and uneven migration can come to a halt. The administration is focusing on infrastructural development and building capacity to promote and showcase local handicrafts, encouraging community integration, and like to make the Ladakh self-reliant. See Fig. 8.5 for an image of rural life.

Tapping Market for Winter Tourism

The Ladakh administration has started organising and promoting winter sports in certain divisions with the rationale that it will evenly spread the tourists round the year and supporting those local communities which reside there in the harsh

Fig. 8.5. The Sankoo Village, Kargil. *Source*: Author.

climatic conditions during the winters. The sports like ice-hockey, ice-skating, ice-climbing, hiking among others are being extensively promoted. Since Ladakh has a prospective market for the winter tourism, the government has found ways to tap that market.

Pilgrimage Tourism and Wellness Tourism

The region is feature with Buddhist culture and various stakeholders have introduced the tour packages focusing on pilgrimage tours and educational tours in a view to propagate the richness of their spiritual heritage and educate the students, researchers among others about the natural and cultural heritage Ladakh is characterised by since long time. In addition, the administration is also in view to launch another dimension of sustainable tourism as wellness tourism by opening yoga retreats as well as health and wellness centres like South India has done.

Organising National and International Events

For showcasing and promoting the local culture, value system and traditions, the administration has unveiled the two carnivals of 13-day Zanskar Youth Festival and 2-day Ladakh's Hemis Festival which has embraced both domestic and foreign tourists at large levels. In addition, to promote one of the dimensions of sustainable tourism, Ladakh has been centre of attraction for first-ever 3-day International Music Festival in Leh where local bands and groups across the country have performed in remembrance of martyrs. Further, the region successfully hosted the World's Highest International Fashion Runway, under Vibrant Ladakh Festival, wherein international models from 14 nations around the globe have participated.

Reliving Restrictions for Tourists

The administration and council, in consultation, have dispensed the restrictions on tourists for visiting the Siachen Base Camp after heavy demand for the same by people of Nubra and associated stakeholders for last two years and civil tourist facility centre is set up near the camp. This will add to the livelihood opportunities in the region. Additionally, the authorities have allowed foreign tourists to stay overnight in Hanle which has Indian Astronomical Observatory, along with the villages of Marsimik and Sko valleys which were earlier open to domestic tourists only. The decision has been taken to increase the influx of tourists and to unveil the unexplored areas of the region.

Ladakh on Cross-roads: Sustainable Tourism vs Ecology Preservation

After being carved out of Jammu & Kahmir, Ladakh has now stuck in choosing between promoting sustainable tourism and preserving the fragile Trans-Himalayan ecology. Tourism in Ladakh has witnessed a consistent rise over the past few decades. However, with each passing year, this growth has imposed an unsustainable burden on the region's natural resources. The desire for an increased number of tourists is juxtaposed with the concerns of environmental experts who are apprehensive about the potential adverse effects on the local ecosystem. To say, there are several incidences where the private vehicles got stuck in lake shore, tourists were deliberating rupturing sand dunes, bikers were doing off-road driving among others, and they have invited lots of criticism on social media indicating the lack of administrative efficiency. Sand dunes are themselves habitat for various species who do ground nesting and damaging sand dunes damages the species also which in turn leads to ecological imbalance. Similarly, off-road driving by tourists disturbs the habitat patterns resulting habitat fragmentation. Additionally, offshore encroachment is damaging rare flora and fauna resulting in shrinking of lakes in the region, specifically in interiors. The major reasons attributed to this dilemma of choosing sustainable tourism vs ecology preservation include the following.

Undue Pressure on Natural Resources

The increasing number of tourists in unbound manner is pressing tremendous pressure on Ladakh's natural resources. Since the region is popularised for its scenic beauty and rich natural heritage, there is no responsibility and accountability when it comes to manage the use of natural resource. Being the mountainous area, people traditionally use dry toilets which are now getting replaced by western toilets and the newer ones are subject to higher installation cost and are not eco-friendly for the region. Further, the higher number of tourists visit, greater quantity of resources will be supplied. To say, the number of guesthouses, hotels, and homestays are rising in uncontrolled manner. The mountains are bombarded for building infrastructure to serve the rising number of tourists which in turn threatening the ecosystem of the region. Also, there are certain products which should not be used in Ladakh as they will be harmful for users and the region as well however they are getting sold over the counter despite knowing that they are not environmentally sustainable. It is well known that natural resources need to be used rationally and any disturbance in ratio of resources to people will create ecological imbalance as it is currently happening in Ladakh.

Excessive Waste Generation

The rapid and unstructured development in Ladakh is creating tonnes of waste which presently cannot be treated properly. The development of certain sub-sectors of industrial and services sector is generating various types of wastes which alarmingly calling for its sustainable disposal. The household waste and sanitation residue has increased with the increase in the number of tourists staying at different places across the region. The people being irresponsible throw packets and packages for several products they use anywhere without considering that this plastic waste can damage the ecosystem of Ladakh like blocking and contaminating the water bodies among others. Also, the initiative has been taken to ban plastic in the region but has not implemented effectively and one can easily find various types of plastic waste here and there at tourist destinations. Further, the construction and demolition activities are undertaken in Ladakh in a rapid and unplanned manner which is generating construction and demolition waste which the administration doesn't know to dispose it off sustainably. Consequently, it is recklessly dumped anywhere leading to ecosystem disturbances in the region. Moreover, people in Ladakh rely on traditional methods of curing diseases using different herbs with medicinal properties. They used to manage their health and wellness themselves and that too efficiently. But the increased visits of tourists, both foreign and domestic, calls for setting up modern medical facilities in an extensive manner and making available modern drugs including other medical paraphernalia so that they can use them which in turn generated biomedical waste that is not even degradable and is utterly unsustainable for the Ladakh as a distinct geographical area to manage this kind of waste. In addition, the region was not open and well-connected with nearby areas which made it electronically backward however rapid introduction of different electronic equipment, appliances and devices generated lot of e-waste in Ladakh. The rising number of tourists carry various electronic equipment and devices like cameras, GPS trackers, chargers, headphones, mobiles, power banks, etc. which adds to the e-waste in the region and depending on the climatic nature and geographic location of the Ladakh, it is very difficult to dispose such kind of waste there. Thus, the higher number of tourists not only generate higher revenues but also higher quantity of wastes thereby challenging the measure to ensure sustainability in Ladakh's tourism industry.

Alarming Levels of Pollution

Ladakh is featured with fragile ecosystems across the divisions and sub-divisions encompassing from Karakoram range, Ladakh Range, and Zanskar Range. This fragility is now more difficult to deal with as the pollution levels for air, water, noise, and light have gone up in recent years in the region. The increased number of tourists lead to increased number of taxis, bikes, and other private vehicles, which deteriorated the air quality of various areas. Since there is no option of using green fuel in most of the vehicles, more importantly in heavy ones, the air pollution is rising in the region. Moreover, the running of vehicles through water streams, throwing garbage in and around the waterfalls, mushrooming camping sites, tampering with agriculture and vegetation, opening flow of household and industrial waste in and around the water bodies has contaminated the water in Ladakh. The region where people happily rely on fulfilling their water needs from water springs and streams are now bound to find ways to filter the water in most of the habitats attracting higher number of tourists. Further, the rising number of tourists and transportation has resulted in noise pollution and Ladakh's ecosystem being highly sensitive is unable to sustain with the same thereby witnessing ecological imbalance affecting the lives of all the living creatures in the region. Additionally, Ladakh is home to India's first Dark Sky Reserve in Hanle and measures are taken to weed out the light pollution there. To the surprise, the lights remain on till late night or midnight at some places while throughout the night at other places in Hanle despite knowing that these lights will overshadow the spectacular view of the sky and milky way. Allowing foreign nationals to stay in Hanle for promoting tourism is going to create sustainability challenges for the administration.

Loopholes in Regulatory Framework

The administration of Union Territory of Ladakh is still lacking a strong regulatory framework to manage the increased influx of tourists. In the name of some appreciable steps, it has levied environmental fee and allowed a campaign for sustainable eco-tourism in the region. The other initiatives taken by the administration in consultation with the Council are focused on promoting tourism extensively along with bringing the untouched areas also in its gamut without any consideration whether the region is capable to carry this extensive tourism. The regulatory aspects taken into account though plan of vision documents include only managing the tourism and environment however there are no strict guidelines how the sustainability will be integrated in region's tourism industry at ground level. To say, the bike providing agencies are not following any guidelines as to which bikes will be made available and how documentation is to be done. One more example is that locals managing guest houses generate permit for tourists which sometimes forge the documents. Further, opening of certain areas like Siachen Base Camp for tourists has its own concerns and can lead to Centre-Territory frictions in the future considering geo-political risks associated with the destination.

Inefficient Information dissemination

Information Dissemination is a significant dimension of tourism sustainability because adoption and integration of sustainable practices in Ladakh's tourism industry depends on what and how well the required information is provided to all the stakeholders so that their responsible participation can be increased. Now with the rapid rise in the number of tourists, it is very difficult for Ladakh's administration to spread awareness to preserve ecological balance in various habitats of the region. The authorities want to increase the participation of local communities but they themself have no idea how to effectively participate so that regions cultural, natural, and spiritual heritage can be promoted in sustainable manner as they are either not given detailed information about the measures taken or there is no awareness among them to capitalise on livelihood opportunities in sustainable way.

Uneven Sectoral Development and Migration Patterns

Since tourism industry has been emerged as the major livelihood provider in Ladakh governing the region's economic development. Tourism industry being a part and parcel of services sector is popularising the sector along with creating forward and backward linkages like construction of places for stay, setting up tours and travels agencies, etc. However, this thrust on services sector is sidelining the other sector that could sustainably contribute to Ladakh's economic development. The younger generation is largely associating itself with this sector and not even paying attention to their studies just like skimming the potential the sector has for the region. This deviation of youth from education and agriculture will gradually affect the vegetation and literacy in an adverse manner which cannot be considered a good practice. In addition, there is haphazard migration happening in the region as people are leaving their original habitat and shifting to areas which are being developed as tourists attracting destinations. In this scenario, some of the interiors of Ladakh are getting deserted and this is against the principles of sustainability.

Climate Change Concerns

The excessive use of natural resources and questioning out the carrying capacity of the region, Ladakh is witnessing climate change issues for past few decades. The entire world is fighting with climate change concerns through action plans and like measures since long, but Ladakh has started experiencing this dangerous phenomenon just after increased influx of tourists which has increased to such a level that the region is incapable to sustain with. In a small span of few years, the region has begun calling for green zone. To exemplify, the average temperature in Ladakh is increased by 3°C in the past three decades. The glaciers are melting with an abnormal rate with some facing existential threat. The region is witnessing water crisis because of rising temperatures and decaying snow covers. The limited vegetation in Ladakh is also in danger due to climate change and can lead to food shortages in future. Further, the climate change is affecting precipitation patterns, extinction of species, health-related issues, and displacement at large. Since, sustainability cannot be incorporated in Ladakh's tourism industry when tourism promotion activities are resulting in climate change concerns. A hockey player and member of Ladakh Women Ice Hockey Foundation mentions:

> I am among the first generation of women hockey players in Ladakh. I've been playing defence in the national team since 2016. Since we don't have a proper artificial ice hockey rink, we are completely dependent on the frozen ponds. And now climate change has created a greater problem for us. When I first started playing, we would get at least two to four months to play in the winter, but now, even in this cold area, winter arrives late and finishes early and so we get only two months of play.

Ladakh stands as an ideal location for the region's beloved winter sport – ice hockey. Initially introduced to eastern Ladakh as a recreational activity by Indian Army personnel around 50 years ago, the sport has gained significant popularity among the locals. Those who have grown up in Ladakh cherish memories from their childhood when playing ice hockey was an integral part of their winter holiday routine. Presently, Ladakh boasts more than 20 ice hockey teams, catering to both men and women. Notably, 18 out of the 20 members in the Indian women's ice hockey team hail from Ladakh. The region hosts renowned winter tournaments like the Lieutenant Governor Cup and the Chief Executive Councillor Cup, attracting teams from various parts of Ladakh. In February, Ladakh hosted the Ice Hockey National Championship 2023. However, over the last few decades, scientists have documented a decline in both snowfall and glacial mass in Ladakh. The increased influx of tourists further complicates the environmental challenges faced by the region.

Serenity Surpassed by Uncertainty and Sustainability

Formerly exclusive to trekkers and bikers, Ladakh was once considered a dream destination. However, it is now evolving into a hub for mass tourism, leading the locals to sense a transformation in their region. Coping with extreme weather is not the sole challenge for residents of high-altitude villages; they are also actively engaged in navigating the shift in lifestyle while striving to preserve their traditions. A small village, nestled in the heart of Hemis High Altitude National Park, is renowned as one of the premier spots globally to witness the elusive snow leopard in its natural habitat, boasting a notable density of these magnificent creatures. Following years of conflict between villagers and wildlife, and thanks to the concerted efforts of the autonomous hill council of Ladakh, this village, situated at an elevation of nearly 5,000 metres, has embraced sustainable tourism as a means of livelihood. By transforming the perceived threat of the snow leopard into an opportunity through the establishment of homestays, Rumbak has become a noteworthy model for other communities. Over the past decade, ecotourism has thrived and expanded in Ladakh due to the increasing number of visitors. However, Rumbak now finds itself heavily reliant on this industry. With the younger generation pursuing education in Leh city and seeking opportunities elsewhere in the country, there is a lack of manpower to tend to the village's cattle and lead them to distant grazing patches in this frigid desert. Additionally, the decline in demand for pashmina has further impacted the village's revenue, presenting a challenge for the community. Similarly, another village providing glimpses of Ladakh's serenity, situated on Manali-Leh Highway, has witnessed considerable loss in livestock in past half a decade though households own small fields, but its location makes it vulnerable to natural calamities as the one it has experienced in 2010. The people there are searching for eco-tourism alternatives with the help of agencies and authorities for earning livelihood, retaining next generations, and marketing their products. However, obtaining and maintaining a balance between sustainable livelihood and uncertainty, being the two sides of region's serenity coin, is really difficult for the villagers.

Need for Tourism Sustainability in Ladakh

The surge in total tourists visiting Ladakh every year has undeniably bolstered Ladakh's economy, but it has also brought to the forefront several challenges related to the conservations of environmental cultural and traditional practices. Ladakh, a cold desert, relies on the melting glaciers for its fresh water supply during the summers. Unfortunately, climate change and global warming have led to accelerated melting rate of these glaciers, diminishing the crucial water supply in Ladakh. This impending water scarcity poses a threat to both the demands of tourism and the essential needs of the local population. The escalating use of motorised vehicles, both two-wheelers and four-wheelers, to facilitate transportation for tourists to various destinations is adversely impacting the air quality in this vegetation-deficient desert. The absence of sufficient green cover exacerbates the pollution's effects. While the tourism industry has emerged as a lucrative avenue for economic gain, it has also triggered a migration of younger generation from interiors to the urban hub of Leh, resulting in the abandonment of sustainable farming practices that once rendered villages self-reliant. With dwindling engagement in farming, villages now rely on external sources for fruits, vegetables, and grains, resorting to chemical fertilisers and pesticides to meet local demands. The unregulated practices and lack of management have significantly contributed to the deterioration of the fragile ecosystem. Given that tourism is the primary economic driver for the region, there is a pressing need to shift focus towards ensuring its sustainability. Iconic tourist destinations like Pangong Lake and Nubra Valley face substantial threats due to mass tourism since they are also incapable to manage different types of wastes generated during the process along with the methods of disposing the same. Ladakh's current infrastructure and sewage treatment capabilities are ill-equipped to handle the escalating volume of waste and sewage generated by the surge in tourist numbers in recent years. The growing concerns extend beyond tourists, as locals also play a significant role in the problem. The economic allure and potential of the region have led to the establishment of commercial ventures on agricultural lands, encroachment on protected areas, disrupting wildlife, and contributing to the proliferation of waste across the landscape.

Incorporation and Assessment of Tourism Sustainability

Ecotourism represents a sustainable approach to travel that nurtures the local environment rather than placing undue stress on it and exploiting natural resources. Given the direct negative impact of conventional tourism on Ladakh's environment, there is a pressing need to shift towards more eco-friendly tourism practices. A fundamental aspect of ecotourism is raising awareness among tourists about the impact of their visits on both the environment and local population. A crucial step is understanding how to downsize this impact, emphasising that ecotourism is most effective when carried out in considerably small and medium tour groups. Moreover, it involves providing fiscal incentives to the locals along with increasing their participation at various levels. Incorporating sustainable practices in tourism industry of Ladakh hinges on fostering an understanding among tourists and travel agencies about the critical importance of preserving natural balance for future generations. To promote ecotourism, it is imperative to view nature as an invaluable component of human survival. For instance, trekking is a popular activity in Ladakh, but many tourists may lack awareness of the region's history and culture. Encouraging the use of homestays during treks not only minimises environmental impact but also allows trekkers to immerse themselves in Ladakhi culture. Ultimately, by embracing ecotourism principles, Ladakh can pave the way for a more sustainable and harmonious coexistence between tourism and its fragile ecosystem through following ways:

Responsible Governance and Tourism Sustainability

The good governance plays a significant role in developing, implementing, and promoting sustainability in tourism sector. A region like Ladakh needs involvements of local governments and local population to launch and making successful any of the eco-tourism promoting initiatives. Responsible governance can lead to responsible tourism via providing eco-friendly products and services to tourists, availing eco-friendly accommodation and stays, promoting traditional business at small and medium levels, rationalising exploration of new tourist destinations along with restricting the development of existing ones in the region. The accountability and transparency dimensions of good governance also help in designing, implementing, and supervising the regulatory framework for developing and practicing sustainable tourism in the region.

Technological Integration and Tourism Sustainability

Utilisation of cutting-edge technologies can help in maintaining tourism sustainability in the region in several ways. Firstly, virtual and augmented reality technologies help tourists to have fascinating experiences remotely without being actually present at the destinations like one experience in the gaming industry. Secondly, drones can be used for supervising the ecologically sensitive areas, thereby assisting in habitat preservation, and satellite imaging can be done to collect and analyse data for promoting sustainable development with least possible human intervention. Thirdly, mobile applications can be used to spread awareness among different stakeholders so that they can take informed decisions, participate in eco-tourism campaigns, know about region's flora and fauna, thereby behaving in a more responsible manner. Furthermore, technological integration doesn't only make travel destinations sustainable but also the travelling practices.

Rationalising Land Use Policy in Mass Tourism Destinations

There is an urgent need for aligning the land use in the region with the carrying capacity in areas attracting greater number of tourists. The administration should categorise tourism destinations into motorable and non-motorable to ensure sustainability. In addition, the vehicles should be categorised before they choose a route to run so that the areas which are not sustainable enough for heavy vehicles can be identified.

Making Value-Chain Efficient through Community Participation

The local population is supposed to be benefitted from the influx of tourists since the Ladakh's economy is tourism-driven in nature, therefore, sustainability practices need to be adopted and integrated through promoting the community-based tourism which in turn will make the habitat, livelihood, and income streams of locals sustainable. The community engagement can be increased by spreading awareness and educating the local people with a view to make them efficient enough to persuade the tourists to stay traditionally, thereby contributing to preservation and conservation of rich natural heritage of the region.

Promoting Theme-based and Responsible Tourism through Capacity Building

The authorities need to promote theme-based tourism in Ladakh including adventure tourism, spiritual tourism, educational tourism, so that decentralised tourism can be incorporated into the industry. This can be achieved through capacity building in considerations with tourism zonation line with the existing potential of various zones wherein the principle of sustainability will be incorporated at the very inception of opening the destinations for tourists.

Implementing Tailor-made CSER Policies for Ladakh

Since Public Private Partnership (PPP) is being encouraged by the government in past few years, it can also be introduced in Ladakh to attain eco-friendliness in various dimensions. The government bodies have their kind of limitations which can be managed through PPP model after adjusting for the potential concerns in regulatory framework for the same. The public authorities should introduce customised polices for Corporate Social and Environmental Responsibility for the region to promote sustainable tourism and eco-friendliness linked with economic prosperity.

Incentivising the Initiatives for Sustainable Tourism

The government launches various plans, polices, programs and campaigns, both single-handedly and in collaboration with various local, national, and international organisations working to promote sustainable tourism, which should be subject to assessment and monitoring at regular intervals so that their performances can be linked to provide fiscal and non-fiscal incentives. The measure taken by the administration in Ladakh for developing and promoting eco-tourism and sustainability in the region at large should be revised as per changing dynamics. The stakeholders should be incentivised for their participation and contribution in adopting and integrating sustainable practices in Ladakh's tourism industry at micro as well as macro levels.

Concluding Remarks

Take Nothing but Memories, Leave nothing but footprints!

– Chief Si ahl

Ladakh, translating to the 'Land of the High Passes', is endearingly referred to as the 'Little Tibet' in India. Characterised by enormity, solitariness, resilience, warm hospitality, spiritual richness, cultural traditions, and scenic beauty, Ladakh evokes a unique blend of experiences. Journeying through Ladakh envelops one in profound serenity and a tranquil sense of peace. However, amidst this tranquillity, a reminder of the region's strategic importance emerges in the form of a notable military presence, reflecting the longstanding territorial disputes involving surrounding nations. However, the inauguration of the Atal Tunnel at Rohtang Pass in October 2020 has ushered in a significant surge in vehicular traffic over the past two years. Remarkably, the traffic influx has experienced an unprecedented fourfold increase compared to the two years preceding the tunnel's opening. This surge in tourist activity, facilitated by the Atal Tunnel, at such a higher altitude, has raised concerns among NGOs and environmentalists. These concerns stem from the potential impact of escalating tourist numbers and the accompanying surge in vehicular traffic on the delicate Trans-Himalayan ecosystem. The concern is further amplified by ongoing initiatives aimed at making the route to Ladakh more accessible and navigable under diverse weather conditions so that more tourists can visit the region. As Ladakh grapples with the delicate balance between preserving its pristine environment and catering to the increasing demands of tourism, thoughtful consideration and sustainable practices become imperative. Therefore, policy crevices should be analysed but comprehensively, master plans should be rolled our but strategically, climate change challenges should be tackled but cautiously, and actions should be taken but inclusively so that Ladakh's beautiful landscapes, spectacular beauty, and breathtaking experiences can be enjoyed forever via maintaining ecological balance, adopting sustainable tourism practices, and synchronising the stakeholders with customised needs of the region.

References

Alcay, A., Montañés, A., Fernández, M. B. S. (2020). Waste generation in Spain. Do Spanish regions exhibit a similar behavior? *Waste Management*, *112*, 66–73. https://doi.org/10.1016/j.wasman.2020.05.029

Brandão, C. D. N., Barbieri, J. C., & Junior, E. R. (2014). Analysis of the social, cultural, economic and environmental impacts of indigenous tourism: a multi-case study of indigenous communities in the Brazilian Amazon. *Sustainable Tourism*, *187*, 175–185. https://repositorio.fgv.br/server/api/core/bitstreams/96bf9c34-1396-4899-93d0-c00cc1f60cd4/content

Ghirardello, L., Walder, M., Rachewiltz M. de., Erschbamer, G. (2022). Cultural sustainability from the local perspective: The example of Transhumance in South Tyrol. *Sustainability*. *14*(15), 9052. https://doi.org/10.3390/su14159052

Ministry of Tourism in (2022). National Stratregy for Sustainable Tourism. https://tourism.gov.in/sites/default/files/2022

NITI Aayog. (2018). Report of Working Group II Sustainable Tourism in the Indian Himalayan Region. scstsenvis.nic.in/WriteReadData/links/Report%20of%20Working%20Group%20II-86489280.pdf

Stone, L. S., & Stone, T. M. (2011). Community-based tourism enterprises: challenges and prospects for community participation, Khama Rhino Sanctuary Trust, Botswana. *Journal of Sustainable Tourism*, *19*(1), 97–114. https://doi.org/10.1080/09669582.2010.508527

Chapter 9

Homestay as a Sustainable Tool for a Greener Future: A Study of Homestays in Madhya Pradesh, India

Brij Mohan Singh[a], Hitesh Tripathi[a], Sharad Srivastava[b] and Gyanendra B. S. Johri[a]

[a]*Department of Tourism Management, Indira Gandhi National Tribal University, Amarkantak, Madhya Pradesh, India*
[b]*Department of Commerce, Indira Gandhi National Tribal University, Amarkantak, Madhya Pradesh, India*

Abstract

This study is an attempt to explore the role of homestays as a sustainable but strategic tool for fostering a greener future, focusing on their prevalence in Madhya Pradesh, India. Homestays contribute to sustainable tourism by providing immersive experiences, supporting local economies, and minimising environmental impacts. The model offers economic benefits to hosts, promotes cultural exchange, and reduces the environmental footprint associated with conventional accommodations. The research analyses the distribution of homestays across different districts in Madhya Pradesh, revealing variations influenced by factors such as urbanisation, cultural appeal, and historical significance. High homestay counts in districts like Bhopal and Indore suggest an urban tourism focus, while lower counts in rural areas present untapped potential. The study discusses the reasons behind the observed disparities, including tourist attractions, infrastructure, and economic development. Despite their positive impact, homestays face challenges like cultural sensitivity, privacy concerns, and

The Need for Sustainable Tourism in an Era of Global Climate Change: Pathway to a Greener Future, 173–194

doi:10.1108/978-1-83608-668-020241026

potential environmental impact. Addressing these challenges is crucial for ensuring the long-term success and sustainability of homestay initiatives in contributing to responsible and meaningful tourism.

Keywords: Homestays; sustainable tourism; environmental impact; greener future; cultural exchange

Introduction

Homestays, as a tool for sustainable tourism, play a pivotal role in fostering a greener future by promoting responsible travel, supporting local communities, and minimising the environmental impact associated with conventional tourism practices. This accommodation model contributes to the broader framework of sustainable tourism in several key ways.

Homestays provide an immersive experience that goes beyond typical tourist accommodations. Travellers have the opportunity to stay with local families, engaging directly with the community's way of life, traditions, and cultural practices. This cultural exchange not only enriches the travel experience but also promotes the preservation of local heritage.

Homestays offer a direct economic benefit to local hosts, channelling income into community pockets rather than large international hotel chains. This economic empowerment enhances the overall well-being of the community, providing opportunities for entrepreneurship and fostering a sense of pride and ownership among locals.

One of the key advantages of homestays is their relatively low environmental impact compared to traditional hotels and resorts. Homestays often involve existing residences, minimising the need for additional construction and reducing the carbon footprint associated with large-scale hospitality infrastructure. Additionally, hosts and guests in homestays are more likely to adopt sustainable practices, such as energy conservation and waste reduction, contributing to overall environmental conservation efforts.

Homestays are a prime example of community-based tourism, where residents actively participate in and benefit from the tourism industry. This approach helps distribute the economic benefits more equitably among community members, ensuring that tourism revenue is dispersed across various sectors and services.

The rise of homestays offers an alternative to the concentration of tourists in crowded urban centres and popular tourist destinations. By dispersing travellers to local neighbourhoods and rural areas, homestays contribute to alleviating the issues associated with over-tourism, such as environmental degradation, congestion, and cultural dilution.

Homestays, especially those designed with sustainability in mind, can incorporate energy-efficient practices and sustainable resource use. Hosts may adopt eco-friendly technologies, such as solar panels or rainwater harvesting systems, contributing to the overall reduction of the environmental footprint associated with accommodation.

Homestays provide a platform for educating both hosts and guests about responsible tourism practices. Hosts can share local conservation efforts, cultural norms, and sustainable living practices, fostering a sense of environmental and cultural stewardship among visitors. Travellers, in turn, gain a deeper understanding of the places they visit and are more likely to adopt sustainable behaviours in their future travels.

Homestays offer a more authentic and personalised travel experience compared to standardised hotel accommodations. This personal touch enhances the overall satisfaction of travellers and fosters a deeper connection between hosts and guests, contributing to the positive impact of tourism on both sides.

Overall, homestays serve as a powerful tool for sustainable tourism, aligning with the principles of economic, social, and environmental sustainability. By choosing homestays, travellers actively contribute to the creation of a greener and more responsible tourism industry, ensuring that their journeys leave a positive and lasting impact on the destinations they explore. Homestays exemplify the potential for travel to be a force for good, supporting local communities and preserving the planet for future generations.

Concept of Homestays

Homestays represent a unique and increasingly popular form of accommodation within the broader hospitality and tourism industry. Rooted in the idea of cultural immersion and local engagement, homestays provide travellers with the opportunity to stay in the homes of residents rather than traditional hotels or resorts. This concept has gained traction globally as travellers seek more authentic and meaningful experiences, fostering a deeper connection with the destinations they visit.

Homestays, at their core, involve the provision of lodging in private homes. Hosts open their residences to guests, offering a more personal and intimate experience compared to standard accommodations. Travellers typically rent a room or part of the host's home, immersing themselves in the daily life and culture of the destination.

A defining feature of homestays is the emphasis on cultural immersion. Travellers have the opportunity to interact closely with local hosts, experiencing the destination from an insider's perspective. This cultural exchange extends beyond a mere stay, fostering a genuine connection between hosts and guests and facilitating a richer understanding of local customs, traditions, and ways of life.

Homestays come in various forms, ranging from urban apartments and suburban houses to rural farmsteads and traditional dwellings. The diversity of host accommodations ensures that travellers can choose experiences that align with their preferences, whether it's a bustling city homestay or a serene retreat in a natural setting.

Hosts in homestays directly benefit from tourism-related income, contributing to the economic well-being of local communities. This economic empowerment often extends beyond individual hosts to local businesses, as guests explore nearby attractions, dine in local restaurants, and purchase goods and services from the community.

Unlike standardised hotel experiences, homestays offer a more personalised and authentic touch. Hosts may share meals with guests, provide insider tips on hidden gems, and create an atmosphere that feels like a home away from home. This personalised approach enhances the overall travel experience, making it memorable and unique.

Homestays are often considered a sustainable form of accommodation. The reuse of existing homes reduces the need for new construction, minimising the environmental impact associated with traditional hotels and resorts. Additionally, hosts and guests in homestays are more likely to engage in sustainable practices, fostering a culture of environmental responsibility within the tourism sector.

Staying in a homestay fosters direct communication and interaction between hosts and guests. This facilitates cross-cultural understanding, breaking down stereotypes and promoting a more inclusive and tolerant global community. The exchange of ideas, perspectives, and stories creates a platform for mutual learning and enrichment.

With the advent of online platforms and booking services, homestays have become more accessible to a global audience. Websites and apps connect hosts with potential guests, allowing for easy bookings, reviews, and communication. Technology has played a crucial role in expanding the reach and popularity of homestays.

Despite their many advantages, homestays come with challenges. Cultural differences, language barriers, and the need for clear communication are considerations for both hosts and guests. Privacy concerns, the maintenance of accommodation quality, and adherence to local regulations are also important factors to address.

The concept of homestays continues to evolve with changing travel preferences and technological advancements. Trends include the integration of sustainability practices, the rise of niche homestay experiences (such as eco-friendly or farm-to-table homestays), and the incorporation of smart technologies for enhanced guest experiences.

Overall, homestays represent a dynamic and impactful concept within the realm of travel and hospitality. They embody the spirit of cultural exchange, sustainability, and personal connection, offering travellers a way to go beyond the superficial aspects of tourism and truly immerse themselves in the local fabric of a destination. As this concept continues to gain momentum, homestays are likely to shape the future of travel, providing a pathway to more meaningful and sustainable experiences for global explorers.

Homestays as a Pathway to a Greener Future

Homestays, as a pathway for a greener future, embody a sustainable and environmentally conscious approach to accommodation within the tourism industry. As the world grapples with the challenges of climate change and seeks ways to minimise ecological footprints, homestays have emerged as a promising alternative that not only provides travellers with unique and authentic experiences but also contributes to the broader goal of building a greener and more sustainable future.

Reduced environmental impact: One of the primary ways in which homestays contribute to a greener future is through their minimal environmental footprint. Unlike large hotels and resorts that often require extensive land development and construction, homestays leverage existing residential structures. This reduces the demand for new buildings, thus curbing deforestation and preserving natural habitats. Additionally, the smaller scale of homestays generally results in lower energy and water consumption, minimising the overall environmental impact associated with accommodation.

Preservation of local ecosystems: Homestays often thrive in locations that are immersed in natural settings, providing travellers with the opportunity to connect with and appreciate local ecosystems. Hosts, who are typically residents of these areas, have a vested interest in preserving the environment. By choosing homestays, travellers indirectly contribute to the conservation of local ecosystems as hosts are motivated to maintain the natural beauty that attracts guests in the first place.

Support for sustainable practices: Many homestays actively adopt sustainable practices, creating a culture of environmental responsibility within the hospitality sector. Hosts may implement energy-efficient measures, such as using solar panels or LED lighting, and adopt waste reduction strategies. These practices not only minimise the carbon footprint associated with accommodation but also set an example for guests who may be inspired to incorporate sustainable behaviours into their lifestyles.

Community empowerment and economic sustainability: Homestays provide a direct economic benefit to local communities, contributing to community empowerment and economic sustainability. Unlike international hotel chains where a significant portion of profits may be funnelled out of the community, homestay earnings stay within the local economy. This localised economic impact helps in building resilient communities that are less vulnerable to external economic shocks.

Cultural conservation and exchange: Homestays foster cultural conservation by providing a platform for the preservation and promotion of local traditions. As guests interact with hosts and engage in day-to-day activities, there is a natural exchange of cultural knowledge. This cultural immersion not only enriches the travel experience but also raises awareness about the importance of safeguarding cultural heritage as an integral part of sustainable development.

Reducing over-tourism and dispersing tourist impact: One of the challenges associated with tourism is the phenomenon of over-tourism, where popular destinations become overcrowded, leading to environmental degradation and a strain on local resources. Homestays offer a solution by dispersing tourist impact to local neighbourhoods and less-travelled areas. This not only reduces the burden on popular destinations but also allows travellers to explore and appreciate the diversity of regions, contributing to a more sustainable and balanced tourism model.

Personal connection to sustainability: Staying in a homestay creates a personal connection to sustainability for both hosts and guests. Hosts, motivated by a desire to provide an authentic experience, often incorporate sustainable practices into their daily lives. Guests, in turn, become more aware of the environmental and cultural nuances of the region they are visiting. This personal connection fosters a sense of responsibility towards sustainable travel, encouraging travellers to make more environmentally conscious choices in their future journeys.

Platforms and technology for sustainable homestays: The integration of technology has played a significant role in promoting sustainable homestays. Online platforms and apps connect hosts with potential guests, making it easier for eco-conscious travellers to find accommodations aligned with their values. These platforms often include features that highlight hosts practising sustainable initiatives, encouraging travellers to choose accommodations with a demonstrated commitment to environmental responsibility.

Education and awareness: Homestays provide an avenue for educating travellers about sustainable practices and the importance of responsible tourism. Hosts, acting as local ambassadors, can share information about local environmental challenges, conservation efforts, and ways in which guests can minimise their impact. This educational aspect contributes to a broader understanding of sustainability and encourages guests to carry these lessons forward into their everyday lives.

Encouraging responsible travel: By emphasising personal connections, cultural exchange, and the preservation of local environments, homestays inherently encourage responsible travel. Travellers who choose homestays are often seeking meaningful experiences that go beyond superficial tourism. This mindset promotes a shift towards more thoughtful and conscientious travel, where individuals become mindful of their impact and strive to leave a positive mark on the destinations they visit.

Overall, homestays serve as a compelling pathway for achieving a greener future in the realm of tourism. Their intrinsic connection to local communities, minimal environmental impact, and promotion of sustainable practices position homestays as a viable and eco-friendly alternative to traditional accommodation options. As travellers increasingly recognise the importance of responsible tourism, the popularity and influence of homestays are likely to grow, contributing to a more sustainable and environmentally conscious global travel industry.

Review of Literature

Over the years, there has been a notable increase in research focused on the homestay concept, particularly as governments leverage it to boost their rural tourism economies. Homestays, as highlighted by Liu (2006), Wang (2007), Ibrahim and Razzaq (2010), and Kayat (2010), align effectively with the promotion and safeguarding of local culture, encouraging active participation from rural communities in the tourism sector. This concept, rooted in creating a home-like atmosphere, fosters a sense of belonging (Lynch, 2003). The dining experience in homestays is often interactive, providing guests with a warm family environment (Gu & Wong, 2006; Richardson, 2004). Apart from offering affordable accommodation, homestays provide frequent opportunities for guests to engage with host families and immerse themselves in the hosts' lifestyles.

In literature, homestays are alternatively referred to as 'home-based enterprise', 'commercial home enterprise', and 'commercial home' (Di Domenico & Lynch, 2007; Lynch, 2005). Utilising residential spaces for homestay businesses represents a middle ground between hosting friends or family and providing a commercial lodging environment like hotels (Kontogeorgopoulos et al., 2015). For less developed countries, homestays serve as a means to generate revenue for local communities, offering guests a unique and authentic experience (Wang, 2007). Hosts benefit not only economically but also by establishing lasting relationships with guests. Homestays, in an indirect way, enable local participation in the nation's tourism industry (Kayat, 2010), contributing to the preservation of authentic local heritage (Wang, 2007).

Foreign tourists, seeking to explore local cultures and interact with local communities, find homestays particularly appealing, making them more attractive in rural areas (Kontogeorgopoulos et al., 2015; Mura, 2015; Wang, 2007). The essence of tradition and local architecture can attract visitors, and homestays often integrate additional activities such as hiking, bird watching, and participation in local festivals to ensure a consistent income base. Factors like technology, infrastructure, hygiene, risk management, cost management, and policies play crucial roles in the success of homestays (Jauhari & Sanjeev, 2010; Sanjeev & Jauhari, 2012).

In places like Thrissur, Kerala, homestays offer diverse income opportunities for local communities through activities like cooking, yoga, Ayurveda, plantation visits, trekking, safaris, and local tours (Paul, 2014). Such initiatives contribute to the promotion of nature-based activities and provide scope for practice-based learning facilitated by locals (Regmi & Walter, 2016). Agyeiwaah (2013) emphasises the essential role of homestays in heritage preservation, stressing the importance of understanding tourists' motivational factors, which can guide policymakers. In summary, the demand for home-like accommodation is significant, with a considerable number of international tourists motivated by socio-cultural factors and the appeal of warm and friendly lodging (Lynch, 2005).

Research Design

This study is conducted purely based on secondary data (Table 9.1). The primary aim of this study is to examine and comprehend the number of homestays in Madhya Pradesh, India, across different districts, as well as to investigate the factors contributing to the observed variations. Additionally, the study seeks to propose homestays as a sustainable accommodation option for a more environmentally friendly future.

The data in Fig. 9.1 represent the number of homestays in different districts of Madhya Pradesh, India. Bhopal has the highest homestay count with 28 homestays. Other districts with notable homestay numbers include Indore (23), Niwari (23), Chhatarpur (21), Jabalpur (8), Mandla (9), Narmadapuram (12), Panna (6), Satna (6), Seoni (6), and Ujjain (10). Several districts such as Dewas, Dhar, Gwalior, Shivpuri, and Sidhi have a lower homestay count, ranging from 2 to 4.

Table 9.1. Total Number of Registered Homestays in Madhya Pradesh (District-wise).

Sl. No.	District	Homestay Count
1	Balaghat	5
2	Bhopal	28
3	Chhatarpur	21
4	Dewas	2
5	Dhar	2
6	Gwalior	4
7	Indore	23
8	Jabalpur	8
9	Khargone	5
10	Mandla	9
11	Narmadapuram	12
12	Niwari	23
13	Panna	6
14	Satna	6
15	Sehore	3
16	Seoni	6
17	Shivpuri	2
18	Sidhi	6
19	Ujjain	10
20	Umaria	3
	Total	184

Source: https://tourism.mp.gov.in/contents?page=registered-homestay&number=2wyUbp51NyrP3L+lLhjGaQ== (as of May 1, 2023).

The data represent a mix of districts from different regions in a particular Indian state of Madhya Pradesh. Bhopal, Indore, and Jabalpur are relatively more urbanised, while districts like Mandla and Niwari might have a more rural or nature-centric appeal.

Higher homestay numbers indicate a potential tourist hotspot or a more developed tourism infrastructure in those districts. Factors such as historical significance, natural attractions, and cultural events may contribute to the popularity of homestays in certain districts. Districts with lower homestay counts could be potential areas for growth in the tourism and hospitality sector. It might be beneficial to explore and promote tourism in these areas.

Overall, this analysis provides a snapshot of the distribution of homestays across various districts, offering insights into the tourism landscape in the region.

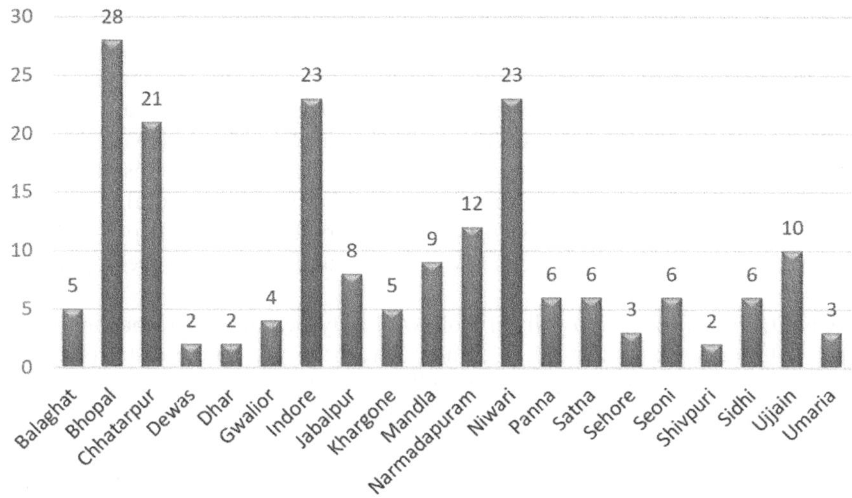

Fig. 9.1. Bar Chart Showing the Number of Registered Homestays in Madhya Pradesh (District-wise). *Source*: Authors' compilation from https://tourism.mp.gov.in/contents?page=registered-homestay&number=2wyUbp51NyrP3L+lLhjGaQ==.

Analysis

There is a notable disparity in the distribution of homestays across districts. Some districts, like Bhopal and Indore, have a significantly higher number of homestays compared to others. This suggests variations in tourism popularity and development.

Urban centres such as Bhopal and Indore seem to attract more homestays, indicating a potential focus on urban tourism. In contrast, districts with lower homestay counts may have a more rural or less-developed tourism infrastructure.

Certain districts, like Bhopal, Indore, and Jabalpur, emerge as potential tourist hotspots, likely due to factors such as historical significance, cultural attractions, or urban amenities.

Some districts with lower homestay counts, such as Mandla and Niwari, may be more focused on nature-centric tourism, given their potentially rural or natural appeal.

The presence of homestays indicates economic opportunities for local communities, offering them a chance to participate in and benefit from the tourism industry.

Districts with lower homestay counts present opportunities for growth and investment in the tourism sector. This could lead to economic development and job creation in these regions.

Discussion

In recent years, the tourism industry has witnessed a paradigm shift, with an increasing number of travellers seeking authentic and immersive experiences.

Homestays, in particular, have emerged as a popular choice for tourists looking to connect with local cultures and communities. The provided data, which outlines the homestay counts in various districts of Madhya Pradesh, India, offer a compelling glimpse into the regional dynamics of tourism in the area.

Bhopal, the capital district, stands out prominently in the data with the highest homestay count of 28. This suggests a thriving tourism and hospitality sector, likely fuelled by a combination of historical landmarks, cultural attractions, and modern amenities. Bhopal's appeal as a potential tourism hotspot is further underscored by its urban character, making it a hub for travellers seeking diverse experiences.

Chhatarpur, Indore, and Niwari follow closely behind, each boasting a significant homestay count of 21, 23, and 23, respectively. These districts exhibit a robust tourism infrastructure, potentially driven by a mix of historical significance, cultural richness, and natural attractions. The diversity in their offerings, from ancient monuments to scenic landscapes, positions them as multifaceted destinations capable of catering to a broad spectrum of tourist interests.

On the flip side, districts like Dewas, Dhar, Gwalior, Shivpuri, and Sidhi present lower homestay counts ranging from 2 to 6. While these districts may not be immediate magnets for tourism, their inclusion in the dataset highlights the nuanced tapestry of the region. The lower counts might indicate a focus on preserving natural landscapes or the need for targeted efforts to promote these areas as emerging tourist destinations.

Mandla and Umaria, with homestay counts of 9 and 3, respectively, exemplify the potential for rural and nature-centric tourism. These districts, possibly endowed with serene landscapes and indigenous cultural experiences, may appeal to travellers seeking a break from urban life. The relatively lower numbers of homestays in these areas open up opportunities for sustainable growth, ensuring that tourism development aligns with the preservation of local ecosystems and traditions.

The overall distribution of homestays across districts hints at a balanced approach to tourism development. The variety in counts showcases a concerted effort to prevent over-concentration in specific areas, promoting a more equitable distribution of economic benefits and reducing the risk of over-tourism in popular destinations.

Economically, the presence of 184 homestays signifies a positive impact on local communities. Tourism has the potential to create jobs, stimulate small businesses, and contribute to the overall economic development of the region. As districts with lower homestay counts represent untapped potential, there is an opportunity for policymakers to implement targeted initiatives to encourage tourism growth in these areas.

Overall, the provided homestay data paints a vivid picture of the regional tourism landscape, offering insights into the diverse offerings and developmental nuances of each district. The numbers tell a story of urban allure, cultural richness, natural beauty, and untapped potential. As the region continues to evolve as a tourism destination, stakeholders have the opportunity to shape policies that foster sustainable growth, preserve cultural heritage, and ensure a balanced and inclusive tourism experience for visitors and locals alike.

Reasons Behind the Disparity in Homestay Counts in the Districts of Madhya Pradesh, India

The observed disparity in homestay counts across the various districts in the provided data can be attributed to a multitude of factors, reflecting the diverse nature of tourism development in the region. Understanding these reasons offers valuable insights into the underlying dynamics shaping the distribution of homestays.

Tourist attractions and historical significance: Districts with higher homestay counts, such as Bhopal, Indore, Chhatarpur, and Niwari, often boast significant tourist attractions and historical landmarks. These districts may be home to iconic monuments, cultural sites, and historical relics that naturally draw visitors, leading to a higher demand for accommodation options like homestays.

Urbanisation and infrastructure: Urban centres like Bhopal and Indore tend to have better-developed infrastructure, including transportation and communication networks. This urban advantage attracts more tourists and encourages the establishment of homestays to cater to their diverse needs. On the other hand, rural districts may face challenges in terms of accessibility and amenities, impacting their ability to attract and accommodate tourists.

Cultural and natural appeal: The cultural and natural appeal of a district significantly influences its attractiveness to tourists. Districts with unique cultural experiences, vibrant traditions, and picturesque landscapes, such as Chhatarpur and Mandla, may naturally attract more visitors seeking authentic and immersive encounters. This appeal translates into a higher demand for homestay accommodations.

Tourism promotion and marketing: Disparities in homestay counts can also be linked to the effectiveness of tourism promotion and marketing strategies. Districts with proactive tourism boards and marketing campaigns are likely to attract more attention from travellers, leading to higher homestay occupancy rates. The visibility of a district in the tourism landscape plays a crucial role in shaping visitor preferences.

Economic development and employment opportunities: The economic development of a district is closely tied to its attractiveness to tourists. More economically developed districts often invest in tourism infrastructure, creating a conducive environment for hospitality businesses, including homestays. Additionally, the prospect of employment opportunities in tourism-related sectors can drive the establishment of accommodations in certain districts.

Government policies and incentives: Government policies and incentives aimed at promoting tourism can significantly impact the distribution of homestays. Some districts may benefit from targeted initiatives, grants, or subsidies that encourage the development of the hospitality sector. Understanding the role of policy in shaping the tourism landscape is crucial for explaining disparities in homestay counts.

Community participation and awareness: In districts where local communities actively participate in and support tourism initiatives, there is a higher likelihood of increased homestay development. Community involvement fosters a welcoming environment for visitors, contributing to positive word-of-mouth and repeat tourism.

Natural resources and environmental considerations: Districts with abundant natural resources, wildlife sanctuaries, or scenic landscapes may attract nature enthusiasts. However, concerns about environmental conservation and sustainable tourism practices might limit the expansion of homestays in these areas to strike a balance between economic development and ecological preservation.

Overall, the disparity in homestay counts reflects the multifaceted nature of tourism development, encompassing cultural, historical, economic, and infrastructural factors. As tourism stakeholders and policymakers work towards fostering sustainable and inclusive growth, understanding these reasons becomes paramount for creating targeted interventions that can address the unique challenges and opportunities in each district.

Economic, Social, and Environmental Impacts of Homestays in Madhya Pradesh

Homestays in Madhya Pradesh have emerged as a significant contributor to the state's economy, social fabric, and environmental sustainability. Understanding the economic, social, and environmental impacts of homestays is essential for policymakers, stakeholders, and communities to leverage the potential benefits and address potential challenges. This article explores these impacts in-depth, highlighting the multi-dimensional nature of homestay tourism in Madhya Pradesh.

Economic Impacts

Homestays play a pivotal role in driving economic growth, job creation, and income generation in Madhya Pradesh's rural areas. By providing alternative sources of livelihood for local communities, homestays contribute to poverty alleviation and economic diversification. The economic impacts of homestays can be observed through various dimensions:

Income generation: Homestays offer rural households an opportunity to earn supplementary income by hosting tourists and providing hospitality services. This additional source of revenue enhances household incomes, improves living standards, and reduces dependency on traditional livelihoods such as agriculture.

Employment opportunities: The operation and management of homestays create employment opportunities for residents, including housekeeping staff, cooks, guides, and artisans. This not only reduces unemployment rates but also fosters skill development and entrepreneurship among rural youth and women.

Local procurement and supply chains: Homestays stimulate local economies by sourcing goods and services from nearby suppliers, including food, handicrafts, and cultural experiences. This strengthens local supply chains, supports small-scale producers, and promotes the circulation of income within the community.

Tourism revenue: The influx of tourists staying at homestays contributes to the overall tourism revenue of Madhya Pradesh, including expenditures on accommodation, food, transportation, and souvenirs. This revenue stream supports infrastructure development, conservation projects, and destination marketing efforts.

Social Impacts

Homestays foster social cohesion, cultural exchange, and community empowerment in Madhya Pradesh, enriching the socio-cultural fabric of rural areas. The social impacts of homestays extend beyond economic transactions to encompass intangible benefits that enhance social capital and cultural heritage preservation.

Cultural exchange: Homestays facilitate meaningful interactions between tourists and local hosts, enabling cultural exchange, cross-cultural learning, and mutual understanding. Visitors have the opportunity to immerse themselves in the customs, traditions, and lifestyle of rural communities, while hosts share their heritage, stories, and hospitality.

Community empowerment: Homestays empower local communities by involving them in tourism decision-making processes, revenue-sharing mechanisms, and capacity-building initiatives. Community participation in homestay development fosters a sense of ownership, pride, and responsibility for preserving cultural identity and natural resources.

Preservation of traditional knowledge: Homestays serve as repositories of traditional knowledge, indigenous practices, and folk wisdom passed down through generations. By showcasing local cuisine, crafts, performing arts, and rituals, homestays contribute to the preservation and revitalisation of intangible cultural heritage in Madhya Pradesh.

Women's empowerment: Homestays often provide opportunities for women to engage in income-generating activities, gain economic independence, and participate in decision-making roles within their households and communities. Women-owned and managed homestays empower female entrepreneurs and contribute to gender equality initiatives.

Environmental Impacts

Homestays have the potential to promote environmental conservation, sustainable resource management, and ecotourism principles in Madhya Pradesh. While tourism activities may exert pressures on natural ecosystems, responsible homestay operations can mitigate environmental impacts and support biodiversity conservation efforts.

Resource efficiency: Homestays encourage resource-efficient practices such as rainwater harvesting, solar energy utilisation, waste reduction, and organic farming techniques. By minimising resource consumption and promoting eco-friendly technologies, homestays reduce their carbon footprint and ecological footprint.

Ecotourism awareness: Homestays serve as platforms for raising awareness about environmental conservation, wildlife protection, and eco-friendly behaviours among tourists and hosts alike. Educational programs, nature walks, and interpretive experiences conducted at homestays promote ecotourism principles and foster a sense of environmental stewardship.

Habitat preservation: Homestays located in ecologically sensitive areas contribute to habitat preservation and biodiversity conservation by promoting responsible tourism practices, supporting wildlife corridors, and minimising habitat fragmentation. Conservation-oriented homestay initiatives collaborate with local conservation organisations and government agencies to protect natural landscapes and endangered species.

Community-based conservation: Homestays engage local communities in participatory conservation initiatives, community-based ecotourism projects, and habitat restoration activities. By involving residents in monitoring, patrolling, and conservation efforts, homestays empower communities to become stewards of their natural heritage and guardians of biodiversity hotspots.

Homestays in Madhya Pradesh have far-reaching economic, social, and environmental impacts that extend beyond mere accommodation provision. By recognising the interconnectedness of these impacts and adopting holistic approaches to sustainable tourism development, policymakers, stakeholders, and communities can harness the transformative potential of homestay tourism to achieve inclusive growth, cultural resilience, and environmental sustainability in rural areas of Madhya Pradesh.

Potential Challenges and Drawbacks Associated with Homestay Initiatives

While homestay initiatives are increasingly popular and contribute positively to sustainable tourism, they also come with their set of challenges and drawbacks. It is essential to recognise and address these aspects to ensure the success and long-term viability of homestay programs.

Cultural sensitivity and cross-cultural communication: Homestays involve cultural exchange, and miscommunication or misunderstandings can arise due to differences in language, customs, and traditions. Both hosts and guests may face challenges in navigating these cultural nuances, potentially leading to discomfort or dissatisfaction.

Privacy concerns for hosts and guests: Privacy is a crucial aspect, and both hosts and guests may find it challenging to strike a balance between providing a welcoming environment and respecting personal space. Hosts might feel invaded, and guests may feel restricted in their activities, potentially leading to a less-than-ideal experience for both parties.

Quality and consistency of accommodations: Maintaining a consistent standard of accommodations across different homestays can be challenging. Discrepancies in cleanliness, amenities, or overall quality may arise, impacting the reputation of the homestay initiative as a whole.

Dependency on seasonal tourism: Homestays in certain regions may heavily rely on seasonal tourism. During peak seasons, hosts may experience high demand, but during off-peak times, there might be a lack of visitors. This can create economic instability for hosts who depend on homestay income.

Infrastructure and accessibility challenges: In some rural or less-developed areas, homestays may face challenges related to basic infrastructure and accessibility. Limited transportation options, unreliable utilities, and inadequate facilities can pose obstacles to both hosts and guests.

Regulatory and legal issues: Homestays may be subject to local regulations, zoning laws, and licencing requirements. Hosts may find it challenging to navigate these regulations, and non-compliance could lead to legal issues. Striking a balance between supporting local economies and adhering to legal frameworks is a persistent challenge.

Economic inequality and gentrification: The economic benefits of homestays might not be evenly distributed among host communities. Certain individuals or areas may benefit more than others, potentially exacerbating economic inequality. Additionally, the influx of tourism can contribute to gentrification, altering the socio-economic landscape of a community.

Environmental impact: While homestays are generally considered more environmentally friendly than large-scale hotels, an increase in the number of homestays can still contribute to environmental degradation. Issues like water usage, waste disposal, and energy consumption need to be addressed to minimise the ecological footprint of homestay initiatives.

Vulnerability to global crises: Homestays, like other sectors of the tourism industry, are vulnerable to global crises such as pandemics, natural disasters, or geopolitical instability. Sudden disruptions can lead to cancellations, financial losses, and uncertainties for both hosts and guests.

Maintaining authenticity and local flavour: As homestays gain popularity, there is a risk of commercialisation and loss of authenticity. Hosts may feel pressure to cater to tourist expectations, potentially diluting the unique cultural and local experience that homestays are meant to provide.

Addressing these challenges requires a collaborative effort from homestay hosts, local communities, policymakers, and travellers. Implementing clear guidelines, providing cultural training, and fostering open communication can help overcome these drawbacks, ensuring that homestay initiatives continue to contribute positively to sustainable tourism.

Policy Implications and Recommendations for Sustainable Homestay Tourism in Madhya Pradesh

As an emerging sector with significant potential to promote rural development, cultural preservation, and environmental conservation, homestay tourism requires supportive policies, regulatory frameworks, and institutional mechanisms to thrive sustainably. Below are the key policy implications and recommendations tailored to the context of Madhya Pradesh, focusing on fostering community participation, enhancing sustainability standards, and promoting destination management.

Community Participation and Empowerment

Policy Implications

Community ownership: Policies should prioritise community ownership and participation in homestay tourism initiatives, empowering residents to actively engage in decision-making processes, benefit-sharing mechanisms, and resource management.

Capacity building: Government agencies, NGOs, and tourism stakeholders should collaborate to design and implement capacity-building programs that equip homestay hosts with the necessary skills, knowledge, and resources to manage their accommodations effectively.

Supportive infrastructure: Policy interventions should focus on improving access to infrastructure, basic amenities, and social services in rural areas, ensuring that homestay communities have adequate support for hosting guests.

Recommendations

Community homestay committees: Establish community-based homestay committees or associations comprising residents, tourism officials, and representatives from civil society organisations to oversee homestay operations, address challenges, and promote cooperation.

Training and skill development: Invest in training workshops, vocational courses, and educational programs tailored to the needs of homestay hosts, covering topics such as hospitality management, cultural sensitivity, marketing techniques, and sustainable tourism practices.

Financial assistance: Provide financial support, grants, and subsidies to aspiring homestay operators for infrastructure development, renovation projects, and marketing initiatives, enabling them to enhance the quality of their accommodations and attract more visitors.

Sustainability Standards and Certification

Policy Implications

Regulatory frameworks: Governments should develop regulatory frameworks and quality standards for homestay accommodations, including guidelines for hygiene, safety, waste management, and environmental sustainability.

Certification programs: Implement certification programs or accreditation schemes that recognise homestays adhering to sustainable practices, encouraging compliance with industry standards and fostering consumer confidence.

Environmental protection: Incorporate environmental impact assessments, biodiversity conservation measures, and waste reduction strategies into homestay development plans, minimising the ecological footprint of tourism activities.

Recommendations

Sustainable tourism guidelines: Develop comprehensive guidelines and best practices manuals for homestay operators, highlighting sustainable tourism principles, responsible hosting practices, and eco-friendly initiatives such as energy conservation, water management, and waste recycling.

Green certification: Introduce a voluntary green certification scheme for homestays that meet specific sustainability criteria, such as eco-friendly infrastructure, use of renewable energy sources, promotion of local biodiversity, and engagement in community conservation projects.

Monitoring and evaluation: Establish mechanisms for monitoring and evaluating the environmental performance of homestay accommodations, conducting periodic audits, and providing feedback to hosts on areas for improvement.

Destination Management and Planning

Policy Implications

Tourism planning: Integrate homestay tourism into broader destination management plans, aligning development strategies with local socio-economic priorities, cultural heritage preservation goals, and environmental conservation objectives.

Zoning regulations: Enact zoning regulations and land-use policies that designate suitable areas for homestay development, taking into account factors such as ecological sensitivity, cultural significance, and community preferences.

Infrastructure development: Invest in infrastructure upgrades, road connectivity improvements, and tourism amenities in homestay destinations, balancing the need for accessibility with the preservation of natural landscapes and cultural authenticity.

Recommendations

Destination branding: Develop cohesive destination branding strategies that highlight the unique cultural identity, natural attractions, and experiential offerings of homestay destinations in Madhya Pradesh, leveraging storytelling, digital marketing, and niche positioning to attract target markets.

Heritage conservation: Support initiatives for heritage conservation, restoration of traditional architecture, and preservation of indigenous crafts in homestay communities, encouraging hosts to showcase local culture and heritage through interpretive experiences, guided tours, and cultural exchanges.

Community-based tourism planning: Foster collaborative partnerships between government agencies, tourism stakeholders, and local communities to co-create tourism development plans, ensuring that homestay initiatives are aligned with community aspirations, livelihood priorities, and sustainable development goals.

Policy implications and recommendations for sustainable homestay tourism in Madhya Pradesh must prioritise community participation, sustainability standards, and destination management strategies to foster inclusive growth, cultural resilience, and environmental stewardship. By adopting a holistic approach that balances economic benefits with social equity and environmental conservation, policymakers can harness the transformative potential of homestay tourism to create a more sustainable and resilient tourism industry in Madhya Pradesh.

Future Prospects and Opportunities for Homestay Tourism in Madhya Pradesh

As one of India's largest states, Madhya Pradesh boasts a rich cultural heritage, diverse natural landscapes, and a burgeoning tourism sector. Leveraging the concept of homestays presents a unique opportunity to enhance visitor experiences, promote sustainable development, and empower rural communities across the state.

Cultural and Heritage Tourism

Madhya Pradesh is renowned for its historical and cultural significance, with attractions such as UNESCO World Heritage Sites, ancient temples, forts, and palaces scattered throughout the region. Homestay tourism offers travellers the chance to immerse themselves in the local culture, traditions, and way of life by staying with host families in rural villages and heritage-rich towns. Future prospects for homestay tourism in Madhya Pradesh include the following:

Heritage homestays: Transforming heritage properties, ancestral homes, and traditional dwellings into homestay accommodations can provide visitors with an authentic glimpse into the state's rich history and architectural heritage. By preserving and showcasing cultural assets, heritage homestays contribute to heritage conservation efforts while generating income for local communities.

Cultural experiences: Offering cultural immersion experiences such as folk performances, craft demonstrations, culinary workshops, and village tours can enrich the tourism experience for visitors and support local artisans, performers, and cultural practitioners. Homestay hosts can serve as ambassadors of Madhya Pradesh's cultural heritage, sharing stories, traditions, and customs with guests.

Ecotourism and Nature-based Tourism

Madhya Pradesh boasts diverse ecosystems, including dense forests, wildlife sanctuaries, national parks, and scenic landscapes, making it an ideal destination for ecotourism and nature-based tourism. Homestay tourism can play a vital role in promoting sustainable nature tourism while fostering conservation awareness and supporting biodiversity conservation efforts. Future prospects for ecotourism homestays in Madhya Pradesh include the following:

Wildlife homestays: Partnering with local communities near wildlife reserves and protected areas to establish wildlife homestays can provide travellers with unique opportunities to observe wildlife in their natural habitats while promoting responsible tourism practices. Homestay operators can collaborate with conservation organisations and forest departments to offer guided wildlife safaris, bird watching tours, and nature walks.

Rural retreats: Creating rural homestay retreats in scenic countryside settings allows visitors to reconnect with nature, experience traditional rural lifestyles, and participate in sustainable farming practices. Rural homestays offer opportunities for activities such as organic farming, nature treks, birdwatching, and cultural exchanges with local farmers and artisans.

Community-based Tourism Development

Empowering local communities to participate in and benefit from tourism development is essential for fostering inclusive growth, reducing poverty, and promoting socio-economic resilience. Homestay tourism presents a grassroots approach to community-based tourism development, enabling rural households to supplement their income, preserve cultural traditions, and enhance livelihood opportunities. Future prospects for community-based homestay tourism in Madhya Pradesh include the following:

Community homestay networks: Establishing community-based homestay networks in collaboration with local self-help groups, cooperatives, and women's organisations can create a sense of solidarity and collective ownership among host families. These networks can pool resources, share best practices, and collectively market their homestay offerings to domestic and international travellers.

Skill development and capacity building: Investing in training programs, skill development initiatives, and capacity-building workshops for homestay hosts can enhance their hospitality skills, business acumen, and cultural competency. Training modules may cover topics such as guest relations, hygiene standards, financial management, and sustainable tourism practices, empowering hosts to deliver high-quality experiences.

Promotion and Marketing Strategies

Effectively promoting and marketing homestay tourism in Madhya Pradesh is essential for attracting visitors, generating demand, and maximising the economic benefits for host communities. Future prospects for promoting homestay tourism in Madhya Pradesh include the following:

Digital marketing platforms: Leveraging digital marketing platforms, social media channels, and online travel agencies to showcase homestay accommodations, highlight unique experiences, and engage with potential travellers. Creating visually appealing content, user-generated testimonials, and virtual tours can capture the attention of adventure seekers, cultural enthusiasts, and nature lovers.

Cultural festivals and events: Organising cultural festivals, rural fairs, and community events that celebrate local traditions, arts, and cuisine can serve as platforms to promote homestay tourism and attract visitors. Collaborating with local authorities, tourism boards, and event organisers can help amplify the visibility of homestay offerings and encourage cultural exchange.

Infrastructure and Connectivity

Improving infrastructure and connectivity in rural areas is crucial for enhancing accessibility, facilitating tourism mobility, and supporting the growth of homestay tourism in Madhya Pradesh. Future prospects for infrastructure development and connectivity enhancements include the following:

Road and transport infrastructure: Upgrading road networks, bridges, and transportation facilities to rural homestay destinations can improve accessibility and connectivity for travellers. Investing in public transportation services, shuttle services, and last-mile connectivity solutions can facilitate seamless travel experiences for visitors.

Digital connectivity: Expanding digital connectivity, internet access, and mobile network coverage in remote rural areas can enable homestay hosts to communicate with guests, manage bookings, and promote their accommodations online. Access to online booking platforms, payment gateways, and communication tools can streamline the reservation process and enhance the visibility of homestay offerings.

Overall, the future prospects and opportunities for homestay tourism in Madhya Pradesh are multifaceted, encompassing cultural enrichment, ecological conservation, community empowerment, and economic development. By embracing sustainable tourism practices, fostering community partnerships, and leveraging the state's natural and cultural assets, Madhya Pradesh can position itself as a leading destination for responsible travellers seeking authentic and immersive experiences. With the right policies, investments, and collaborations, homestay tourism has the potential to contribute significantly to the state's socio-economic growth while preserving its heritage and natural heritage for future generations.

Conclusion

In conclusion, the exploration of homestays as a sustainable tool for a greener future in Madhya Pradesh, India, reveals a nuanced and dynamic landscape within the tourism industry. The provided data on homestay counts across different districts offers valuable insights into the varied patterns of tourism development, economic impact, and potential areas for growth. As homestays gain prominence in the broader context of sustainable tourism, their significance extends beyond providing unique accommodation experiences to fostering community engagement, preserving local heritage, and minimising environmental impact.

Homestays, deeply rooted in the principles of cultural immersion, create a bridge between hosts and guests, facilitating meaningful exchanges that go beyond the surface of traditional tourism. The data showcases a diverse distribution of homestays, with districts like Bhopal and Indore standing out as urban hubs with rich historical and cultural appeal, while others, such as Mandla and Niwari, offer a potential haven for nature-centric and rural tourism. This diversity not only caters to a range of tourist preferences but also reflects the multifaceted nature of the Madhya Pradesh region.

The disparities in homestay counts underscore the influence of factors such as historical significance, urbanisation, cultural appeal, and effective tourism promotion. Urban centres with well-developed infrastructure and cultural attractions tend to attract more homestays, while rural areas may face challenges related to accessibility and amenities. The presence of 184 registered homestays across the region signifies a positive economic impact, creating opportunities for local communities and businesses.

The regional analysis prompts discussions on the potential for sustainable growth in districts with lower homestay counts. Mandla and Umaria, with their nature-centric appeal, emerge as areas ripe for development, offering a counterbalance to the concentration of tourism in urban centres. Policymakers and stakeholders can leverage this information to implement targeted initiatives that promote tourism, create employment opportunities, and ensure a more equitable distribution of economic benefits.

Homestays serve as a testament to the evolving landscape of travel preferences, where authenticity, cultural exchange, and sustainable practices take precedence. The success of homestays lies not only in providing travellers with unique and personalised experiences but also in their role as catalysts for economic empowerment at the community level. By directly benefiting local hosts and businesses, homestays contribute to a more inclusive and community-driven tourism model.

Despite their many advantages, homestays come with challenges and considerations. Cultural sensitivity, privacy concerns, and the need for consistent quality are factors that require attention. Additionally, the potential vulnerability of homestays to global crises underscores the importance of resilience and adaptability in the tourism sector.

In navigating these challenges, there is an opportunity for collaboration between hosts, local communities, policymakers, and travellers. Clear guidelines, cultural training, and a commitment to responsible tourism practices can help address these drawbacks, ensuring that the positive impact of homestays on sustainability and community development remains intact.

As the tourism industry continues to evolve, homestays exemplify a pathway to a greener future, where responsible travel choices align with economic growth, cultural preservation, and environmental conservation. Travellers, by choosing homestays, actively participate in shaping a tourism industry that leaves a positive and lasting impact on the destinations they explore. As Madhya Pradesh embraces the potential for sustainable tourism, the journey towards a greener future unfolds, guided by the principles of authenticity, community engagement, and environmental stewardship. In this journey, homestays stand as beacons of responsible tourism, illuminating a pathway towards a more sustainable and harmonious coexistence between travellers and the places they visit.

Bibliography

Agyeiwaah, E. (2013). International tourists' motivations for choosing homestay in the Kumasi Metropolis of Ghana. *Anatolia*, *24*(3), 405–409.

Chakraborty, B. (2020). Homestay as a reliable promotional tool for cultural tourism and security in Indian context. *IJSSTH*, (21), 5.

Choudhary, N., & Manda, V. K. (2022). Neo tourism: A paradigm shift in the era of tourism. *Recent trends in tourism and hospitality* (pp 1–17). Eureka Publications.

Dash, S. P. (2022). Exploring the prospects of homestays in Indian tourism and hospitality industry as an alternate preference for stay. *International Journal of Built Environment and Sustainability*, *9*(3), 1–10.

Di Domenico, M., & Lynch, P. A. (2007). Host/guest encounters in the commercial home. *Leisure Studies*, *26*(3), 321–338.

Dutta, P. K. (2012). Guidelines for promotion of homestays in Arunachal Pradesh. *Tezpur: Western Arunachal Landscape programme, WWE.*

Government of Madhya Pradesh, Madhya Pradesh Tourism Board. (2023, May 1). https://tourism.mp.gov.in/contents?page=registered-homestay&number=2wyUbp51NyrP3L+lLhjGaQ==

Gu, M., & Wong, P. P. (2006). Residents' perception of tourism impacts: A case study of homestay operators in Dachangshan Dao, North-East China. *Tourism Geographies*, *8*(3), 253–273.

Ibrahim, Y., & Razzaq, A. R. A. (2010). Homestay program and rural community development in Malaysia. *Journal of Ritsumeikan Social Sciences and Humanities*, *2*(1), 7–24.

Jauhari, V., & Sanjeev, G. M. (2010). Managing customer experience for spiritual and cultural tourism: An overview. *Worldwide Hospitality and Tourism Themes*, *2*(5), 467–476.

Kayat, K. (2010). The nature of cultural contribution of a community-based homestay programme. *Tourismos: An International Multidisciplinary Journal of Tourism*, *5*(2), 145–159.

Kontogeorgopoulos, N., Churyen, A., & Duangsaeng, V. (2015). Homestay tourism and the commercialization of the rural home in Thailand. *Asia Pacific Journal of Tourism Research*, *20*(1), 29–50.

Kulshreshtha, S., & Kulshrestha, R. (2019). The emerging importance of "homestays" in the Indian hospitality sector. *Worldwide Hospitality and Tourism Themes*, *11*(4), 458–466.

Liu, A. (2006). Tourism in rural areas: Kedah, Malaysia. *Tourism Management*, *27*(5), 878–889.

Lynch, P. A. (2003). *Conceptual relationships between hospitality and space in the homestay sector* (Doctoral dissertation, Queen Margaret University College).

Lynch, P. A. (2005). The commercial home enterprise and host: A United Kingdom perspective. *International Journal of Hospitality Management*, *24*(4), 533–553.

Mura, P. (2015). Perceptions of authenticity in a Malaysian homestay – A narrative analysis. *Tourism Management*, *51*, 225–233.

Paul, H. K. (2014). Homestays for the development of tourism in Thrissur district. *CLEAR International Journal of Research in Commerce & Management*, *5*(5), 64–67.

Pasanchay, K., & Schott, C. (2021). Community-based tourism homestays' capacity to advance the Sustainable Development Goals: A holistic sustainable livelihood perspective. *Tourism Management Perspectives*, *37*, 100784.

Regmi, K. D., & Walter, P. G. (2016). Conceptualising host learning in community-based ecotourism homestays. *Journal of Ecotourism*, *15*(1), 51–63.

Richardson, K. (2004, October). Homestay: Opening a world of opportunity. In *Australian International Education Conference* (pp. 5–8).

Sahoo, S. S., & Bg, M. M. (2019). Marketing strategy for homestay as an alternative tourism product–A study on local community involvement in rural Odisha. *Think India Journal*, *22*(21), 859–875.

Sanjeev, G. M., & Jauhari, V. (2012). The emerging strategic and financial issues in the Indian hospitality industry: An overview. *Worldwide Hospitality and Tourism Themes*, *4*(5), 403–409.

Singh, A., Bathla, G., & Raheja, K. (2022). *Current and futuristic trends in accommodation sector* (p. 334). Eureka Publications.

Singh, T. V. (1991). The development of tourism in the mountain environment: The problem of sustainability. *Tourism Recreation Research*, *16*(2), 3–12.

Wang, Y. (2007). Customized authenticity begins at home. *Annals of Tourism Research*, *34*(3), 789–804.

Chapter 10

The Role of Digital Tourism Platforms in Sustainable Tourism Development

Rajinder Kaur[a], *Amrita Saini*[b] *and Sandeep Singh*[c]

[a]*University School of Business, Chandigarh University, Punjab, India*
[b]*Chandigarh Business School of Administration, Chandigarh Group of Colleges, Punjab, India*
[c]*Punjabi University, Punjab, India*

Abstract

To overcome the problem of over tourism, depletion of natural resources and various adverse effects on local communities, sustainable tourism has emerged with the different eco-friendly approaches. Without the usage of online resources, green tourism cannot be promoted. Hence, this chapter demonstrates the significance of digital media platforms to promote sustainable tourism. The chapter is based on descriptive study and primarily used the secondary sources of data. The data is primarily supported by credible, previously published works, such as journal articles and working papers. The primary focus is placed on the importance of digital media platforms in promoting sustainable tourism. The findings demonstrated that with the assistance of digital technology, tourist destinations can move forward on the path of sustainability as it contributed positively for the sustainability of tourist destinations. The adopting digital technology has become essential for advancing environmentally friendly travel since it leads to enhancement in destination marketing, improvement in the management of tourist resources as well as enrich the visitor experience.

Keywords: Sustainable tourism; digital media; online platforms; eco-friendly; tourism development

The Need for Sustainable Tourism in an Era of Global Climate Change: Pathway to a Greener Future, 195–214

doi:10.1108/978-1-83608-668-020241028

1. Introduction

As per the World Tourism Organisation (WTO), 2005, 'Tourism is a social, cultural and economic phenomenon which entails the movement of people to countries or places outside their usual environment for personal, business and/ or professional purposes' Being the largest sector globally, travel and tourism accounting for 10.3% of the world economy and expected to grow to 500 billion US dollars by 2029. This industry has not slowed down over the last ten years; rather, it has become more significant and a significant source of foreign exchange (Menon et al., 2021). Among the world's biggest and fastest-growing sectors, with a substantial impact on the world economy, aids in the improvement of infrastructure, reduction of poverty by generating more employment opportunities all over the world (Fang et al., 2016). Both developed and developing nations view the tourism sector as a catalyst for economic expansion.

The top 10 nations for foreign tourist arrivals (FTAs) in India in 2021, according to the Ministry of Tourism of India, are the United States, Bangladesh, the United Kingdom, Canada, Nepal, Afghanistan, Australia, Germany, Portugal, and France. The percentage shares of the USA's FTAs in 2017, 2018, 2019, 2020, and 2021 were 13.72, 13.80, 13.83, 14.36, and 28.15, respectively. Maldives (82.8%) had the highest positive growth rate in India's free trade agreements (FTAs) in 2021 compared to 2020, followed by Nigeria (44.4), Nepal (28.7%), the United States of America (9.1%), Yemen Arab Republic (6%), Kenya (4.6%), and Portugal (3.4%).

2. Sustainable Tourism

In the contemporary global economy, sustainability has emerged as a key topic of interest for a wide range of stakeholders from various business sectors. The sustainable business model is starting to gain traction. These days, businesses build their sustainable business models with customers, stakeholders, and shareholders in mind, with an emphasis on creating value for all of these parties in addition to the environment and society (Baldassarre et al., 2017).

With the high growth tourism sector, a number of natural, social, and economic issues are also brought about, such as over tourism, the exhaustion of natural resources, and detrimental consequences on communities located nearby (Baloch et al., 2023; Sisneros-Kidd et al., 2019). In response to all these types of challenges, sustainable tourism has emerged with a new shift from traditional tourism practices to eco-friendly tourism options which indicates how to plan for and manage tourism while integrating the economy, society, and the environment. (Connell et al., 2009). A generally recognised definition of sustainable tourism development has not yet been established by empirical literature (Asmelash & Kumar, 2019), however it has been defined by the World Tourism Organisation's (WTO) as a 'development that meets the needs of present tourists and host regions while protecting and enhancing opportunity for the future'. However, a number of important issues in this field are still unanswered as the empirical literature continues to examine a number of fundamental questions for the growth of sustainable tourism. It is also the method by which individuals sustain change in a state of steadiness in which investments, resources, institutional reform, and technological advancement all serve to meet and exceed the needs and ambitions of people both now and in the future. 2017 has been designated as the International Year of Sustainable Tourism for Development by the UN. The purpose of this declaration made during the 70th UN General Assembly was to raise awareness of the role that the tourism industry plays in promoting the economic, social, and environmental pillars of sustainability. Sustainable tourism promotes indigenous cultures, preserves the long-term ecological balance, and uses tourism to generate jobs for the local populace in towns and villages (Aynalem et al., 2016). For instance, guests receive an ideal area where they are able to relax and take in the beautiful surroundings and organically grown food from nearby farms (Kauppila et al., 2009). Visitors learn about sustainability and witness its attainment.

2.1 Ecotourism Versus Sustainable Tourism

Ecotourism is highly related with the education of visitors with regard to environment and making their involvement in the activities such as conservation and cultural, whereas sustainable tourism has a goal to create travel options with little effect and advantages for destinations and their communities (Jokela & Minoia, 2021).

Although ecotourism ought to be sustainable at all times but it is not similar as all forms of sustainable travel (Brown & Vergragt, 2008). For instance, we can travel by train to a lodge that uses sustainable energy. Even though this is a sustainable alternative, it's not like we are getting an education about the environment and our neighbourhood.

2.2 Sustainable Practices by the Tourism Companies

- *Eco-friendly Accommodation*: exploring and supporting stay home that uses renewable energy sources, adheres to green building guidelines, and conserves water.
- *Community-Based Tourism*: researching tourism-dependent local communities that actively engage in the industry and reap its benefits, making sure that financial gains are shared properly.
- *Biodiversity Conservation*: looking into programmes like marine conservation areas, wildlife sanctuaries, and sustainable forestry techniques that aim to save and maintain regional ecosystems.
- *Cultural Preservation*: exploring initiatives and regulations aimed at preserving and advancing regional customs, heritage sites, and cultures in order to guarantee that tourism both honours and advances cultural diversity.
- *Responsible Wildlife Tourism*: analysing methods that put conservation and animal welfare first, abstaining from actions that injure or exploit wildlife for the amusement of visitors.
- *Sustainable Transportation*: looking up the use of carbon-reducing transportation strategies, such as bicycle sharing programmes, public transportation, and electric cars.
- *Waste Management*: seeking into efficient waste management and recycling initiatives in popular tourist locations, along with initiatives to reduce the use of single-use plastics in the sector.

- *Certification Programs*: investigating and endorsing certification schemes like Green Key or Earth Check that honour and incentivise companies that use sustainable practices.

Non-sustainable Tourism: Tourism practices that negatively affect the environment, society, and economy are referred to as mass tourism. It frequently puts immediate financial gain ahead of long-term sustainability (Higgins, 2018). If not handled properly, non-sustainable tourism can result in social unrest, cultural exploitation, and environmental damage.
Non-Sustainable Practices: The excessive growth and Exploitation: Examining instances in which the quick expansion of tourism results in overdevelopment, placing strain on regional ecosystems and resources (Hoffman et al., 2022).

- *Cultural Insensitivity*: analysing situations in which tourism leads to the deterioration of regional cultures or to tourists acting disrespectfully, which affects the destination's identity.
- *Impact of Mass Tourism*: examining the detrimental impacts of mass tourism, such as resource and infrastructure strain, increased pollution, and crowding.
- *Destroying Natural Habitats*: investigating instances in which tourism exacerbates environmental problems like deforestation, deteriorating coral reefs, or upsetting wildlife migration patterns.
- *Absence of Local Involvement*: examining situations in which tourism profits are concentrated in the hands of a small number of people, resulting in social and economic inequality, and local communities are left out.
- *Excessive Use of Energy and Water*: examining situations in which tourism-related activities put a strain on local resources by consuming excessive amounts of water and energy.
- *Inadequate Waste Management*: investigating areas where insufficient waste management techniques lead to pollution and environmental deterioration, impacting nearby populations and ecosystems. Examining instances of irresponsible wildlife tourism, such as when visitors feed animals or interfere with their natural behaviours, in order to cause harm to wildlife or disturb their natural habitat.

3. The Goals of Sustainable Tourism

Twelve primary goals for sustainable tourism have been set up in 2005 with the efforts of the World Tourism Organisation and the United Nations Environment Programme (UNEP). These will contribute to showing how sustainable tourism could assist with some of the greatest worldwide issues that the industry is facing at the moment:

(i) *Economic Viability*: ensuring the long-term viability, competitiveness, and accomplishment of tourism locations and businesses.
(ii) *Local Prosperity*: it is a sort of guarantee for the tourism-related activities' contribution to the financial upliftment of local communities with taking into consideration the considerable portion of the revenue.

(iii) *Employment Superiority*: boosting the quantity of regional jobs stimulated by the tourism sector and guarantee that returns and working conditions are reasonable and safe provided without and discrimination among employees.
(iv) *Social Equity*: making sure that tourism will encourage social and economic advantages that are dispersed to the nearby communities, enhancing every aspect of life.
(v) *Visitor Realisation*: facilitating visitors have a safe and enjoyable trip without discriminating against them on account of their sexuality, race, age, gender, or other characteristic.
(vi) *Local Control*: permitting neighbourhoods to take part in the process of planning and making decisions related to tourism in their communities.
(vii) *Community Wellbeing*: raising the standard of living and well-being of the surrounding communities, to make sure about the availability of resources and are not become the victim of any sort of mistreatment or taken wrong advantage of in the name of tourism.
(viii) *Cultural Richness*: honouring the customs of host communities, authenticity, cultural history, and individuality.
(ix) *Physical Integrity*: maintain all landscapes as well as making sure that tourism does not become the reason of physical or aesthetic injury and degradation.
(x) *Biological Diversity*: to preserve species, distinctive ecosystems, and natural regions by keeping an eye out for harm and taking appropriate action.
(xi) *Resource Efficiency*: to build and run tourism buildings and services using as many sustainable and renewable resources as possible.
(xii) *Environmental Purity*: keeping trash free the land, water and air from pollution, and other waste that may be left behind by the tourists or tourism-related enterprises.

4. Sustainable Tourism Development

Different types of tourist destinations such as mass tourism and niche tourism segments took advantages from the development of rules and policies of sustainable tourism (Hughes, 1995). To guarantee the enduring viability of tourism growth, an appropriate equilibrium between these three aspects needs to be determined. All features like environmental, sociocultural and economic of tourism development are all covered by sustainability principles (Liu & Wall, 2006). So, sustainable tourism ensures the basic rules such as:

(i) Ensure optimal utilisation of the environmental resources that are mandatory for the enlargement of the tourism sector, supporting the conservation of important organic procedures, biodiversity, and the ecological heritage.
(ii) Uphold the host communities' cultural values and built the living traditional heritage; encourage intercultural tolerance and understanding; and respect their sociocultural authenticity.
(iii) Ensure that long-term, sustainable economic activities, such as societal services for host societies and steady employment and revenue-earning opportunities, are provided to all parties involved in a fair manner, helping to reduce poverty.

There is a dire need of robust political leadership as well as informed participation from the end of each and every concerned stakeholder to build the engagement and trust (Brokaj, 2014). In order to achieve sustainable tourism, it is imperative that ongoing impact assessments be conducted and that preventive and/or corrective measures be taken as needed. Sustainable tourism should raise visitors' awareness of sustainability issues and promote the adoption of sustainable travel habits in addition to offering them a meaningful experience and preserving a high degree of satisfaction.

5. India Sustainable Tourism Market Outlook (2022–2032)

The total revenue of the India sustainable tourism industry will continue to increase from its estimated US$26.01 million in 2022 to US$151.88 million with 2032, with a fast annualised growth rate (CAGR) of 19.3% from 2022 to 2032. Between 2% and 5% of the world market for sustainable tourism is made up of sales from the Indian sector. As per the projections of future market insights, Projected size of the Indian Sustainable Tourism Market for the year 2032 is US$151.88 Millions and the Value-driven CAGR for India's Sustainable Tourism Market from 2022 to 2032 is estimated as 19.3% of entire tourism market in the world. Apart from this, Indian sustainable tourism market recorded with the share of 5% to 9% in the year 2021 in the global market.

5.1 An Outlook for the Sustainable Tourism Market in India, 2017–2021: A Comparison with 2022–2032 Forecasts

It is believed that the tourism sector is a people-oriented one that creates jobs that support the expansion of local companies. India's need for sustainable tourism is rising as the country places greater emphasis on economic development, cultural and traditional practices, and resource preservation. These are the main drivers increasing India's market share for sustainable tourism. The size of the India sustainable tourism market is anticipated to be determined by several key factors over the course of the forecast period, such as the country's natural environment, cultural traditions, heritage, and many more. India is encircled by tropical vegetation in the south and the northern Himalayan peaks. There are many different types of natural habitats in India, including endless stretches of coastline with stunning beaches and mangroves (Adger et al., 2005). Consequently, eco-friendly travel is gaining popularity in India. Sustainable tourism leads to well-developed infrastructure and attractive tourists destinations due to its huge potential to grow with rapid pace. Apart from this, as sustainable tourism can create long-term growth patterns and leverage the country's success in the services sector, it is becoming increasingly important in India.

In contrast to the industrial sector, which is expected to generate 45 jobs for every million rupees spent, India's travel and tourism industry is anticipated to create 78 jobs. It is one of the biggest service sectors in India, along with construction. besides the fact that tourism employs a diverse range of workers, from unskilled labourers to experts, but it also provides greater job and MSME trade opportunities for women. State governments and federal government has launched a large number of spending plans to boost the historical and cultural sector. Besides this, other several kinds of opportunities have been built for the ecological protection and development of natural landscape for generating the interest of foreign investors to thrive the sustainable tourism sector of India. However, it has been projected that these factors will pay a significant role to 19.3% in demand for sustainable tourism in India over the course of the projection period.

5.2 Significant Trends Advancing the Sustainable Tourism Industry in India

The commitment of the tourism sector to sustainable, inclusive growth is particularly important for livelihoods. Owing to its enormous diversity, India is a well-liked travel destination worldwide. Therefore, it is essential to guarantee the long-term viability of the tourism sector. The majority of nations view sustainable tourism as a key trend for the growth of the travel and tourism sector.

Moreover, one of the things propelling the expansion of India's sustainable tourism market is the country's vastness, which offers breath-taking and serene landscapes in addition to cultural and historic landmarks. In essence, it is a nation that offers travel destinations for every kind of traveller.

Element driving the need for sustainable tourism in India is the rise in disposable income, which has fuelled travellers' desire to visit stunning and eco-friendly locations. Travellers searching for affordable, environmentally friendly vacation spots should consider India. Together, these elements account for the growth of the sustainable tourism market in India between 2022 and 2032.

5.3 India's Sustainable Tourism Industry to Benefit from Industrialisation and Urbanisation

The global trends of industrialisation and development have led to the growth of the sustainable tourism business in India. In addition to the rural areas experiencing the effects of urbanisation, this movement in India's sustainable tourism sector has created new employment opportunities and income brackets. India's need for sustainable tourism is expected to increase as it addresses issues related to equal growth. The 'return to roots' tourism theme is also associated with concerns about climate change, improved accessibility, and greater interest in heritage and culture. The increasing availability of quality time visits for participatory settings in tourism experiences across the nation is contributing to the growing popularity of sustainable tourism in India. Additionally, community-owned, culturally expressive, and environmentally sustainable enduring ways of living, art, and legacy are driving the size of the India sustainable tourism business.

At the same time, local communities gain from the meaningful relationship that occurs between guests and hosts. As attitudes and mind-sets change, a sense of local pride and gratitude for tourists is ingrained. Because of this, when a visitor connects with the modern, they are faced with India's vernacular traditions, which fuels the country's need for sustainable tourism. India is a multi-interest, year-round destination due to its vast geographical area and diversity of habitats, which is another element expected to fuel the expansion of the country's sustainable tourism industry.

6. GSTC Destination Criteria

Many guidelines and specifications for sustainable tourism from every continent are taken into account by the GSTC Destination Criteria (GSTC-D), which are based on decades of prior work and international experience. Throughout the development process, they were widely consulted in a number of languages in both developed and developing countries worldwide. Best practices, certification standards, indicators, and criteria from different geopolitical and cultural contexts around the world in the travel and tourism industry are incorporated when applicable. The practicality, relevance, and applicability of potential indicators to a broad range of destination types were assessed.

They were field tested all over the world. The ISEAL Alliance, an international organisation that provides guidelines for the development and implementation of sustainability standards across all industries, and ISO codes of conduct were all taken into account in the process of developing the Criteria. The GSTC Destination Criteria v2.0 represents the initial version of the criteria. The GSTC-D v2's performance indicators are meant to provide guidance when assessing adherence to the Criteria. By applying the criteria, a destination can support both the 17 Sustainable Development Goals and the 2030 Agenda for Sustainable Development. The 17 SDGs that are most closely associated with each of the Criteria are identified, one or more of them.

The following are a few examples of how tourism management groups anticipate using the criteria:

- Provide a solid framework for places looking to improve their sustainability.
- Help clients select trustworthy, eco-friendly vacation spots.
- Provide as a standard by which the media may identify places and enlighten the public about their sustainability.
- Assist certification and other optional destination level initiatives in making sure their requirements align with a generally acknowledged baseline.
- Provide programs in the public, non-profit, and commercial sectors with a foundation for creating sustainable tourism standards.
- Act as fundamental standards for organisations that provide education and training, like universities and hotel schools.

6.1 Sustainable Tourism and the GSTC Criteria

The GSTC Criteria serve as the international standards for sustainability in travel and tourism. The public is educated and made aware of through the application of the criteria. They are used in the processes of measurement, assessment, and policy-making, and they provide the basis for certification. They are the result of a global effort to establish a shared language regarding sustainable tourism.

They can be divided into four groups: sustainable management (D), cultural effects (C), environmental effects (D), and social and economic effects (A). These recommendations are the result of decades' worth of prior work by global industry experts. In both developed and developing nations, they were heavily consulted throughout the development process. They represent our objective of reaching a global understanding of sustainable tourism.

The standards-setting code of the ISEAL Alliance was followed in the development of the criteria. An international group called the ISEAL Alliance offers guidelines for managing sustainability standards in all sectors of the economy. Based on applicable ISO standards, that code was created. Lastly, the GSTC Criteria provide a starting point for businesses, governments, and travel destinations. Every tourist destination has its own distinct culture, ambiance, traditions, and laws. Consequently, the purpose of the criteria is to be customised to the specific circumstances of each location and activity, adding additional criteria as needed.

There are three sets of standards

(i) **Industry Standards for GSTC** = refers to the sustainable management of the private sector travel business, with a current concentration on hotels and tour operators.
(ii) **GSTC Destination Standards** = has to do with managing tourism destinations sustainably.
(iii) **GSTC MICE Standards** = has to do with managing venues, event planners, and events and exhibits sustainably.

6.2 GSTC in India

Sustainable tourism, sometimes known as ecotourism in India, is gaining importance. In 2020, Booking.com performed research which revealed that 96% of Indian passengers considered sustainable travel to be significant. Additionally, 76% of them expressed a greater resolve to make sustainable travel decisions in the future. Moreover, 73% of visitors are choosing more environmentally friendly options in their daily lives as a result of seeing the effects that tourism has on the environment while on holiday. Even while these results are encouraging, there are still challenges. Of Indian tourists, 39% are unsure of where to go for eco-friendly travel options, and 54% believe there aren't enough eco-friendly travel options accessible. Thus, more information about eco-friendly travel agencies in India is required. In India, the GSTC Criteria are already in use and have been gaining traction lately.

7. Role of Technology in Promoting Sustainable Tourism

The relationships between the accessibility of information about tourism services, the influence of information technologies, and their online sale-purchase process had an effect on the digitalisation of tourism (Mbaiwa, 2003). Airlines, hotels, travel agencies, and other tourism-related businesses typically concentrate less on the aspects related to service delivery and physical location against higher stress is given on creating a competitive edge with the help of travel services' online component (Cuomo et al., 2021).

Furthermore, the relationship between supply and demand in the travel industry has completely changed as a result of technological advancement, which also has a significant impact on consumers' travel habits. Digitalisation and new technologies have had a big impact recently, producing a tonne of data, information, and knowledge (Minghetti & Buhalis, 2010). Digitalisation has revolutionised tourism initiatives, products, experiences, business ecosystems, destinations, and ways of interacting with people, according to Al-Mulali et al. (2021). It has also changed how people interact (Jingen Liang & Elliot, 2021). All things considered, there ought to be greater focus on the ways in which adopting new technology might influence a more sustainable path for tourism development. Through the utilisation of technology, the tourism sector can strive to lessen its impact on the environment, aid local communities, and conserve cultural heritage.

Rather than being a distinguishing characteristic, technology should be seen as a tool to support sustainable tourism initiatives. An all-encompassing strategy that prioritises both traditional sustainable practices and technical breakthroughs will eventually result in a prosperous and ethical tourist sector. Even the most established markets are being disrupted by recent technology advancements, which can also lead to robust and strategically agile operations.

There are many methods through which tourism and travel industry can drive transformative change by making sustainable technology solutions such as:

(i) Embracing digital technology solutions by using less paper.
(ii) Making investments in eco-friendly technology.
(iii) Utilise data analytics to streamline operations.
(iv) Collaborate with suppliers who place a high priority on sustainability.
(v) More use of digital technology to foster connections between tourists and the community.

7.1 Embracing Digital Technology Solutions by Using Less Paper

Reducing paper use and, whenever possible, implementing digital solutions are two ways to help a travel agency become more sustainable. It might entail implementing digital communication and teamwork tools, as well as employing electronic tickets and e-vouchers. Travel agencies can use digital solutions in a number of ways to prepare for a sustainable future:

- *Online Booking Systems*: travel agencies can put in place online booking systems that let clients make reservations and pay for their trips online, doing away with the need for paper tickets and other hard copies.
- *Virtual Tours*: travel agencies can provide clients with virtual tours of their locations, enabling them to explore and discover the area before they ever leave their homes.
- *E-Newsletters*: travel agencies are able to notify their clientele via e-newsletters about upcoming trips, exclusive deals, and other information. It lessens the requirement for marketing materials printed on paper.

- *Social Media Marketing*: travel agencies can connect and interact with consumers on social media by posting news, images, and other content that encourages environmentally friendly travel.

7.2 Making Investments in Eco-Friendly Technology

Water conservation systems, energy-efficient appliances and lighting, and renewable energy sources are examples of green technology that travel agencies can invest in.

Travel agencies can invest in green technology in a number of ways, including:

- *Invest in Energy-Efficient Transportation*: for their transportation needs, travel agencies can purchase electric or hybrid cars. They will ultimately save money on fuel in addition to lessening their carbon footprint.
- *Use Renewable Energy Sources*: to lessen their dependency on fossil fuels, travel agencies can invest in solar or wind turbines to produce their own electricity.
- *Implement Green Initiatives Into Their Operations*: to lessen their impact on the environment, travel agencies can invest in recycling programmes, water conservation measures, and environmentally friendly waste management systems.
- *Encourage Eco-Tourism*: travel agencies can make investments in environmentally friendly lodging and programmes that support sustainability and protect natural areas.
- *Invest in Carbon Offset Programmes*: by funding initiatives that lower greenhouse gas emissions or raise carbon sinks, travel agencies can offset their carbon emissions.
- All things considered, spending money on green technology can not only lessen its negative effects on the environment but also enhance its reputation and draw in eco-aware clients.

7.3 Utilise Data Analytics to Streamline Operations

Travel companies can use data analytics to streamline their processes and make better decisions about how they can minimise their environmental impact in several ways.

- *Finding Patterns and Trends in Travel Demand*: travel agencies can improve their operations and lessen their environmental effect by using data analytics to identify patterns and trends in travel demand. Travel agencies might concentrate on providing more eco-tourism options, for instance, if data indicates that there is a high demand for eco-tourism.
- *Examining The Environmentally Harmful Effects of Travel*: data analytics can assist travel agencies in determining how much of an environmental impact their operations have and in understanding that impact. They can use data, for instance, to find the most fuel-efficient routes and means of transportation or to find ways to use less energy.
- *Finding Opportunities for Sustainability*: by examining data on the environmental effects of various travel destinations and lodging options, data analytics can

assist travel agencies in finding opportunities for sustainability. They can use it to make better-informed choices about where to sell travel packages and how to lessen their influence on the environment.
- *Assessing the Effectiveness of Sustainability Initiatives*: travel agencies can use data analytics to track the effectiveness of their sustainability initiatives and pinpoint areas for improvement. They can use data, for instance, to monitor how their initiatives are affecting their carbon footprint and pinpoint areas in which they can improve even more.

7.4 Collaborate with Suppliers Who Place a High Priority on Sustainability

Prioritising sustainability in their operations and forming partnerships with suppliers are two more ways that travel businesses can effect radical change. It might include lodging establishments with environmentally conscious rules and procedures or transportation companies that drive fuel-efficient cars.

The following measures can be taken to collaborate with travel technology providers who place a high priority on sustainability:

- *Research Possible Partners*: seek out providers of travel technology who have demonstrated a commitment to environmental responsibility and sustainability. It could entail reading the mission statement of the partner company, looking up any sustainability certifications or honours they have won, and reading customer testimonials or case studies.
- *Determining Top Priorities for Sustainability*: establish which sustainability priorities are most critical to the company. It could entail cutting back on waste, conserving resources, helping out local communities, and lowering greenhouse gas emissions.
- *Communicate the Sustainability Targets*: A business should be confident about the sustainability priorities when approaching possible partners. It will assist in making sure the supplier can support business' goals for sustainability and is a suitable fit for business's needs.
- *Negotiate Sustainability Terms*: sustainability terms and conditions with the possible partner must be ensured during the negotiation process. This could entail adopting eco-friendly products, cutting down on energy use, or helping out in the community.
- *Checking Out and Evaluating the Partnership*: after completing the process of making partnerships, a business needs to keep an eye on and evaluate how it will affect sustainability. It could entail establishing sustainability goals and monitoring the partner's progress towards reaching them, reviewing their performance on a regular basis, and making necessary adjustments.

By taking these actions, sustainability aspiring organisation can make sure that the travel technology providers are dedicated to promoting the long-term viability and health of the travel sector and that they prioritise sustainability.

7.5 More Use of Digital Technology to Foster Connections Between Tourists and the Community

Travellers can engage with local communities and support sustainable tourism initiatives through the use of technology. Communities may benefit economically from this and it can aid in the preservation of regional cultures.

- *Online Resources and Applications*: travellers can utilise online resources and applications to establish connections with local businesses, organisations, and communities that are dedicated to sustainable tourism practices. Travellers can use these platforms, for instance, to make reservations for lodging, tours, and activities with regional businesses dedicated to sustainable practices, like eco-friendly lodging and conservation-focused tours.
- *Social Media*: travellers can engage with local communities and organisations promoting sustainable tourism by using social media platforms. Travellers can discover sustainable tourism initiatives in the area and receive recommendations for sustainable places to dine, stay, and explore by, for instance, following local tourism boards and organisations on social media.
- *Virtual and Augmented Reality*: travellers can learn about local cultures and communities in a more interactive and captivating way by taking advantage of immersive experiences offered by virtual and augmented reality technology. Travellers can use virtual reality, for instance, to explore a local village virtually or gain insight into the history and culture of a community by following the guide's perspective.

7.6 Digital Technology Uses for Sustainability in Aviation Sector

Air travel is a primary cause of climate change, the airline industry has a big environmental impact. But there are actions you can take to improve the industry's sustainability:

- *Utilisation of More Fuel-Efficient Aircraft*: by purchasing newer, more fuel-efficient aircraft, airlines can lower their carbon emissions. The amount of fuel used by these aircraft per passenger lowers the flight's carbon emissions.
- *Carbon Offset Programmes*: a lot of airlines give their passengers the choice to use them to offset their carbon emissions. Through these programmes, customers can invest in clean energy or carbon reduction projects in order to offset the carbon emissions from their flight for a fee.
- *Reducing Waste*: by utilising more environmentally friendly materials for in-flight amenities like recycled paper goods and biodegradable cutlery, airlines can cut down on waste.
- *Investing in Renewable Energy*: to power their operations, some airlines have started to invest in renewable energy sources. This covers producing electricity with solar power or biofuels.
- *Promoting Alternate Forms of Transportation*: for shorter distance travel, airlines can also persuade passengers to think about using buses or trains as an alternative to flying.

8. Five Eco-Friendly Technologies to Encourage Sustainable Travel

The travel business is shifting towards more environmentally friendly tourism practices, which means that technology plays a bigger and bigger role now. In addition to being crucial for the preservation of the environment, this change is also necessary for the tourism industry to continue operating profitably (Mehta et al., 2023).

8.1 Solutions for Renewable Energy

Any sustainability program's holy grail is renewable energy, and the tourism industry is no exception. Although it is too soon to declare that it is revolutionising sustainable tourism, there are indications that more and more tourist locations are utilising renewable and natural energy sources:

- *Solar Power*: a lot of hotels and resorts are using solar panels to provide sustainable energy, which lowers energy expenses and greenhouse gas emissions.
- *Wind Energy*: with their ability to provide a sustainable power source with little impact on the environment, wind turbines are gaining popularity as a tourism option for isolated and coastal locations.
- *Hydropower*: some eco-lodges and rural tourist destinations use small-scale hydropower systems as a dependable, sustainable energy source that blends in with the surroundings.

8.2 Technologies for Water Conservation

Traveling to sunny beaches isn't the only thing that comes to mind when someone mentions tourism, but it is frequently the first thing that comes to mind. The issue is that water scarcity frequently blights the same locations that sunseekers 'seek'. a problem made worse by the increase in visitors. The following water-saving techniques are assisting in addressing this problem:

- *Intelligent Irrigation Systems*: these systems minimise water waste by optimising water usage for landscaping in resorts and popular destinations.
- *Water Recycling*: to clean and reuse water for non-potable uses, many tourism facilities now employ water recycling techniques, reducing the demand on nearby water resources.
- *Low-flow Fixtures*: installing low-flow showers, toilets, and faucets in hotels and lodges is an easy yet efficient method to reduce water usage without sacrificing the visitor experience.

8.3 Innovative Waste Management

Sustainable tourism relies on efficient waste management to protect the environment and improve the visitor experience in general. Not only may innovative ideas in this field help the environment, but they also improve the well-being of the local population.

The following are some significant innovations:

- *Composting Initiatives*: to cut down on organic waste and create useful compost for regional usage, numerous tourist locations have put in place composting initiatives.
- *Recycling Activities*: recycling initiatives are important in minimising landfill waste and raising visitors' awareness of environmental issues, especially in popular tourist locations.
- *Biodegradable Products*: by switching to eco-friendly and biodegradable products, hotels and resorts lessen the impact of single-use items and cut down on their overall waste output.

8.4 Green Transportation

By definition, tourism is dependent on transportation. Although there has been progress in the area of flying sustainability, much more work has to be done to effectively address this issue. But after we arrive at our location and start exploring the neighbourhood, the sustainability of the energy-efficient transportation we use is affecting tourism (Bala & Singh, 2023).

Among the advancements that make exploring distinctive, entertaining, and sustainable are:

- *Electric Vehicles (EVs)*: to cut down on noise and pollution, a lot of tourist spots are introducing EVs, from buses to rental automobiles.
- *Bike Sharing Programs*: for an environmentally and health-conscious travel option, many cities and tourist destinations now offer bike-sharing programs. Bicycles provide an emission-free means of transportation.
- Modernising public transportation can help minimise the need for private automobiles, which in turn reduces the total carbon footprint of tourism. Cleaner, more effective choices are available for public transportation.

8.5 Sustainable Construction Materials and Methods

The infrastructure of tourism is changing, with smart technologies and sustainable building materials drastically lowering environmental footprints (Bala et al., 2023). These methods incorporate cutting-edge technologies for effective operation in addition to employing environmentally acceptable components.

- *Eco-Friendly Construction*: new tourist developments have a lessening effect on the environment when they use materials like bamboo, recyclable steel, and green insulation.
- *Smart Building Technologies*: cutting-edge automation and energy management systems are becoming extensively used. This includes working together with producers of door access control systems to improve security and reduce energy use.
- *Green Walls and Roofs*: these live architectural elements improve air quality, act as insulation, and lessen urban heat islands, all of which help make tourist facilities more sustainable.

9. The Relationship Among Three: Tourism, Technology, and Sustainable Development

The four phases of ICT development are apparent to the tourism industry: opportunity, interruption, entanglement, and encroachment. These stages paved the way for the development of automated reservation systems, global distribution networks, travel agencies' websites, expedited and straightforward reservation procedures, and price competition. ICT started to become a noticeable part of many consumers' daily lives during these phases. We take the help of social media and other innovations like Airbnb, ICT, etc. ICT arrangements are extensively used in the tourism sector to support vital tourism functions that benefit both suppliers and guests, like accommodation, transportation, and communication. Their adoption has been impacted by the changing competitive environment.

The United Nations has established the Sustainable Development Goals (SDGs), which are 17 objectives including topics like equality, poverty reduction, climate change, and environmental preservation. As per the UN, tourism has the potential to contribute, directly or indirectly, to all of the goals and can play a significant role in delivering sustainable solutions for people, the planet, prosperity, and peace. The 17 SDGs for tourism can be broadly categorised into four areas: environment (Goals 6, 13, 14, 15), economy (Goals 1, 7, 8, 9, 10, 11, 12), governance (Goal 17), and social-cultural (Goals 2, 3, 4, 5, 16). In turn, ICT developments are frequently seen as promoting the SDGs and enhancing society.

Three technologically enabled goals and how they support the SDGs are described below:

9.1. Platforms for the Sharing Economy

Technology led sources similar to Airbnb and Couchsurfing has a great future for sustainable practices with SDGs for the better outcomes.

These sources carry the utmost importance for each and every segment of the society, either it is related to the people from developing nations, women entrepreneurs, or small/rural firms' access to the market. They have also built and improved the things like cultural learning, cross-cultural friendships, and economic sustainability. It is possible to argue that platforms create jobs, make underutilised resources accessible, and have the potential to lower resource usage.

9.2. Instruments for DMOs

There is a huge possibility to enhance the sustainable tourism with the help of ICTs driven destination management system. Several applications can be used to manage the information for providing maximum satisfaction to customers, motivating society participation as well as proper energy consumption management with its impacts. Furthermore, ICTs also play a significant role for Destination Management Organisations (DMOs) for the development of sustainable tourism.

These include destination management systems (DMS), virtual tourism, carbon dioxide calculators, location-based services (LBS), community informatics

(CI), and environment management information systems (EMIS). ICTs offer new distribution channels and the potential to improve stakeholder interaction and communication. They are crucial to the collection, processing, organisation, and distribution of information. They are back reporting, observation, and impact measurement.

It has been observed that among the several key tools for the help of destination managers, Destination Management System (DMS) plays a significant role. It can be utilised for the various activities such as managing the information and resources, making aware the educators as well as marketing of eco-friendly tourists destinations. Environment Management Information Systems (EMIS) support tourist ability to plan the things and to take better decisions by disseminating helpful information on the effects of tourism at the destination and monitoring emissions and waste management. Location Based Services (LBS) contribute to deliver the information on tourists' particular locations to assist destination management in educating visitors about sites and attractions to visit as well as sensitive locations, appropriate tourist behaviour, and most environment friendly choices in the destination.

The use of community informatics (CI) can be advantageous for community engagement, heritage and tradition preservation, interpretation, community cohesion, and tourism education. It entails involving the local community in decisions about the development and planning of tourism at the destination. Conversely, virtual tourism (VT) offers a novel approach to interacting with products and services associated with travel. It can slow the deterioration of attractions and have a positive environmental impact by reducing the number of visitors and spreading knowledge about the site. As a last example of a product innovation, a carbon calculator tells travellers about their carbon footprint both before and during their trip.

9.3. Local Communities' Empowerment

ICTs can significantly affect the neighbourhood both directly and indirectly. The literature on tourism has placed significant emphasis on local communities as a vital resource for the development of sustainable tourism, underscoring the significance of their engagement and participation. Utilising the natural and cultural assets of the towns, tourism plays a significant role in poverty alleviation and community development by providing employment opportunities for women. Small enterprises can manage and advertise their businesses and reservations with the use of ICTs. It provides advantages like skill development, improved access to education, exposure to the outside world, boosted confidence, and increased community involvement in addition to economic benefits for the residents. For instance, the study discovered that ICTs were crucial to Malaysian community development. More chances for education and ICT training were made available to the community through the implementation of ICTS in homestay lodgings. The neighbourhood gained knowledge about how to use the several online booking services for lodging. Because of the success of the ICT-enabled firm, communities could have better access to higher education and hospitality skills.

Additionally, the residents were able to educate themselves about environmental protection thanks to their internet access. ICT and socioeconomic development are related. The socioeconomic development of people, as well as larger organisations and nations, is impacted by business operations and services like online education and e-commerce, which are made possible by computing resources, the Internet, mobile phones, GPS, and other devices. Impacts at the individual level include life quality, income, and education. ICT at the organisational level can improve a company's ability to compete internationally as well as factors like political freedom and the nation's output.

9.4. COVID-19 Virtual Tourism Market

Through digital experiences, people can virtually 'travel' to locations or visit tourist attractions without ever leaving their homes. This is known as virtual tourism. It makes use of a variety of technological elements, including narration, audio, video, graphics, and virtual reality software. Interactive presentations featuring informed and entertaining tour guides can be recorded or streamed live for virtual tourism experiences.

The goal of virtual tourism is to replicate a close-to-real touring experience of a certain location or tourist site. Virtual tourism can be used to plan a future in-person vacation or to get a taste of what a destination or tourist site has to offer. While in-person travel is still recovering from the COVID-19 pandemic, major players in the VR technology space such as Google, GoMeta, Valve, Ximmerse, Samsung Electronics, Microsoft Corporation, Sony Interactive Entertainment, and Facebook's Oculus VR LLC are driving significant progress in the global virtual tourism sector.

The intention is not to supplant in-person travel but rather to expand into new markets, such as school groups and older adults with discretionary means who are unable to travel because of health issues or limited mobility. Additionally, industry Research Data states that 'the global virtual tourism market is predicted to reach the valuation of US$24.10 billion by 2027, having been valued US$5 billion in 2021'.

10. Conclusion

Currently, efforts and practices related to sustainable tourism are being prioritised by nations across the globe. The primary obstacle for implementing the sustainable tourism, however, is the lack of clarity surrounding its meaning, which has been mirrored in inadequate tourism practices. Clarity in both concept and practice is necessary to ensure that the assessment of sustainable tourism is founded on reliable criteria. In the previous few years, there has been a strong interest in ensuring sustainable development that minimises the adverse impact on the surroundings, particularly in the tourism industry, which involves intricate interactions with the natural system. The tourism industry is a complex ecosystem with a wide range of players, including electronic platforms, travel firms, tour operators, lodging providers, destination management companies, tourists attractions, transportation activities, and physical tourist offices as well as online service providers. Additionally, this industry is the significant player in the EU economy and is a major employer. It has been observed from the past studies that Digital technology integration in sustainable tourism is a complicated and complex issue that requires collaboration from a variety of stakeholders, including technology providers, local governments, tour operators, and communities.

Through the use of cutting-edge technology, digitalisation may continue to raise the bar for travel experiences to the point where they may support the sustainable development goals of tourism. The study has demonstrated a wide range of technological innovations that have a major effect on the hospitality and travel industry. These comprise technologies for business management, including blockchain, data analytics, automation and sophisticated robotics, cloud-based computing, smartphone and tablet technology, and data analytics. At the last, it is concluded that digitalisation is important and key resource to innovate business models for the global recognition of tourism industry. Apart from this, technological and digital environment trends are pushing tourism businesses to adopt new, creative business models and procedures, as well as to become more adaptable. These developments also bring with them challenges that must be recognised and addressed in light of market and business strategy. Investing in cutting-edge tools or procedures serves to foster creativity, sustainable business practices, and readiness for market shifts that will affect the company. As a result, travel agencies are starting to understand how important they are and that investing in digital tools is a long-term commitment.

References

Adger, W. N., Arnell, N. W., & Tompkins, E. L. (2005). Successful adaptation to climate change across scales. *Global Environmental Change*, *15*(2), 77–86.

Al-Mulali, U., Solarin, S. A., Andargoli, A. E., & Gholipour, H. F. (2021). Digital adoption and its impact on tourism arrivals and receipts. *Anatolia*, *32*(2), 337–339.

Asmelash, A. G., & Kumar, S. (2019). The structural relationship between tourist satisfaction and sustainable heritage tourism development in Tigrai, Ethiopia. *Heliyon*, *5*(3), e01335.

Aynalem, S., Birhanu, K., & Tesefay, S. (2016). Employment opportunities and challenges in tourism and hospitality sectors. *Journal of Tourism & Hospitality*, *5*(6), 1–5.

Bala, R., & Singh, S. (2023). Examining the antecedents and consequences of corporate commitment towards climate change. In K. Sood, S. Grima, P. Young, E. Ozen, & B. Balusamy (Eds.), *The impact of climate change and sustainability standards on the insurance market* (pp. 269–278). Scrivener.

Bala, R., Singh, S., & Sharma, K. K. (2023). Relationship between environmental knowledge, environmental sensitivity, environmental attitude and environmental behavioural intention – A segmented mediation approach. *Management of Environmental Quality: An International Journal*, *34*(1), 119–136.

Baldassarre, B., Calabretta, G., Bocken, N. M. P., & Jaskiewicz, T. (2017). Bridging sustainable business model innovation and user-driven innovation: A process for sustainable value proposition design. *Journal of Cleaner Production*, *147*, 175–186.

Baloch, Q. B., Shah, S. N., Iqbal, N., Sheeraz, M., Asadullah, M., Mahar, S., & Khan, A. U. (2023). Impact of tourism development upon environmental sustainability: A suggested framework for sustainable ecotourism. *Environmental Science and Pollution Research*, *30*(3), 5917–5930.

Brokaj, R. (2014). Local governments role in the sustainable tourism development of a destination. *European Scientific Journal, 10*(31), 87–99.

Brown, H. S., & Vergragt, P. J. (2008). Bounded socio-technical experiments as agents of systemic change: The case of a zero-energy residential building. *Technological Forecasting and Social Change, 75*(1), 107–130.

Connell, J., Page, S. J., & Bentley, T. (2009). Towards sustainable tourism planning in New Zealand: Monitoring local government planning under the Resource Management Act. *Tourism Management, 30*(6), 867–877.

Cuomo, M. T., Tortora, D., Foroudi, P., Giordano, A., Festa, G., & Metallo, G. (2021). Digital transformation and tourist experience co-design: Big social data for planning cultural tourism. *Technological Forecasting and Social Change, 162*, 120345.

Fang, B., Ye, Q., & Law, R. (2016). Effect of sharing economy on tourism industry employment. *Annals of Tourism Research, 57*, 264–267.

Higgins-Desbiolles, F. (2018). Sustainable tourism: Sustaining tourism or something more? *Tourism Management Perspectives, 25*, 157–160

Hoffmann, F. J., Braesemann, F., & Teubner, T. (2022). Measuring sustainable tourism with online platform data. *EPJ Data Science, 11*(1), 41.

Hughes, G. (1995). The cultural construction of sustainable tourism. *Tourism Management, 16*(1), 49–59.

Jingen Liang, L., & Elliot, S. (2021). A systematic review of augmented reality tourism research: What is now and what is next? *Tourism and Hospitality Research, 21*(1), 15–30.

Jokela, S., & Minoia, P. (2021). Tourism platforms. In C. P. Krieg & R. Toivanen (Eds.), *Situating sustainability: A handbook of contexts and concepts* (pp. 223–237). Helsinki University Press.

Kauppila, P., Saarinen, J., & Leinonen, R. (2009). Sustainable tourism planning and regional development in peripheries: A nordic view. *Scandinavian Journal of Hospitality and Tourism, 9*(4), 424–435.

Mbaiwa, J. E. (2003). The socio-economic and environmental impacts of tourism development on the Okavango Delta, north-western Botswana. *Journal of Arid Environments, 54*(2), 447–467.

Mehta, P., Kaur, A., Singh, S., & Mehta, M. D. (2023). "Sustainable attitude" – A modest notion creating a tremendous difference in the glamourous fast fashion world: Investigating moderating effects. *Society and Business Review, 18*(4), 549–571.

Menon, S., Bhatt, S., & Sharma, S. (2021). A study on envisioning Indian tourism–Through cultural tourism and sustainable digitalization. *Cogent Social Sciences, 7*(1), 1903149.

Minghetti, V., & Buhalis, D. (2010). Digital divide in tourism. *Journal of Travel Research, 49*(3), 267–281.

Liu, A., & Wall, G. (2006). Planning tourism employment: A developing country perspective. *Tourism Management, 27*(1), 159–170.

Sisneros-Kidd, A. M., Monz, C., Hausner, V., Schmidt, J., & Clark, D. (2019). Nature-based tourism, resource dependence, and resilience of Arctic communities: Framing complex issues in a changing environment. *Journal of Sustainable Tourism, 27*(8), 1259–1276.

Weblinks

www.unwto.org
www.tourism.gov.in
www.futuremarketinsights.com

Chapter 11

Tech-Driven Horizons: Pioneering Sustainable Tourism for Environmental Resilience and SDGs Triumph

Mahua Basu

St. Xavier's College (Autonomous), Kolkata, India

Abstract

Currently, the tourism industry stands at a crossroads and is in desperate need of a paradigm shift toward sustainable practices. The tourism sector is interwoven by culture, economy, technology, and environmental awareness. Exploring trends and challenges could help us to better understand the transformative forces required to tailor the tourism industries, pivotal to employment, economy, and culture. Technology can act as a catalyst and offer tools and solutions to minimize the ecological footprint of tourism activities. A new era in tourism ushered with the advent of technology that changed the industry's landscape and offered innovative solutions to address and tackle challenges in multiple dimensions. Embracing such advancements like artificial intelligence can foster responsible tourism and force the world toward a positive change. This paper aims to explore the intersection of technology and sustainable tourism and envisions a future where technological advances will pave the way for environmental resilience and triumph over the SDGs. It also emphasizes the need for a participatory approach and collaboration amongst various stakeholders to harness the full potential of technology toward a resilient future.

Keywods: Sustainable tourism; technology; sustainable development; artificial intelligence; SDGs

The Need for Sustainable Tourism in an Era of Global Climate Change: Pathway to a Greener Future, 215–247

doi:10.1108/978-1-83608-668-020241032

1. Introduction

Tourism is a uniquely intangible and perishable product. It is vulnerable to natural, economic, and political events (Rahmadian et al., 2022). It is a socio-cultural and economic event that involves people heading to a destination, either for personal or business activity (Escobar & Margherita, 2021). Tourism has extraordinarily boomed in recent decades. Further, aviation has opened up travel from domestic to international arenas. Before the COVID-19 pandemic, the number of worldwide visits had more than doubled since 2000. In 2022, travel and tourism contributed to around 7.7 trillion US dollars globally. Rise in living standards especially in the North block, reducing, extended haul travel costs, increase in holiday entitlements, demography transition, and consumer demand has led to significant growth in tourism with more and more international visits to the global south (Goodwin, 1998).

With years increasing number of tourists started recognizing the detrimental impacts on the places of their trips and almost 80% of all travelers consider sustainable tourism to be the need of the hour.

The tourism setting is a dynamic tapestry interwoven by culture, economy, technology, and environmental awareness. At this juncture in time, it seems imperative to delve into the multifaceted condition of global tourism. Exploring trends and challenges could help us to better understand the transformative forces required to tailor the tourism industries, pivotal to employment, economy, and culture. The year 2022 had 963 million international tourists exhibiting a 111% growth in comparison to 2021. The five most visited places are France, Spain, the USA, Turkey, and Italy. Some tourist destinations encountered unexpected consequences; some places have developed unconsciously without any strategic planning and integrated effort. Hence integrated tourism seems to be valuable in attaining sustainability (Lisi & Esposito, 2015).

1.1 Setting the Stage

Tourism is marketed globally but consumed at the production point. Internationally dominated tour companies are based in tourist-originating nations, but the established entrepreneurs reign nationally in the destinations. Opportunities of local people need to be maximized at the level of destinations mostly in the developing countries, in the way of increasing foreign exchange income, employment, and funding to preserve natural and cultural heritage. Tourism growth promotes new destinations search that will affect more and more communities (Goodwin, 1998). Tourism should foster a constructive impact on the destinations. It should create an environment that beckons travelers and offer memorable journeys and unique experiences. Building this foundation is crucial to shaping the overall tourism experience. This can influence both the tourists and the communities explored.

1.1.1. Infrastructure and Accessibility

(i) *Robust Transportation Networks*: an efficient transportation network saves traveling time adding appeal to tourist spots. Well-connected airports, railways, roadways, and waterways improve ease of access and trouble-free and cheap public transportation adds to the advantage. Singapore's Changi Airport offers top-tier transportation facilities and is ranked one of the world's best.

(ii) *Accommodation Options*: a diverse range from budget-friendly to expensive accommodation options can also attract a variety of people from diverse societal backgrounds. A Booking.com survey reveals that 57% of travelers prioritize lodging for enhancing their traveling experience.

1.1.2. Cultural Significance and Authentic Experiences

(i) *Cultural Preservation*: responsible tourism practice safeguarding cultural assets. Many tourists are attracted to the authenticity of local culture, traditions, and heritage. Kyoto in Japan is known for preserving historically relevant places and traditions that create a cultural vibe.

(ii) *Unique Experiences*: in many places, local businesses are encouraged to offer authentic interactions that can set the destination a class apart from others. It is observed that the majority of tourists choose destinations based on culture and nature.

1.1.3. Safety and Security

(i) *Infrastructure for Safety*: the rapid availability and implementation of effective emergency response systems to ensure visitors' safety, like in New Zealand, makes it one of the most sought tourist places. Transparent communication raises confidence in tourists.

(ii) *Health and Sanitation Measures*: many governments implement health protocols to address concerns like COVID-19. According to the World Travel and Tourism Council (WTTC), 80% of tourists look for hygiene and safety protocols when deciding to travel.

1.1.4. Marketing and Promotion

(i) *Strategic Destination Branding*: the compelling campaign "Inspired by Iceland" successfully showcased, narrated, and highlighted Iceland's natural beauty and culture. Effective marketing is most often boosted by using digital platforms.

(ii) *Collaborative Partnerships*: more than 50% of tourists explore novel destinations through online traveling platforms and agencies adding to digital marketing. Strategic collaboration with traveling agencies, influencers, and local businesses can foster the tourism industry into a thriving venture.

1.2 Need for Sustainable Tourism

Sustainability in tourism is presently crucial in the era marked by environmental challenges and hence an increasing need for awareness and responsible tourism practices.

1.2.1. Preserving Natural Resources

(i) *Conservation of Biodiversity*: biodiversity and wildlife conservation enable us to safeguard our natural habitat with varied flora and fauna. This ensures our coming generation to enjoy and experience the enormous diversity on this planet. In many places of the world, protected areas and wildlife corridors have been designated to protect biodiversity and nurture ecotourism.

(ii) *Minimizing Environmental Footprint*: the tourism sector contributed to nearly 5% of the GHG emissions worldwide in 2019 (UNWTO). Sustainable tourism on the other hand endeavors to lower GHG emissions, restrict waste generation, and implement energy-efficient practices to minimize the detrimental impacts of traveling on the environment.

1.2.2. Empowering Local Communities

(i) *Socioeconomic Benefits*: sustainable tourism empowers local and native people, creates employment opportunities, and supports local businesses. It acts as a positive catalyst for bringing socioeconomic changes along with the preservation of indigenous culture. Communities get engaged in a more meaningful way. Hence it is pivotal for poverty alleviation by providing means of income. Bhutan for instance emphasizes high-value, low-impact tourism ensuring a considerable proportion of revenue to return to the local population to assist in their education, healthcare, and infrastructural development.

1.2.3. Economic Resilience and Diversification

Sustainable tourism diversifies revenue streams by less depending on one type of industry and contributes to economic stability. Maldives created a resilient tourism industry by introducing eco-friendly practices like commitment and enhanced focus on marine conservation.

2. The Ever-Changing Dynamics

The dynamics of tourism evolve contiguously and are dependent on a multitude of factors that may range from technological development to large-scale events as well as changes in customer preferences. The stakeholders need to understand these altering dynamics to innovate, cope, and thrive in the dynamic environment.

2.1 Pre-Pandemic Resurgence

COVID-19 shockwaves have been tumultuous for the tourism industry both globally and locally. However, the industry is slowly bouncing back giving evident signs of recovery. Of late travel restrictions have eased, vacation efforts are on the move and pent-up wanderlust is fueling a resurrection in travel demand.

2.2 Post-COVID Recovery

2.2.1. Easing Travel Restrictions

This is crucial for augmenting tourism and its revival. Globally governments of different nations have introduced different measures to facilitate traveling and attract tourists. Many countries have shifted to visa-free or visa-on-arrival policies to simplify tourist entry. To streamline tourist health status verification, many governments have introduced digital health certificates or passports. Digital COVID certificate enables people to prove their vaccination status to cater to smooth cross-border travel. Countries like Maldives implemented negative Polymerase Chain Reaction (PCR) test results as proof to allow travelers and also to manage public health concerns. Vaccinated tourists were often given incentives like Iceland has done for such visitors and need to fulfill a few testing and quarantine requirements. Entry procedures are also expedited by introducing online health declaration forms and a pre-registration process. This also minimizes arrival contact.

The establishment of travel corridors or bubbles is yet another step. Australia and New Zealand have enforced travel bubbles to allow residents to travel without quarantine requirements between these two nations. In some places, alternative quarantine options or reduced quarantine periods are introduced to attract more tourists, for example "Phuket Sandbox" program of Thailand. Targeted marketing campaigns related to ease and safety measures have been launched by both the governments and tourism authorities, such as "Dubai Assured" by Dubai. Air travel has become more seamless through the Open Skies Agreement that permits airlines to operate and transit freely to enhance connectivity and lower airfares.

2.2.2. Technological Integration

Tourism is heavily dependent on technological advancements that impact a person's journey as well as the operational efficiency of the business. Sustainable tourism and technology are closely knotted. It gathers invaluable data on traveler's behavior, likes, and dislikes that help governments and tour companies to decide about policies, rules, infrastructure, and other facilities thus fostering a most sustainable tourism sector along with safeguarding the well-being of local people (Nicola-Gavrilă, 2023). Hotels, aviation, and tourism companies use digital technology as millions of Europeans plan summer vacations. Employing useful, attractive, and data-driven innovation that ensures stress-free vacations can determine the mode of future traveling (Zientarski & Such-Pyrgiel, n.d.).

Travelers often use apps to research a given destination. The travel agencies, Expedia Group, and Booking.com website enable booking online tickets, hotel reservations, renting cars, cruise trips, etc. Airbnb acts as a broker, charging a commission through each booking of accommodations, short-term housing, cabins, and beach houses. Google Maps and Waze-like satellite navigation apps use the global positioning system (GPS) and help tourists to explore and navigate new places by foretelling travel times and route details. Several apps are available that provide insights into local attractions like restaurants, heritage sites, and handicrafts. Digital wallets like Apple Pay, Google Pay, and Paytm, ease out transaction mode ensuring secure and contactless payments. Applications like Opera and Cloudbeds use Property Management Systems for front desk activities like reservation, check-ins, and check-out management. The Customer Relationship Management software assists in keeping tourist profiles and tracking choices to offer personalized services. The Central Reservation System ensures real-time availability and pricing consistency of hotels. Similarly, Open Table like restaurant management and marketing platforms permit travelers to table bookings for dining in advance.

In the present day, technology redefines the tourism business and its success. Table 11.1 shows the interface of technology integration. The details are discussed later in this chapter.

3. Sustainable Tourism

Tourism involves commuting to a destination, lodging, relaxing, entertainment, food, and shopping, either for recreation, or business or family trips (Manzoor et al., 2019). The year 2017 was announced as the *'International Year of Sustainable Tourism for Development'* by the United Nations. Mostly, the notion of sustainable tourism was developed intending to lower the impact of mass tourism. Different terms emerged over the past years like eco-tourism, green tourism, responsible tourism, and sustainable tourism (Roblek et al., 2021).

Table 11.1. The Interface of Technology Integration Toward Improving Sustainable Tourism Business.

S. No.	Benefits	Benefit Routes
1	Enhanced customer experience	Online booking and planning of tours Personalization of travel experience
2	Efficient operations	Automated check-ins at airports to hotel Data analytics to understand customer behavior, preferences, and trends
3	Connectivity and communication	global connectivity Social media platforms and their influence
4	Innovative travel experiences	Virtual and augmented reality IoT and wearable
5	Safety and security	Biometrics and facial recognition Emergency response systems
6	Environmental sustainability	Green Technologies and eco-friendly practices
7	Economic impact	More e-Commerce and online transactions
8	Adaptation to trends	Remote work and digital nomadism

3.1 Paradigm Shift and Environmental Imperatives

The dimensions of sustainable tourism are far-reaching. Not only does it need to achieve the economic objective but also environmental (including climate resilience), and social objectives as well with an emphasis on the remediation of environmental, social, and economic impacts (Vázquez Loaiza et al., 2019). In the case of environmental protection, special emphasis should be given to lakes, coastal regions, and protected areas. Hotels, golf courses, restaurants, and aquatic sports can damage coastal tourism leaving a long-term negative impact on the gross domestic product (GDP) and occupation. Local workers are generally less skilled, and hence ill-paid; expatriates often take managerial work (Kashem et al., 2022).

Sustainable tourism depends on three pillars – environmental, social, and economic that underline its perspective. While the environment pillar restores the ecological functions, conserves biodiversity, and natural heritage, the socio-cultural pillar considers and reveres the socio-cultural genuineness of host communities, conserves their cultural heritage & traditional values, and facilitates intercultural understanding and forbearance. The economic pillar ensures feasible lasting economic growth that can offer social and economic recompense to its stakeholders such as job security, revenue generation prospects, and easing of poverty (see Fig. 11.1). Also, this should provide high tourist contentment and meaningful experience (Escobar & Margherita, 2021).

3.1.1. Impact of Climate Change

Transportation, accommodation, and activities form the key elements of tourism and emit GHGs. The tourism industry contributes to 5% of the carbon footprint worldwide. Holiday travels and pilgrimage tours contribute much but the distance traveled, duration of stay, and transport mode also serve as the factors for carbon emission (Filimonau et al., 2014). The bulk of the emissions in tourism is due to transport. The United Nations World Tourism Organization (UNWTO), United Nations Environment Program (UNEP), and World Meteorological Organization (WMO) published a statement on tourism and climate change in 2007. Though aviation contributes 17% of all tourist trips globally, inter-regional travel between Europe, America, Asia, the Pacific, the Middle East, and Africa accounts for 92% of tourist trips by air route (Peeters & Dubois, 2010). Climate change considerably alters the characteristic features of these places bringing erosion of coasts. The quality of beaches falls and hence loses its attractiveness leading to a fall in economy. Here stricter measures are needed to stop such deterioration. The climate also affects the mountain areas. The stakeholders are compelled to adopt the principle of sustainability in the specter of looming climate change. Globally, aviation consumes more than 5 million barrels of oil everyday aviation adding a considerable proportion of global anthropogenic CO_2 emissions, around 2–2.5% (Grote et al., 2014), with a growing trend in the future. Fuel costs are one of the greatest burdens to the airlines amounting to 32% of the global airline operating expenses in 2014 (Pagoni & Psaraki-Kalouptsidi, 2016). Non-CO_2 emissions like water vapor, NO_x, soot, and SO_x also affect the climate (Grote et al., 2014).

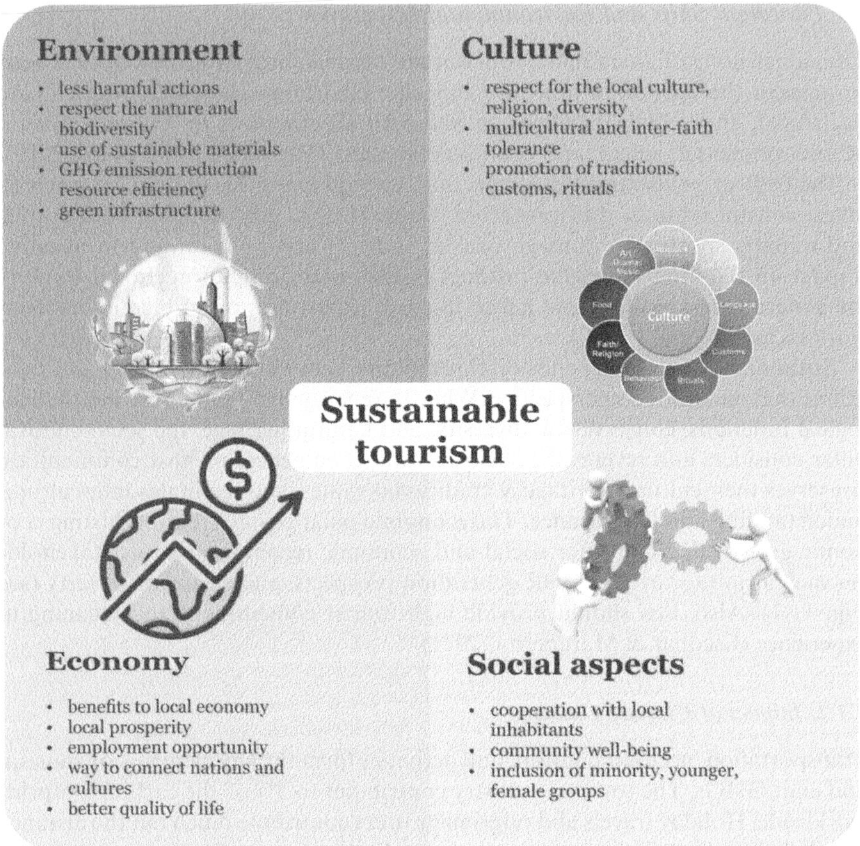

Fig. 11.1. The Pillars of Sustainable Tourism – Environmental, Economic and Socio-Cultural Aspects.

Controlling speed to achieve the most economical flight time can help the airplane to optimize fuel consumption. This can save carbon emissions by 1% on average. However, there are time-dependent costs as well such as, passenger compensation in case of delay (Edwards et al., 2016). Especially, long-haul flights on older and less efficient planes. Ground transportations with the help of cars and buses also play crucial roles. The International Civil Aviation Organization (ICAO) aimed for a 2% improvement in fuel efficiency per year, to stabilize global net CO_2 emissions at 2020 levels (Edwards et al., 2016). Technological and operational efficiency along with alternative fuel use in aviation can be the long-term strategies to meet climate goals. Apart there exist market-based measures such as voluntary carbon offsetting, environmental charges at airports, cap-and-trade policies, and others. The 2012 introduced European Emissions Trading Scheme of the EU is the biggest cap-and-trade policy (Pagoni & Psaraki-Kalouptsidi, 2016). The Air Transport Action Group (ATAG) achieved an improvement of average fuel efficiency of 2.1% per year since 2010. An increase in the number of travelers and an increase in the number of trips of long-haul tourism will increase GHG emissions in the near future. To mitigate such impacts a change in behavior and shifts in transport modes becomes pertinent.

3.1.2. Eco-Friendly Practices

Tourism causes overcrowding, and environmental damage and brings changes in local culture (Bozkurt & Şeker, 2023). The guide to incorporate sustainable development considerations in tourism was published as *'Sustainable Tourism Development: Guide for Local Planners'* in 1993 by the World Trade Organization (WTO); and then *'Agenda 21 for the Travel and Tourism Industry'* in 1997 by WTO (Dodds & Graci, 2012). Sustainable tourism along with environmental protection is dependent greatly on political will to prioritize tourism projects in tune with relevant international environmental conventions along with funding. Regulations in the Exclusive Economic Zone (EEZ), United Nations Convention on the Law of the Sea (UNCLOS), Rio Summits from 1992 till date the World Summit on Sustainable Development (WSSD), and many more are specifically important to be enforced. Incentives may be useful strategies for alleviating poverty, empowerment, and employment generation. Degradation of habitats and ecosystems, depletion of natural resources, deterioration of air, water, and soil quality, wildlife must be restricted, imparting education and knowledge on sustainable resource utilization like energy, food, water, transport, and communication (Roblek et al., 2021).

The issues of economic transformations, poverty, inequalities, and job opportunities should be addressed; security, recreation, adventures, development, and services should be properly managed (Roblek et al., 2021). UNWTO predicts that the tourism industry will engage 1.8 billion people by 2030 (Escobar & Margherita, 2021). Development of rural tourism in rural environments like agro-tourism, vineyard tourism, adventure tourism, and others places more focus on eco-tourism. Here, landscape plays the center of attraction, such as vegetation, and architecture uniqueness. The task is not that easy, as stakeholders confront many challenges like managing divisions within the communities, other risks, and vulnerabilities (Roblek et al., 2021).

Irrespective of management strategies, the networking of social capital becomes imperative. In case of improper social orientation, market needs may not be met through local resources thus blocking sustainability in community tourism. Coordination and partnership between stakeholders like governments, tourism companies, and visitors are integral to such development. A conflict always exists between economic growth and environmental protection but the two need to be made harmonious but carefully calibration to safeguard sustainability in the use of resources, becoming energy adequate, along with continuous upgradation and development to strike an environmental balance in tourism (Roblek et al., 2021). It is worth mentioning that the UNESCO World Heritage sites exhibit unique cultural, historical, and natural assets invaluable to the history of mankind. The World Heritage sites reach 1,157 in 167 nations as of January 2023. About 39 of these have mixed properties, 900 of these sites have cultural significance, and 218 with natural significance. Here tourism is integral in assisting these places in the way of conservation and also socio-economic development (Bozkurt & Şeker, 2023).

Environmentally responsible tourism should be able to bring innovations in hotel infrastructure, build employee awareness by introducing waste and wastewater treatment, noise control and soundproofing, enhance visual impact, and others. Slow tourism is a new type of sustainable tourism that emphasizes leisure and restful style to accomplish life values and enhance wellbeing (Roblek et al., 2021). Table 11.2 highlights some of the key sustainable tourism approaches.

There is an increasing trend for tourist inflow in the developing nations. With a variety of natural resources and climatic assortment, South Asia serves a distinct region attractive to tourists. According to the World Travel and Tourism Council (WTTC), Pakistan generated US 19.4 billion in 2017 which was 6.9% of its GDP (Manzoor et al., 2019). Bhutan is much sought after by eco-travellers as it is the only carbon-neutral nation. Spain uses tourism as one of the tools of local development for rural mass (Vazquez Vicente et al., 2021).

Sustainable development focuses on innovating new business ways, boosting employment along assisting conserving landscapes. It is imperative to emphasize on integration and transformation of cultural heritage into tourism products; thus converting an asset into a resource. In most cases, such a conversion seems complicated due to a lack of association between various players involved in conservation and tourism (Roblek et al., 2021). Hence sustainable tourism plays a crucial in achieving the 2030 SDGs (UNWTO, 2017). The contribution to the achievement of SDGs can be direct and indirect particularly when it comes to Goal 8, Goal 11, Goal 12, Goal 13, and Goal 14 (WTO, 2005). The success of

Table 11.2. Approaches to Sustainable Tourism.

Approaches to Sustainable Tourism	Field of Application
Environment Impact Assessment (EIA)	Evaluating the environmental impact of a tourism project through measuring carbon emissions, energy and water consumption, waste generation, etc.
Management of sustainable destination	Resource optimization and allocation, protecting ecosystems, and measuring visitor flow
Transportation	Traffic management to decongest, optimizing travel routes, and promoting green transport
Collaboration and innovation	Data sharing and adopting best practices
Consumer awareness and education	Transparency and authenticity of information and promote responsible choices

sustainable tourism requires a unified response of stakeholders, visitors, and tour service agents, proper concepts and strategies, and efficient implementation in all spheres such as health & hygiene, employment, etc., to overcome the environmental, social, and economic challenges in tourism (Kashem et al., 2022).

3.2 Achieving Sustainable Development Goals (SDGs) Through Tourism (see Fig. 11.2)

Both tourism and technology help to achieve the UNSDGs. According to the UNSDGs, sustainable tourism can contribute to five main areas mainly: sustainable economic growth, alleviating poverty and generating employment, environmental

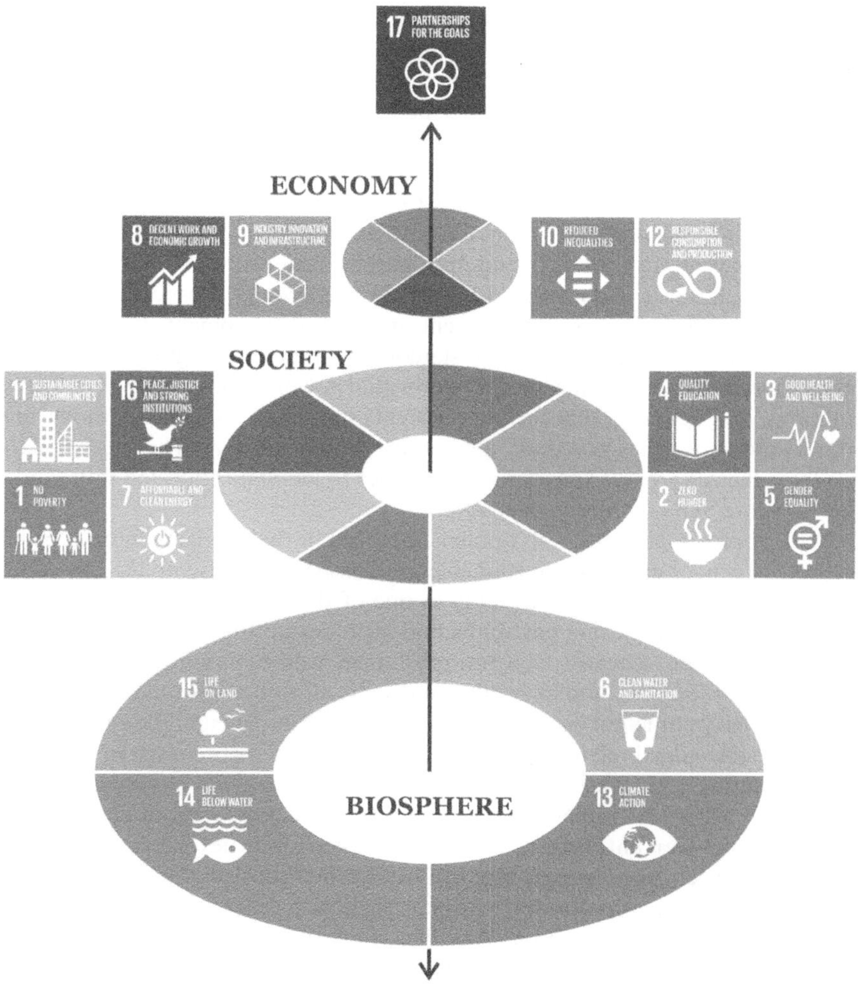

Fig. 11.2. Interconnectedness Between Sustainable Tourism and the SDGs: Sustainable Tourism Can Contribute to all the 17 SDGs.

protection, resource efficiency, and climate change, cultural values and heritage, peace and security, though it can touch upon all the 17 SDGs. Tourism affects the national and global environment, economy, and society. Along with other organizations, the UNWTO catered to sustainable tourism by creating the Global Code of Ethics for Tourism (GCET), the International Network of Sustainable Tourism Observations (INSTO), the Steering Committee on Tourism and Development, the Sustainable Tourism-Eliminating Poverty (ST-EP) Initiative, and so on. Tourism is found to possess immense potential to achieve the SDGs (Tourism and the Sustainable Development Goals – Journey to 2030).

SDG 1 (no poverty) and SDG 2 (zero hunger), can be met by creating job opportunities and promoting food production from local supply chains. Direct involvement of local people in community-based tourism fosters revenue generation that can in turn be invested in community development projects like education, health, infrastructure, and traditional crafts, contributing to reducing poverty. Agro-tourism can offer unique touring experiences connecting tourists with local agricultural practices; consolidating local food production systems, sustainable fishery, and sustainable seafood consumption, prioritize environmental stewardship, and promote food security, steps leading toward SDG 2. Initiatives focused on health results and positive outcomes for public health. Priority to both host and tourists boosts mental health and well-being. Adoption of safety protocols, vaccination, proper sanitation and hygiene practices, and facilities in the hotels help meet SDG 3 (good health and well-being). Responsible tourism paves the way for education enrichment and skill development. Cross-cultural exchange can break the monotony of stereotypes and promote global citizenship. Often tour operators provide interpretation guides, workshops, audio-visual presentations, and activities that facilitate insights into local history, traditions, and conservation strategies practiced, giving experiential learning. The tourist sector can introduce awareness programs and courses to facilitate educated tour guides that promote SDG 4 (quality education).

With approximately 58% of women employed in the tourism sector, it caters to SDG 5 (gender equality). Sustainable tourism pattern is exemplified by resistance to incorporating gender equality as one of the core principles (Eger et al., 2022). Reports state strong environmental attitudes and behavior of women as compared to men. Greater recognition of men–women interdependency, conflicts, value differences, and environmentalism is all to boost women's participation in the environmental decision-making process. Tourism and the environment share a double-edged relationship. Environment and tourists coexist in harmony, complementing each other, whereas anthropogenic activities strengthen the environment in symbiosis rather than harming it. On the contrary, in competition, the conflict between anthropogenic activity and economic development strives to dominate over environment leading to uncontrolled tourism, land degradation, marine and freshwater pollution, transport pollution, untreated sewage disposal, etc., and ultimately degradation. Way back in 1995, the UNEP underlined the call for education and guidance regarding ecotourism and other tourism forms. Tourism is affected locally and globally as it rests on human relationships as considered a leisure journey. Age, gender, ethnicity, and nationality can be used to draw relations between hosts and guests. Tourism ideologically shifted women to integrate them with the outside world. Studies reveal that women are inherently concerned about conservation. An important substitute for women is the commoditization of domestic work for provisioning tourist services. In Greece, women are involved through rural tourism, forming agro-tourism cooperatives, and running nature-based accommodations to offer the use of local products and preserve tradition and cultural heritage. Women thus earn and contribute to their families bringing economic independence to them. Women, who were conventionally disadvantaged and excluded from decision-making, now become independent, knowledgeable, and stakeholders (Skanavis & Sakellari, 2008). Responsible water usage, and its conservation, and community involvement in various such practices can help tourism industries contribute to SDG 6 (Clean Water and Sanitation). Hotels can be accountable for efficient water management, conserving water (like rainwater harvesting), and reducing water wastage by installing efficient faucets and wastewater plants, conducting timely checks for pollutants, pathogens, and water standards, and also solid waste management. Revenue generated can be reinvested in constructing safe water sources and sanitation facilities. Undoubtedly, the tourism sector is energy intensive. The entire industry can transition toward clean and sustainable energy resources. By promoting renewable energy sources like solar, wind, tidal, wave, and small hydropower schemes (wherever feasible), thus transiting from fossil fuel use, tourism can green the energy requirements and promote SDF 7 (Affordable and Clean Energy). Implementing smart technologies can optimize energy consumption but with efficient heating, cooling, and lighting facilities.

The tourism industries have great potential to create job opportunities, drive economic development, and contribute to achieving SDG 8 (Decent Work and Economic Growth). Sustainable tourism promotes local hiring, skill development, capacity building and benefits the host community, and creates sustainable livelihoods. According to the International Labour Organization (ILO), decent work rests on ethical labor practices that in turn depend on employment creation, social protection, decent wages, equal opportunities, rights at work, and social dialogue. Decent work goes beyond employment and also considers its quality. Socially responsible tourism focuses on a non-discriminatory work atmosphere, reasonable remuneration, fine work-life balance, employee participation along social dialogue. Women, migrants, and youth get better opportunities in the tourism sector than in other sectors (Winchenbach et al., 2019). Sustainability advocates for low-impact infrastructure, efficient waste management, energy-efficient building, sustainable tourism practices, and others. Eco-friendly transport infrastructure is the key aspect of tourism. Smart transport systems and networks, smart energy management, and digitalized services in tourism sectors altered business models, improved efficiency, and promoted SDG 9 (Industry, Innovation, and Infrastructure). SDG 10 (Reduced Inequalities) can be promoted by considering inclusive growth, equal economic opportunities, empowering the marginalized, and community engagement. Guided tours, local crafts, and cultural performances can improve the lives of the local communities to bridge the financial and social differences.

Planning and developing tourist spots is another crucial aspect. Sustainable urban development and preserving the cultural and natural heritage of potential sites will attract tourists enabling the tourism industry to contribute to SDG 11 (Sustainable Cities and Communities). Responsible visitor behavior to reduce environmental impact and choosing sustainable options in all spheres promotes sustainable lifestyles. Jordan promotes tourism in underserved regions and poverty pockets to promote development. Italy focuses preservation of cultural and natural heritage in rural and hilly setups; Thailand facilitates rural community welfare through tourism (Tourism and the Sustainable Development Goals – Journey to 2030). Building sustainable supply chains lowering waste generation, conserving water, and restricting carbon emissions in tourism promotes SDG 12 (Responsible Consumption and Production). An increase in domestic recycling, reduced dependency on plastics, spreading relevant information and awareness, initiating eco-certifications in various aspects, and adopting an eco-friendly lifestyle, can all promote SDG 12. Cyprus tourism is founded on its "National Action Plan for a Green Economy." This program guarantees resource efficiency along with sustainable production and consumption. The government promotes Ecolabel, Eco-management and Audit Scheme (EMAS), and the Green Key to launching hotels (Tourism and the Sustainable Development Goals – Journey to 2030).

Promoting low carbon practices, mitigating GHG emissions, and fostering climate resilience is imperative. The link between tourism and climate change is complex. Climate change is influencing investment, planning, and operations. Increasing climate change will increase food and water insecurity, displace more than a billion people, jeopardize health, change the ecosystem, increase wildlife extinction, augment transportation costs, cause sea-level rise, onset climatic hazards, and compromise economic and social development, all detrimental to sustainable tourism (Scott, 2021).

French tourism inclusive of accommodation and devices exhibits 9–10% of its global warming potential (GWP). Throwing responsibility is debatable in the case of global tourism, whether it should be the emitting nations or the receiving nations. Flights and maritime transport pose ethical issues in greenhouse gas (GHG) emission evaluation (Dubois & Ceron, 2006). Traveling produces around 8% of global carbon emissions. Even if enthusiasm for voluntary changes in consumers is seen, most people are not willing to cut off international travel. Consumers getting tired of energy conservation and emission cuts at home and office become more vulnerable to rebounds while on vacation (Hall et al., 2013).

The first challenge to the tourism industry is a transition to a low-carbon economy. Extensive decarbonization to cut the sector's emissions as well as reach net-zero emissions is necessary. The 2003 *Djerba Declaration on Tourism and Climate Change* identifies travel and tourism to be a major sector responsible for climate change. Carbon emissions from traveling have increased from 5% in 2005 to around 8% globally. Since international shipping and aviation are excluded from individual countries' emission pledges in the Paris Agreement, their emissions will increase massively (Scott, 2021). Innovative technologies facilitate smart transport operations, and smart energy management, to reduce carbon footprint and mitigate climate change promoting SDG 13 (Climate Action) (Nicola-Gavrilă, 2023).

The International Energy Agency (IEA) has developed a route for net-zero emissions by 2050. Can electrification of short-haul flights and hydrogen and synthetic fuels for longer flights have the potential, can be a research agenda (Scott, 2021)?

The global tourism industry pledged at the United Nations Framework Convention on Climate Change (UNFCCC) COP 26 in Glasgow to achieve a net-zero target by 2030 with 50% reduction by 2030. The hotel industry together can take initiatives potentially resulting in reducing emissions by 48 million tons of CO_2e. Table 11.3 gives some of the decarbonization initiatives. Many hotels are using energy-efficient lighting and this sector has eliminated 39 million tons CO_2e.

Table 11.3. Notable Decarbonization Initiatives Toward Offsetting Carbon Footprint and Mitigating Climate Change (Dickert, 2023).

Decarbonization Initiatives	**Potential GHG Emissions Reduction (Tons CO_2e)**
Retrofit efficient HVAC system	35–130
Installation of energy-efficient appliances	8–30
Installation of double-pane windows	7–25
Installation of sun shades on windows	1–25
Installation of low-flow fixtures	4–19
Using energy-efficient lights	3–9
Default turning off mini bars	4–8
Opting out of daily towel and linen changes	3–7
Installation of more insulators	1–7
Using reduced laundry temperature	2–4
Vehicle convoy electrification	3–4
Recycling wastes	2–3
Installation of smart lift software	1–3
Installing keycard switches in guest rooms	1–2
Installing motion sensors for lights in the corridors	0.4–2
Installing an efficient pool pump	0.8–1
Placing a pool cover	0.7–1
Installing dual-mode flush toilets	0.3–1
Introducing paperless procedures	~ 0.7
Restricting food wasting	~ 0.7
Eliminating disposables	0.4–0.6
Harvesting rainwater	0.021–0.033
Developing native gardens	0.008–0.013
Installing smart irrigation	0.06–0.013

Maritime and coastal tourism agencies should endeavor to protect underwater life, develop coastal areas, and practice responsible fishing to uphold SDG 14 (Life Below Water). Whale watching, dolphin excursions, and scuba diving can be permitted under strict regulations to ensure minimal disturbance. Often boats, cruise operators, and tourism companies are to meet the eco-certification criteria to promote aquatic life conservation. Bangladesh and several other nations mentioned the role of tourism in progressing blue economy with priority given to coastal and marine tourism. Curacao promotes scientific tourism in the coral reef system and invests in ocean-based industries (Tourism and the Sustainable Development Goals – Journey to 2030). Similarly, tourism should give particular attention while constructing and designing tours to protect protected areas and conserve biodiversity contributing to SDG 15 (Life on Land). Hiking, bird watching, and nature walks should be promoted more to reduce ecological footprints. Both visitors and hosts can take part in afforestation and reforestation practices to contribute toward the rehabilitation of degraded lands and habitats. Community engagement and empowerment boost sustainable tourism through cross-cultural understanding to reduce misunderstandings, foster partnerships and collaborations, that can create robust, responsible institutions, enforce ethical guidelines, develop respect for local customs and traditions, and overall foster inclusive governance and SDG 16 (Peace, Justice, and Strong Institutions). Multistakeholder partnerships through collaboration between governments, policymakers, businesses, NGOs, and the community enhance the effectiveness of tourism projects by integrating capacity-building, knowledge sharing, training, and leveraging technology from a variety of sources. These global networks connect diverse stakeholders and encourage responsible investments to prioritize environmental and social sustainability and SDG 17.

4. Challenges on the Horizon in Pioneering Sustainable Tourism

4.1 Overtourism: Balancing Demand and Preservation

Tourism is an economic boon, but over-tourism poses threats. Destination crowding indicates popularity and fame (Jacobsen et al., 2019). Over-tourism is a relatively new term. However, it received global attention in 2017 due to media reporting local complaints that disrupted resident's lives. Protests were shown with signature and graffiti *Tourist go home*. One of the major factors contributing to over-tourism is affordable travel and good connectivity. Around 615 million visitors traveled to Europe in 2016 (WTO, 2017). Online reservations and Home away also added to over-crowding. Some cities like London and Paris have proper infrastructure to accommodate large volumes of tourists. For small- and medium-sized cities, overcrowding often alienates residents because of increased rents, noise, etc. Long queues can affect tourist experiences. Over-crowding impacts commuting, food, energy consumption, waste management, water supply, and others (Stanchev, 2018).

4.2 Health and Safety Concerns

Previously, developing nations were less chosen owing to their living conditions like increased pollution, and irresponsible behavior that jeopardized health and sanitation status. Diseases can be transmitted from the visitors to the residents and vice versa. New infectious diseases are a great concern. Accidents may cause deaths or injuries and disability (Bauer, 1999).

Lately, the tourism industry was the first to get hit by the COVID-19 pandemic which witnessed a fall of global revenue by 63% in 2020 internationally amounting to a loss of US 538 billion dollars. International tourism rose to US 1.47 trillion dollars in 2019. Consumers are now more sensitive and aware of future pandemics and other socio-economic and environmental issues, bringing socio-cognitive mindfulness to the forefront of tourism. COVID-19 lesson elicited the need of this industry to transform from traditional unsustainable business models to non-conventional, sustainable, and resilient business models with newer and innovative concepts of tourism infrastructure. Thus this industry is expected to undergo restructuring, optimization, and digitization (Roblek et al., 2021).

Tourism monoculture practice negatively impacts the destination, over dependence on tourism makes the area economically vulnerable. Often it generates goods and services demand not produced locally and hence sourced from outside; particularly problematic in rural areas. Visitors cannot access local people when they are confined in hotels, coaches, safaris, museums, etc. Their only point of interaction is the entry and exit points. The same is the case for cruise ship passengers with all-inclusive packages. On the better side, it offers additional livelihood options for localities and diversifies the local economy (Goodwin, 1998).

Sometimes tourism sector involves a precarious nature of work with uncertainty (insecurity and irregularity) and predictability transferring risk from employer to worker. Worker falls under the employer's obligation for social security, pension benefits, health insurance, etc. Such work is usually called temporary casual or part-time. While people in the many developed nations choose part-time options for getting flexibility, those in the South have limited choices and become hostage to employer's whims. Many of the hospitality and table companies exploit their employees with overwork and underpayment coupled with rare options for promotion. Underpaid jobs are mostly part-time, seasonal, or low-skilled like washing, room cleaning, etc. that are given to women or migrants. Women also work as waitresses or housekeepers, with lowly paid wages, and take up multiple jobs to meet their ends (Ioannides et al., 2021).

Since birth men and women are subjected to gender restrictions. According to a United Nations report, women comprise 70% of the total population suffering from low living standards. Generally, women work more than men but receive only 10% of the global income and possess less than 1% of the global assets. More women are uneducated than men. Women face inequality in all spheres and things must change (Ertac & Tanova, 2020).

4.3 Regional Dynamics

Tourism trips can center around recreation and leisure, visiting friends and relatives, health and medical grounds, pilgrimage, shopping, transit purposes, or education apart from business trips. Regional dynamics refers to unique patterns, trends, and factors shaping the tourism industry in specific geographical regions. A blend of cultural, economic, environmental, and political factors influences the regional dynamics. Australia and New Zealand are famous for nature tourism. Divers and nature enthusiasts often select the Great Barrier Reef. Cities like Paris, Rome, Venice, Barcelona, Athens, Petra, and Istanbul attract millions of tourists owing to their historical landmarks, museums, and cultural uniqueness. Mallorca in Spain, Amsterdam in the Netherlands, and Santorini in Greece are some of the most chosen places. In Southeast Asia, Thailand, Vietnam, and Indonesia are known for their natural beauty and beaches. Boracay, a small island is known for its clear waters, coral reef, and white sands. Luxury tourism with extravagant shopping and high-end accommodations is available in Middle East countries like UAE, Qatar, and Oman. Kenya, Tanzania, and South Africa attract people to wildlife safaris and eco-tourism. Galapagos, Serengeti, Kruger, and Masai Mara are very popular places for animal lovers. People indulge in sun, sand, and water activities in the Bahamas, Jamaica, and Dominican Republic. Barbados and Cozumel offer cruise tourism with sports. Trekkers select the Himalayas and Alps.

Tourists are changing their preferences, hence destinations are changing. Customers look for unique experiences, places with cultural richness, scenic beauty, and authenticity. Hence off-beat destinations are taking over the conventional ones. Moreover, tourism is quite susceptible to external shocks, like geo-political tensions, hazards, and disasters like floods, earthquakes, volcanoes, and forest fires, as well as epidemics and pandemics.

4.4 The Need for Resilient Tourism Models

According to WTO, 1996:

> Sustainable tourism development meets the needs of the present tourists and host regions while protecting and enhancing the opportunity for the future. It is envisaged as leading to management of all resources in such a way that economic, social and aesthetic needs can be fulfilled while maintaining cultural integrity essential ecological processes, biological diversity and life support systems.

Projects that can withstand challenges, adapt to changes, and ensure enduring environmental, social, and economic viability can turn into resilient tourism models. They always give priority to the conservation of environmental components and resources, social accountability, and economic stability. Resilient tourism focuses on local communities, wilderness safaris to conserve biodiversity, climate resilience, green buildings, green schools for sustainable education, digital platforms, and circular economy, and integrates them into policies and programs.

5. Technological Solutions to Sustainable Tourism Challenges

Currently, tourism is not only sustainable but also smart. *Smart tourism* is the totality of all incorporated efforts at the target place that fuels smart technology to attain environmental and socio-economic sustainability driven by novelty and competitiveness. The term "smart" gained popularity and describes technological advancements along with socio-economic progress and relies on open data, big data, sensors, and novel means of connection between man and machines, multi-devices, and a plexus of information exchange. It stems from traditional as well as e-tourism (Escobar & Margherita, 2021). In this context, e-tourism faces many challenges from the very start owing to technical issues that take place due to interoperability (Lisi & Esposito, 2015). Smart tourism entails several components and layers like smart destinations, smart businesses, and smart experiences. Smart tourism relies on collecting huge amounts of data, storing them intelligently, processing them, combining them, analyzing them, and then utilizing big data to inform business innovation, operations, and services. The rapid surge of information communication technology (ICT) has popularized computing, storage, and communication through cloud computing, mobile devices, open data, and big data mining. Currently, AI, IoT, robotics, social media platforms, cyber security, 3D printers, and BlockChain have boosted the tourism sector to develop toward digital transformation.

Four key components of smart destinations identified by Gretzel, Sigala, Xiang, and Koo are smart mobility, smart environment, smart economy, and smart governance (Gretzel et al., 2015; Nicola-Gavrilă, 2023). With mobile technology, tourists can have real-time conversations with service providers to plan their trips (Nicola-Gavrilă, 2023; Xiang et al., 2017). Smart destinations guarantee sustainable development by applying smart city principles to both urban and rural places and taking both residents and tourists into consideration. Smart destinations support mobility and ensure resource availability & distribution, quality of life, and sustainability. A smart business ecosystem assists in swapping over touristic resources and the co-creation of the tourism experience. The business component encompasses public–private partnership, vigorously interconnected stakeholders, digital core business processes, and organizational quickness. The government turns into a provider of infrastructure and data and becomes more open, focusing on technology. Here, consumers also create value and monitor thereby taking part in governance. Smart experience comprises tech-mediated tourism experiences, further enhanced through personalization, context awareness, and real-time monitoring (Tsaih & Hsu, 2018). According to Gretzel et al. (2015), there are five layers of smart tourism. They are (i) a physical layer; (ii) a smart technology layer; (iii) a data layer; (iv) a business layer; and (v) an experience layer. While the physical layer comprises transport, resources, and service infrastructures; the smart technology layer comprises business solutions and customer applications. The data layer constitutes data storage, big data, and data mining applications and the business layer constitutes innovations founded on available technology and data sources. The 5th layer (experience layer) is a technology and data-enhanced consumption of experiences. Smart tech which forms the foundation of smart tourism, involves a correlation between the physical and digital world. The physical layer allows the technology layer to develop the data layer. Various data types are combined in the data layer to facilitate the formation of the business layer. Services and applications produced by this layer permit the formation of better touristic experiences and sustainable goals in a destination. Thus experience layer takes both visitors and residents of the destination into consideration (Escobar & Margherita, 2021).

5.1 Applications of Artificial Intelligence

Artificial intelligence (AI) acts as a transformative force and revolutionizes the industry with sophistication, redefining the lifestyle and work (Zientarski & Such-Pyrgiel, n.d.). In times of increased demand and its fluctuations, AI applications make complexities and service availability easier, though they can potentially intimidate one's confidentially and security or mislead privacy reports (Kashem et al., 2022). The "chatbots" are used to assist in online ticket booking, retaining the privacy of customer's information. Chatbots can easily and effectively handle grievances and fast queries from passengers. The confidential issue is handled by *blockchain* technology, preventing hacking or leaking. *Blockchain* technology increases performance, enhances competitive advantages, and also perks up customer satisfaction. It can improve virtual reality for the subsequent day tourism promotion. It can assist in ensuring genuine source and food lineage. It can exhibit the nature of peer-to-peer decentralized association with tourists. The IT provider for the air transport industry, SITA Company uses this technology to secure an integrated automatic fingerprint identification system. Thus customers can use wearable or mobile devices in place of carrying identification documents like passports and driving licenses. Blockchain and AI enable fast reservations providing secure ledgers and evading intermediary expenditures, certain taxes, paper records, and manual communication via a smart contract. Peer-to-peer trade without any intermediary becomes possible without any human error, duplicate bookings, or manhandling of luggage. Hotels can reduce expenses using robotic services in place of chefs, bartenders, servers, etc. With the increasing number of loss of baggage and integrating blockchain technology into luggage systems, travelers gain confidence by being able to track their baggage on their phones. Travelers can use cryptocurrency mode of payment in tourism by applying AI and blockchain technologies, creating a versatile economy, and providing secure financial transactions to international passengers (see Fig. 11.3). Such security and openness can cause investors and tourists to flow. The use of a decentralized ledger offers transparency and security. In fossil-fuel-intensive nations, blockchain may lower energy trading transaction costs. The application of AI and blockchain augments efficiency increases revenue collection, and saves money that can be invested elsewhere. Its implementation can reduce the effect of future crises like pandemics. Facial recognition patterns used AI and blockchain during COVID-19 to permit speedy check-in (Kashem et al., 2022).

Artificial intelligence improves the overall consumer experience by increasing efficiency along with personalization. AI is capable of analyzing enormous

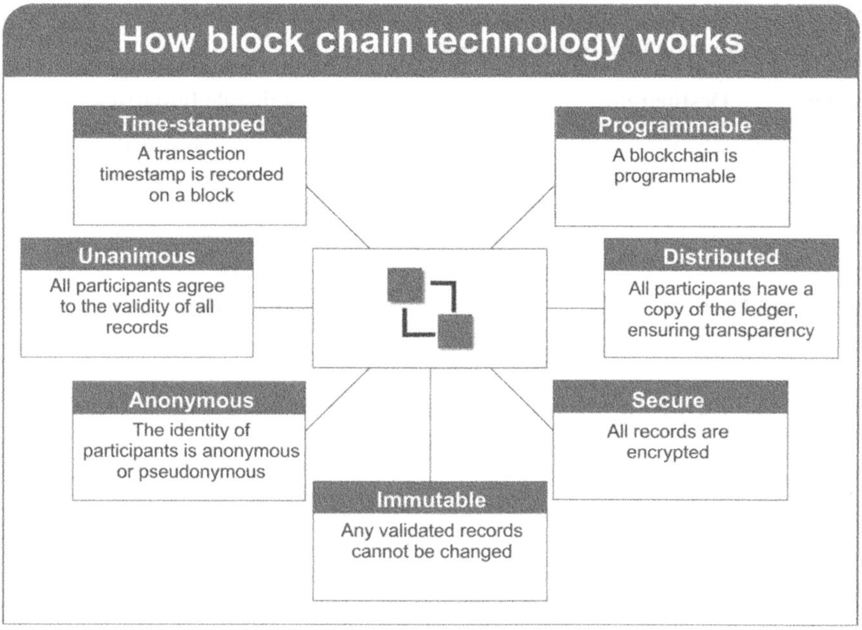

Fig. 11.3. Application of Blockchain Technology in the Tourism Industry.

amounts of data and images quite rapidly, precisely, and consistently. Artificial intelligence can recognize unique features not immediately apparent to tourists. Besides it enables the identification of the most popular sites and helps in planning sustainable tourism strategies (see Table 11.4). Analysis of visitors' data and likings, Artificial intelligence adds personalization of tourism experiences (Bozkurt & Şeker, 2023) (see Fig. 11.4). AI-driven chat rooms are now used by airline companies like Lufthansa, KLM, and Air New Zealand that allow conversations allowing passengers for personalized booking of their tickets. The introduction of a voice assistant Alex by the tour company Virgin Holidays allows customers to interact with a smart speaker and book for holidays. Several airports use autonomous vehicles, uses facial recognition software that is efficient, secure, and confirms customer identity (Zientarski & Such-Pyrgiel, n.d.).

Artificial Neural Networks (ANN) seem promising for analyzing and managing classification, conservation, and sustainable development in places like the WHS. ANN designed with optimization algorithms like swarm optimizations can be used to analyze and optimize aspects like the impact of the rise in sea level on cultural heritage sites (Bozkurt & Şeker, 2023). *Radial Basis Function (RBF)* based on ANN has 3-layered structures with an input layer, hidden layer, and output layer that can estimate any incessant function to arbitrary accuracy and needs few training samples. *Multilayer Perceptron (MLP)*, is another type of ANN, that can serve especially useful in classification tasks that need high-dimensional input data like image categorization, speech identification, and natural language processing (Bozkurt & Şeker, 2023).

Table 11.4. Various Cities Embarking on Technology Integration to Improve Tourism to Smart and Sustainable Tourism.

S. No.	Destinations	Smart and Sustainable Initiatives
1	Seoul, South Korea	Provides fastest internet speed; awarded as the most wired city in the world; sets a good example by having national, metropolitan, and tourism-specific open data that simulates tourism app development.
2	El Hierro, Spain (island)	With Wi-Fi networks and WiMAX; 26 free points of access, open public places that give advantages to both residents and tourists.
3	Madrid, Spain	Intend tourist inflow with “Madrid 21 Destinos” for promoting less visited places. Use sensors and beacons to monitor tourists, intend to increase destination competitiveness.
4	Las Palmas	Placed IoT sensors in large shopping areas to facilitate the shopping tourism section. This is to catch the visitors who acquire goods outside their residence and this factor determines their travel decision.
5	Zagreb, Croatia	Focus on environmental objectives; their ‘Freewa Project’ aims to conserve potable water sources with a web platform. This enables locating free potable water locations along with reusable water bottles throughout the city using a mobile app.
6	Aix-en-Provence, France	The project ‘Aix Living Paces’ collects data from numerous street sensors; these data help the local start-ups to analyze and offer solutions to residents and tourists in terms of air quality and ease the flow of people.
7	Helsingborg, Sweden	Promotes green tourism and mobility by giving the tourists to use the cycling infrastructure with the help of the city’s interactive online map.
8	Copenhagen, Denmark	‘Miljokajakken’ or the Environment Kayak is an environmental initiative toward the cleanest seawater with the opportunity to swim around and watch the city from a new perspective. The tourists exhibit their responsible behavior by renting a kayak free in exchange for their positive impact during their visit and collecting rubbish while kayaking.

(*Continued*)

Table 11.4. (*Continued*)

S. No.	Destinations	Smart and Sustainable Initiatives
9	Helsinki, Finland	To become carbon neutral by 2030, many hotels offer digital carbon footprint calculators to make visitors conscious of their impact on the visited place.
10	Venice, Italy	Encourage tourists to behave responsibly with reverence for nature, landscapes, and residents through the "#EnjoyRespectVenezia" program. Sensors fitted throughout the city. Tourists have day-to-day forecasts on the number of expected visitors that allow them to choose their place of visit. Offers incentives on less visited places to attract visitors flow.

5.2 Big Data for Resource Management

Since the tourism sector thrives on information, it needs dynamic, updated, timely, and pertinent information that assists in decision-making processes. This is where *Big data* can provide updated and incalculably informed inferences concerning behavior and human actions that can improve the tourism sector (Xu et al., 2020). Big data can alternatively serve to measure sustainability. The UNWTO promoted the exploration of big data in 2017 in the field of tourism statistics. But such can be challenging and hence the stakeholders are required to be cautious in matters of precision, legitimacy, and dependability (UNWTO, 2017). On the other hand, *open data* provides free access, use, and modification and can be shared by anyone irrespective of purposes (Escobar & Margherita, 2021).

5.3 Information Extraction

In the era of big data, huge volumes of data in the form of audio, video, images, text, and animation are produced. They are increasingly posing challenges owing to their high volume and variety of unstructured, structured, or semi-structured machine-readable data. Information Extraction (IE) refers to the application of information extraction tools for automated retrieval of specific structured information linked to preferred topics from bodies of text documents, databases, websites, or multiple sources. Various IE tools are proposed to extract structured and useful information from the unstructured data. Unstructured data comes from a variety of sources, has multiple formats, lacks standardization and schemes, and is completely heterogeneous. The IE systems are based on Natural Language Processing (NLP), language modeling, and structure extraction techniques. Travel blogs serve as a rich source to model human traveling behavior and characterize tourist destinations. The huge bulk of unstructured data requires extensive processing to be transformed into knowledge.

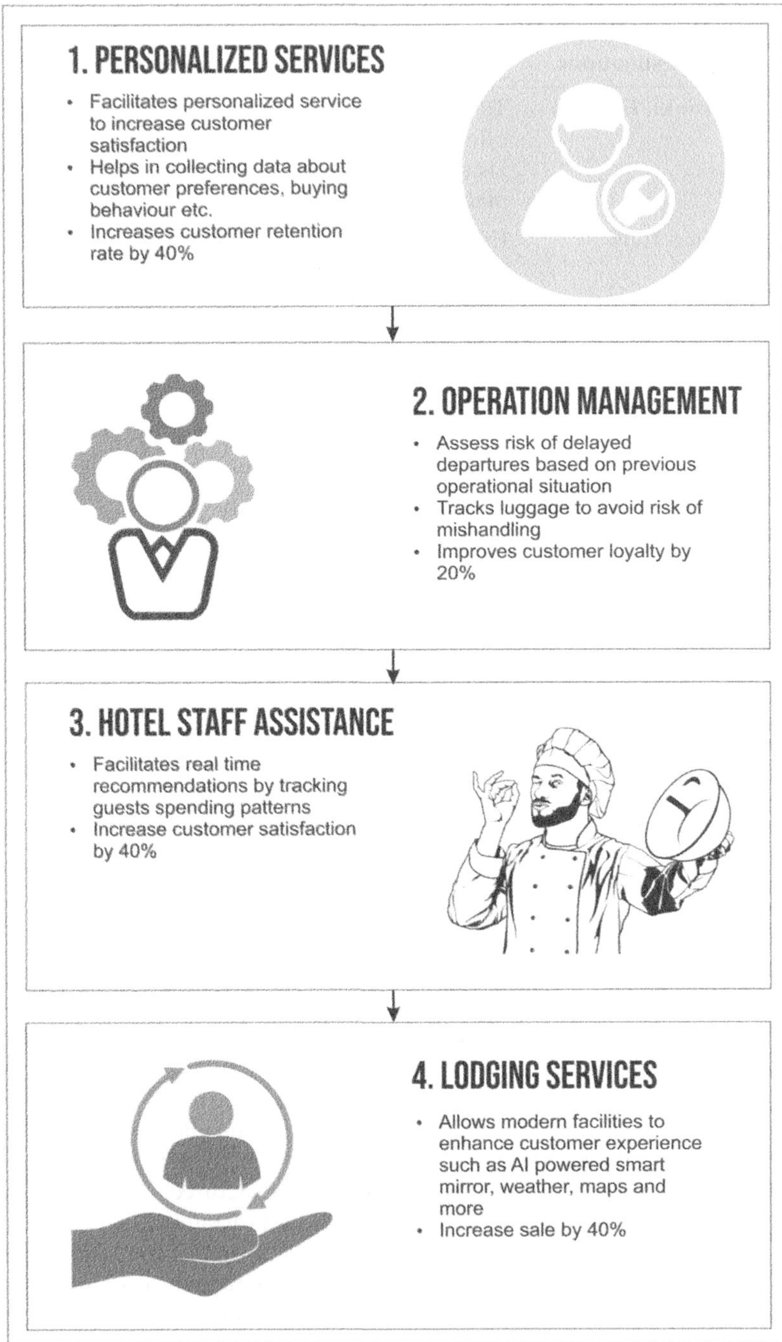

Fig. 11.4. Application of Artificial Intelligence in the Tourism Industry.

MonkeyLearn is a machine learning-based text analytics platform that helps the tourism business understand its strengths and lacuna for improvement. MonkeyLearn's natural language processing capabilities can be applied to analyze sentiments in customer reviews by collecting data in one place, extracting, and sorting information about complaints, satisfaction, etc., in a simple and structured manner. Scrapy is a Python web crawling and scraping tool assisting in real-time monitoring of tour industry competitors, optimizing pricing strategies, analyzing preferences and patterns, and permitting an understanding of market trends in an organized way. Rosoka uses NLP and develops insights by extracting important entities from unstructured texts in over 200 languages. It performs geospatial analysis and provides the users with coordinates. Tools like Wit.ai, Beautiful Soup, and Aylien empower the tourism industry by automating data collection and developing insights to improve the decision-making process.

Information Extraction helps to handle an enormous amount of data through *Web Information Extraction (WIE)* that uses HTML/XML labels and layered format accessible in an online text to develop a less linguistic approach. WIE-ONTOUR is a wrapper-based WIE tool implemented in Java that can help travelers with lodgings like hotels and B&Bs available on the website. This tool is also able to calculate the distance between the extracted housing from the place of interest (Lisi & Esposito, 2015).

5.4 Cloud Services and the Internet of Things (IoT)

Multiple tourism applications can be formed by cloud services. Information analysis, collection of data, and automated systems can be created by the Internet of Things (IoT) through using sensors, chips, and actuators. Both cloud services and the IoT were supported by the end-user internet service system (Escobar & Margherita, 2021). The IoT sensors supply real-time data enabling streamlining operations as well as meeting client satisfaction. It can be used for optimizing air traffic, improving fuel management, and distributing aprons. Both staff and passengers can avail real-time status on flight status. Passengers get useful information and notifications on their smartphones. Existing data analysis can recognize potential problems likely to happen and enable mitigation (Zientarski & Such-Pyrgiel, n.d.). In the post-pandemic times, *IoT* can provide affordable systems that can reduce disease dissemination, through mask detection, contact tracing, and control of atmospheric quality.

5.5 Augmented Reality

Augmented reality (AR) permits the overlaying of 3D animated objects and multimedia content on realistic images enabling for example the 3D historical reconstructions by which we can relive the past at historically important places coupled with self-guided sightseeing and exploration (Roblek et al., 2021). Heathrow Airport, introduced an AR children's application that permits them to open Mr. Men and Little Miss through a digital badge scan, installed at the airport (Zientarski & Such-Pyrgiel, n.d.). In smart tourism, the destination is knowledge-based, and all stakeholders are getting systematic and efficient access to information in contrast to traditional tourism (Escobar & Margherita, 2021).

5.6 Social Media

A huge amount of real-time data can be shared in various formats. *Social media* enables fast virility of data, promoting improvement, amplification, and transformation. Technologies also foster fast processing, visualization, and analysis that help in decision-making as well as making strategies (Rahmadian et al., 2022). In the present day, social media is quite obvious for sharing pictures, experiences, and opinions (Van der Zee & Bertocchi, 2018). Website development and social media systems encourage visitors to share their touring experiences in the digital world (Cong et al., 2014). Not only do social media platforms permit tourists to get information but also to share their points of view and suggestions. Such information can be both positive and negative and may comprise accommodation, places, attractions, and other services. Tripadvisor, TravelPod, Tripblog, and Weibo promote and use other services to reach a wider community. The digital footprints of the tourists turn into precious data sources for marketing, and business, and provide a vision of surroundings, cultures, customs, and communities.

Many companies like Thomas Cook, replaced travel brochures with Virtual Reality (VR) giving pre-holiday experiences to future travelers. They claim that such an application has increased their Royal Caribbean flights by 45% (Zientarski & Such-Pyrgiel, n.d.). The launching of the Virtual Tourism Observatory offers to collect information & data for analyzing the current trends in tourism. EU's "eye@RIS3" records the emerging picture of introducing smart specialization in European regions (Zientarski & Such-Pyrgiel, n.d.).

5.7 Virtual Communities and Virtual Tours

Virtual Communities and Virtual Tours can also help in the preservation of threatened destinations affected by tourist activities. The posts that are provided with embedded geolocations help to recognize the flow of visitors. Geo-tagged pictures assist in analyzing tourists' spatial networks linked to cultural heritages, ecosystem services, and conservation sites. Photos with GPS coordinates make identification of the place of interest easy.

6. The Road Ahead: Collaborative Strategies

6.1 Collaborative Solutions

The tourism sector is an amalgamation of mutually dependent service providers. Informal governance dominates hospitality in case of family-related dynamics; destination management is featured by formal governance. Destinations are primarily influenced by tourist activities. Governance gives important guidelines to gear this industry. Of late, social and environmental perspectives of governance receive much consideration apart from economic performance. As networking is very important in this industry, governance tends to coordinate destination spots besides organizing. Good governance addresses multiple dimensions like authenticity, transparency, efficiency, and participation. In the case of tourism, good governance requires multistakeholder participation (both public and private) together (see Fig. 11.5) to share expertise and resources. The governance needs a holistic approach to citizen participation and cooperation with stakeholders (Bichler & Lösch, 2019).

6.2 Digital Nomadism and Remote Work Trends

The remote work trend is on the rise post-pandemic. Workers and employees are expected to commute to their workplaces, but a digital nomad rejects these traditional ways of living and working. The term "digital nomads" was coined by Makimoto and Manners in 1997 (Poulaki et al., 2023), and refers to people who travel while working. They move away from the daily schedule, and reject routine timetable of work, travel, and the office; rather they work on a laptop in distant places. Pandemics resulted in worldwide lockdowns in institutions, offices,

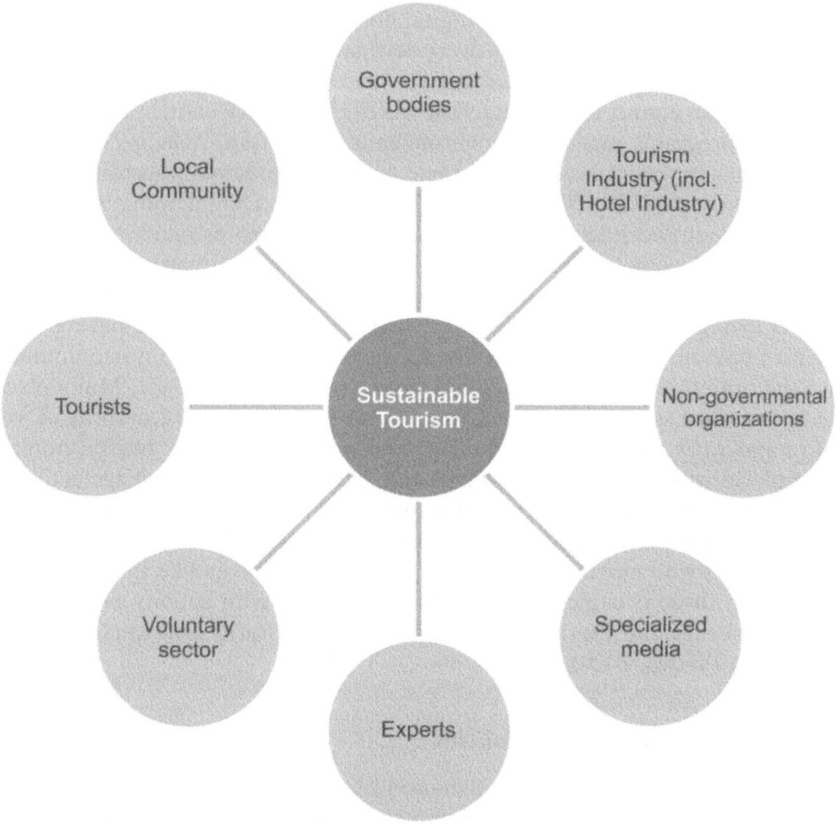

Fig. 11.5. Participatory Approach for Sustainable Tourism.

cultural norms, and social obligations; office gatherings, and meetings disappearing suddenly. This in turn opened up dialogues that challenged the conventional 5 days work 8-hour work, and the core practice of 'remote work' became the mainstream. Several countries issued remote work visas in 2020 permitting people to work in their province during lockdown. More and more workers imagined the likelihood of remote working (Cook, 2023). The evolution of work, technological development, and traveling work in harmony opened the opportunity for exceedingly specialized employees so that they can practice their professional activities anywhere in the world (Mouratidis, 2018).

People working remotely in their towns or places are termed digital nomads! And how often should they travel to be labeled as remote workers or digital nomads? A Digital nomad is defined as those who *use digital technologies to work remotely, they have the ability to work and travel simultaneously, have autonomy over frequency and choice of location, and visit at least three locations a year that are not their own or a friend's or family home*. Research findings put forward 5 types of digital nomads as in the illustration (Cook, 2023) (see Fig. 11.6).

Digital nomadism, a present-day trend in the information-communication network society, evolved due to digitization and mobility. Greece has become an attractive destination as they favor digital nomads. "Slowmadism" refers to slow travel with a feel of local culture in every destination. Workers build a community, make friends, and unwind the facets of the place as much as possible. They often establish deeper connections with fellow travelers and the localities and often volunteer for programs meant for environmental protection, reducing the carbon footprint that in reality is one of the main principles underlying sustainable tourism (Poulaki et al., 2023). They promote local business. Digital nomads are on the rise since remote work has taken a shift from trend to mainstream. Not only has digitization helped in fastening services in the travel and tourism sectors, but also tourists who can use this technology to boost their work efficiently and fruitfully while enjoying tourist places. They are a global tourist audience having lesser expenses without an office, enjoy a better quality of life due to no commuting, have the option of working hours, are free to pursue hobbies, etc. Many places like Barbados, Estonia, Georgia, Croatia, and Anguilla give 12 months of visa to live and work to pull digital nomads and promote tourism. Digital nomads prefer places with less expense, less pollution, and natural beauty (Vagena, 2021). The Indonesian government created a tourism digitization program in 2018 to attract these people to tourist spots. They categorized three types of nomads: Glam Packers who maximize the use of tourist social media like Instagram, and Lux Packer, those occupied in intermedia operation, with activity presented by online platforms, and Flash Packer who settles temporarily and work in a place (Prabawa & Pertiwi, 2020).

6.3 Formulating Sustainable Policies

Governance comprises the implementation of policies and before this comes to its formulation. Good policies need enormous research and input from people who implement such policies at the grassroots level or impact level. Developing

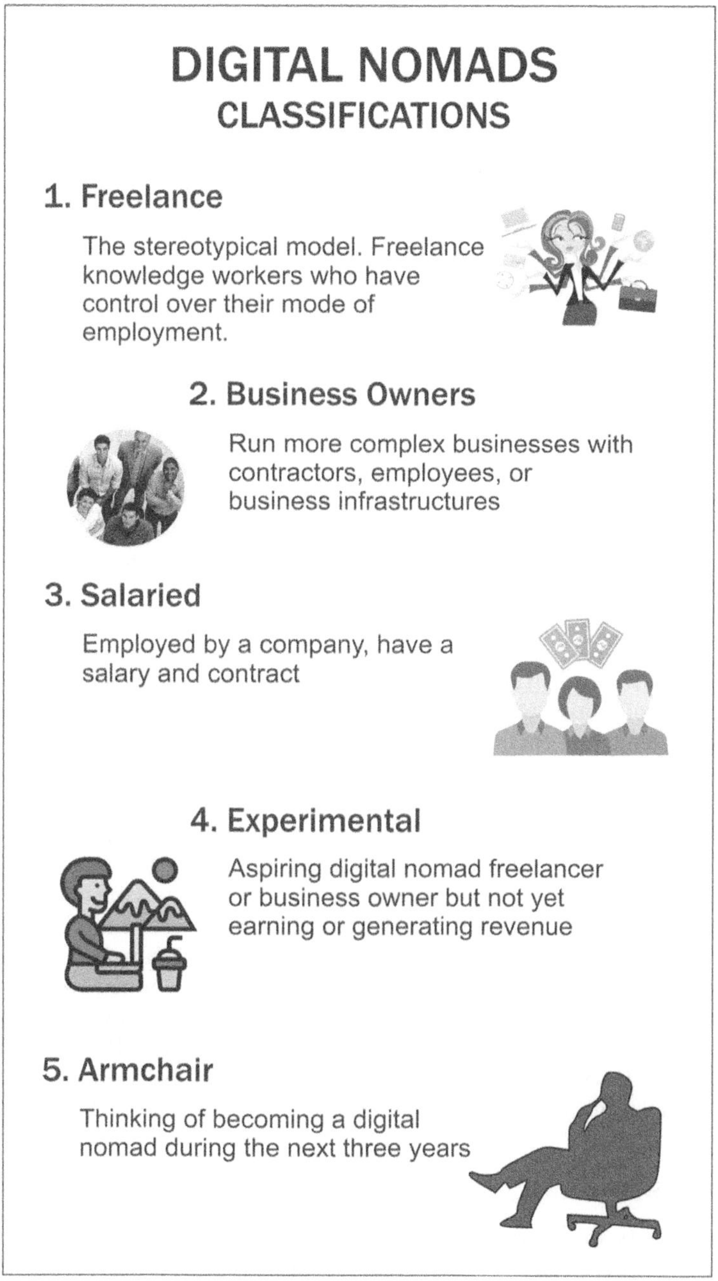

Fig. 11.6. Types of Digital Nomads.

sustainable policy is a continuous process that offers with chance of improvement. While formulating, the guidelines of the United Nations conferences and summits related to sustainable development and tourism along with their outcomes needed to be complied with. It comprises a plan review, followed by formal and legal adoption of the plan. The next step involves the integration of plans and policies into the public and private sector programs, continuous monitoring measured through tourist satisfaction, project development, marketing success, revenue generated, etc., further adjusting the plans and policies with formal and periodic revision (Wong et al., 2010).

6.4 Certifications

Seven thousand tourism products are certified globally of which six thousand are in Europe. The EU Ecolabel Award Scheme developed environmental standards for tourist accommodation (Font, 2002). Certification refers to the process of providing documented assurance by which a product, process, or service is assessed for confirmation against a standard. The government of Costa Rica started the Certification of Sustainable Tourism (CST) for companies complying with green policies and comprising a 5-step scale assessment for sustainable development (Holub, 2015). With the benefits of certification programs in Costa Rica for sustainable tourism, worldwide approaches were initiated.

The *Blue Flag Program* by the Foundation for Environmental Education (FEE) implemented in 1987 is meant to integrate sustainable tourism with the management of coasts, with safe and clean beaches and marinas, along with voluntary engagement in environmental activities. The basic criteria for beaches and marinas are in tune with the EU Urban Waste Water Directive. Additionally, beaches must comply with the EU bathing Water Directive (Toth, 2000). Blue Flag has stepped outside Europe to the Caribbean and South Africa; and turned out the single most popular certification program (Font, 2002).

Hotels can impact the environment by way of energy consumption, water consumption, waste generation, etc. Two categories of certifications are in practice; environmental certifications for buildings and environmental certifications for hotel operations. The Global Sustainable Tourism Council recognizes several certification schemes like Earth Check, Green Globe, and Green Key; while Leadership in Energy and Environmental Design (LEED), Building Research Establishment Environmental Assessment Methodology (BREEAM), and Excellence in Design for Greater Efficiencies (EDGE) are building certifications.

7. Conclusion

The multiplicity of issues in sustainable development can inevitably be transferred to the concept of sustainable tourism and there exists multiple tailor-made definitions and applications; freedom to choose varying shades of green (White et al., 2006).

Tourism offers a promising future for employment. Adopting dignity should be the core value of sustainable tourism jobs. Psychological dimensions can add to broader societal implications (Winchenbach et al., 2019). Studies on Northern Cyprus women entrepreneurs show that such women have better autonomy and self-belief with psychological empowerment that helps not only them to flourish but also benefit the society and economy. Hence tourism can act as a route to integrate women into social and economic development (Ertac & Tanova, 2020). Most importantly, the workers in different sub-sectors should be recognized as human beings with basic human rights, they need to be valued in both monetary and non-monetary terms (Winchenbach et al., 2019). Many tour companies shifted their HR management policies in tune with CSR and improved their working environment and hiring practices. Their daily operations engage sustainable practices by encouraging customers to recycle, reuse, water saving, and energy conserving practices. Sustainable tourism goes beyond employment generation and upholds labor rights, a safe working place, security of jobs, and equal opportunities, ensuring human and worker rights in a threat-free atmosphere (Ioannides et al., 2021).

In the present day, the travel and tourism industry stands at a crossroads. Our choices made today shall tailor the future of this industry, persuading economies, and reinforcing cultural identities and uniqueness and environment. Tourism has an enormous amount of potential to strike a harmonious balance between societal well-being, and economic development along with safeguarding the environment. But obviously, the significance of sustainable tourism is still underscored. By prioritizing sustainability, upgrading and implementing technological advances and development, and fostering partnerships among all stakeholders, the tourism industry can march toward a resilient, equitable, and enriching future. Businesses, destination spots, governments, and travelers can adopt responsible practices for a collective contribution toward a sustainable global tourism industry.

Acknowledgments

The author is thankful and acknowledges the contribution of Dejus for the illustrations.

References

Bauer, I. (1999). The impact of tourism in developing countries on the health of the local host communities: The need for more research. *Journal of Tourism Studies*, *10*(1), 2–17.

Bichler, B. F., & Lösch, M. (2019). Collaborative governance in tourism: Empirical insights into a community-oriented destination. *Sustainability*, *11*(23), 6673.

Bozkurt, A., & Şeker, F. (2023). Harmonizing heritage and artificial neural networks: The role of sustainable tourism in UNESCO world heritage sites. *Sustainability*, *15*(17), 13031.

Cong, L., Wu, B., Morrison, A. M., Shu, H., & Wang, M. (2014). Analysis of wildlife tourism experiences with endangered species: An exploratory study of encounters with giant pandas in Chengdu, China. *Tourism Management*, *40*, 300–310.

Cook, D. (2023). What is a digital nomad? Definition and taxonomy in the era of mainstream remote work. *World Leisure Journal*, 1–20.

Dickert, C. (2023, April 21). What does a net zero hotel look like? *Visual capitalist*. https://www.visualcapitalist.com/sp/what-does-a-net-zero-hotel-look-like/

Dodds, R., & Graci, S. (2012). *Sustainable tourism in island destinations*. Routledge.

Dubois, G., & Ceron, J.-P. (2006). Tourism and climate change: Proposals for a research agenda. *Journal of Sustainable Tourism, 14*(4), 399–415.

Edwards, H. A., Dixon-Hardy, D., & Wadud, Z. (2016). Aircraft cost index and the future of carbon emissions from air travel. *Applied Energy, 164*, 553–562.

Eger, C., Munar, A. M., & Hsu, C. (2022). Gender and tourism sustainability. *Journal of Sustainable Tourism, 30*(7), 1459–1475.

Ertac, M., & Tanova, C. (2020). Flourishing women through sustainable tourism entrepreneurship. *Sustainability, 12*(14), 5643.

Escobar, S. D., & Margherita, E. G. (2021). Outcomes of smart tourism applications on-site for a sustainable tourism: Evidence from empirical studies *The Big Data-Driven Digital Economy: Artificial and Computational Intelligence* (pp. 271–283). Springer.

Filimonau, V., Dickinson, J., & Robbins, D. (2014). The carbon impact of short-haul tourism: A case study of UK travel to Southern France using life cycle analysis. *Journal of Cleaner Production, 64*, 628–638.

Font, X. (2002). *Critical review of certification and accreditation in sustainable tourism governance*. Leeds Metropolitan University.

Goodwin, H. (1998). Background paper for the workshop on sustainable tourism and poverty elimination in preparation for the 1999 session of the commission on sustainable development. DFID and DTER.

Gretzel, U., Sigala, M., Xiang, Z., & Koo, C. (2015). Smart tourism: Foundations and developments. *Electronic Markets, 25*, 179–188.

Grote, M., Williams, I., & Preston, J. (2014). Direct carbon dioxide emissions from civil aircraft. *Atmospheric Environment, 95*, 214–224.

Hall, C. M., Scott, D., & Gössling, S. (2013). The primacy of climate change for sustainable international tourism. *Sustainable Development, 21*(2), 112–121.

Holub, M. (2015). Ecotourism certification programs: Standards and benefits. *Baltic Journal of Economic Studies, 1*(1), 67–74.

Ioannides, D., Gyimóthy, S., & James, L. (2021). From liminal labor to decent work: A human-centered perspective on sustainable tourism employment. *Sustainability, 13*(2), 851.

Jacobsen, J. K. S., Iversen, N. M., & Hem, L. E. (2019). Hotspot crowding and over-tourism: Antecedents of destination attractiveness. *Annals of Tourism Research, 76*, 53–66.

Kashem, M. A., Shamsuddoha, M., Nasir, T., & Chowdhury, A. A. (2022). The role of artificial intelligence and blockchain technologies in sustainable tourism in the Middle East. *Worldwide Hospitality and Tourism Themes, 15*(2), 178–191.

Lisi, F. A., & Esposito, F. (2015). *An AI application to integrated tourism planning*. Paper presented at the Congress of the Italian Association for Artificial Intelligence. Springer

Manzoor, F., Wei, L., Asif, M., Haq, M. Z. U., & Rehman, H. U. (2019). The contribution of sustainable tourism to economic growth and employment in Pakistan. *International Journal of Environmental Research and Public Health, 16*(19), 3785.

Mouratidis, G. (2018). *Digital nomads: Travel, remote work and alternative lifestyles*. Lund University.

Nicola-Gavrilă, L. (2023). Empirical approaches regarding iterdependency between technology and sustainable tourism. *Journal of Environmental Management and Tourism (JEMT), 14*(4 (68)), 2140–2147.

Pagoni, I., & Psaraki-Kalouptsidi, V. (2016). The impact of carbon emission fees on passenger demand and air fares: A game theoretic approach. *Journal of Air Transport Management, 55*, 41–51.

Peeters, P., & Dubois, G. (2010). Tourism travel under climate change mitigation constraints. *Journal of Transport Geography*, *18*(3), 447–457.

Poulaki, I., Mavragani, E., Kaziani, A., & Chatzimichali, E. (2023). Digital nomads: Advances in hospitality and destination attractiveness. *Tourism and Hospitality*, *4*(3), 483–498.

Prabawa, I., & Pertiwi, P. R. (2020). The digital nomad tourist motivation in Bali: Exploratory research based on push and pull theory. *Athens Journal of Tourism*, *7*(3), 161–174.

Rahmadian, E., Feitosa, D., & Zwitter, A. (2022). A systematic literature review on the use of big data for sustainable tourism. *Current Issues in Tourism*, *25*(11), 1711–1730.

Roblek, V., Drpić, D., Meško, M., & Milojica, V. (2021). Evolution of sustainable tourism concepts. *Sustainability*, *13*(22), 12829.

Scott, D. (2021). Sustainable tourism and the grand challenge of climate change. *Sustainability*, *13*(4), 1966.

Skanavis, C., & Sakellari, M. M. (2008). Gender and sustainable tourism: Women's participation in the environmental decision-making process. *European Journal of Tourism Research*, *1*(2), 78–93.

Stanchev, R. (2018). *The most affected European destinations by over-tourism*. Universitat de les Illes Balears.

Toth, R. (2000). Implementing a worldwide sustainable tourism certification system. *Tourism and the Sustainable Development Goals – Journey to 2030*. RB Toth Associates.

Tsaih, R.-H., & Hsu, C. C. (2018). *Artificial intelligence in smart tourism: A conceptual framework*. ICEB 2018 Proceedings. Guangxi Normal University.

UNWTO, U. (2017). Tourism and the sustainable development goals: Journey to 2030. *Highlights*.

Vagena, A. (2021). Digital nomads and tourism industry. *Academia Letters*. Article 765. https://doi.org/10.20935/AL765

Van der Zee, E., & Bertocchi, D. (2018). Finding patterns in urban tourist behaviour: A social network analysis approach based on TripAdvisor reviews. *Information Technology & Tourism*, *20*(1–4), 153–180.

Vázquez Loaiza, J. P., Pérez-Torres, A., & Díaz Contreras, K. M. (2019). Semantic icons: A sentiment analysis as a contribution to sustainable tourism. *Sustainability*, *11*(17), 4655.

Vazquez Vicente, G., Martin Barroso, V., & Blanco Jimenez, F. J. (2021). Sustainable tourism, economic growth and employment—The case of the wine routes of Spain. *Sustainability*, *13*(13), 7164.

White, V., McCrum, G., Blackstock, K., & Scott, A. (2006). *Indicators and sustainable tourism: Literature review*. Macaulay Institute.

Winchenbach, A., Hanna, P., & Miller, G. (2019). Rethinking decent work: The value of dignity in tourism employment. *Journal of Sustainable Tourism*, *27*(7), 1026–1043.

Wong, E. P., Mistilis, N., & Dwyer, L. (2010). Understanding ASEAN tourism collaboration—The preconditions and policy framework formulation. *International Journal of Tourism Research*, *12*(3), 291–302.

WTO. (2005). *Making tourism more sustainable: A guide for policy makers (English version)*. World Tourism Organization.

Xiang, Z., Du, Q., Ma, Y., & Fan, W. (2017). A comparative analysis of major online review platforms: Implications for social media analytics in hospitality and tourism. *Tourism Management*, *58*, 51–65.

Xu, F., Nash, N., & Whitmarsh, L. (2020). Big data or small data? A methodological review of sustainable tourism. *Journal of Sustainable Tourism*, *28*(2), 144–163.

Zienarski, P., & Such-pyrgiel, M. (2019). Sustainable tourism development in the era of digital transformation. *OPUS International Journal of Society Researches*, *10*(17), 2157–2167.

Chapter 12

Harnessing Artificial Intelligence for Sustainable Tourism Practices

Pranjal Jain[a], *Pooja Jain*[b] *and Anju Jain*[c]

[a]*Lovely Professional University, India*
[b]*Google, USA*
[c]*University of Delhi, India*

Abstract

The 28th meeting of the Conference of Parties (COP) to the United Nations Framework Convention on Climate Change (UNFCCC) brought together over 70,000 delegates to address climate change, carbon emissions reduction, and sustainable development. The tourism industry, contributing 8% of global carbon emissions, poses significant environmental challenges. The surge in tourism post-COVID restrictions and aviation emissions doubling by 2050 exacerbate concerns. Literature indicates a complex relationship between tourism, economic growth, and environmental impact. Utilizing the STIRPAT model, our study examines sustainable tourism intensity across G20 countries. The analysis reveals positive correlations between tourist arrivals, energy consumption, urbanization, and carbon emissions. Integrating AI, our research explores predictive modeling, smart energy management, and personalized recommendations to mitigate carbon emissions. We call for concerted action to leverage AI for greener, more sustainable tourism practices, advocating for a shared vision of tourism that safeguards the planet while enriching traveler experiences.

Keywords: Sustainable tourism; artificial intelligence; carbon emission; STIRPAT model; community engagement

The 28th COP to UNFCCC (2023) convened with a focus on discussions surrounding climate change mitigation, carbon emission reduction, and the imperative to

The Need for Sustainable Tourism in an Era of Global Climate Change: Pathway to a Greener Future, 249–267

doi:10.1108/978-1-83608-668-020241034

reverse environmentally damaging trends triggered by anthropogenic activities. Central to these discussions is the ambitious goal of achieving net climate neutrality, while striving to limit the global temperature increase well below 2°C, preferably below 1.5°C, and transitioning toward clean energy sources. This collective effort aims to ensure responsible stewardship of our planet and secure a sustainable future for generations to come. Amidst these discussions, the tourism industry has emerged as a significant contributor to global carbon emissions, accounting for approximately 8% of the total (Lenzen et al., 2018). This figure has been escalating rapidly in recent years, with current emission rates reaching around 5 gigatons of carbon dioxide annually. Projections suggest that the sector could release a staggering 200 gigatons of carbon dioxide equivalent between 2022 and 2050, underscoring the urgent need for sustainable tourism practices. A report by the UN World Tourism Organization (UNWTO, 2022) highlights these concerns, highlighting a notable surge in tourism activity, including the adoption of "revenge tourism" practices following the lifting of COVID-induced restrictions. This resurgence in travel poses significant challenges to environmental sustainability, exacerbating carbon emissions within the tourism sector (Shi et al., 2020).

Studies conducted by Lee et al. (2021) shed light on the concerning trend of escalating carbon emissions from the aviation industry. Their research indicates that between 1960 and 2018, carbon emissions from aviation soared to a staggering 1,034 million tons. Even more alarming is the projection that these emissions are expected to double by 2050, with fuel consumption projected to triple. This trajectory poses formidable challenges to the long-term sustainability of international tourism (Zhang & Liu, 2019). For instance, the exponential growth of the aviation industry and its associated carbon emissions are closely intertwined with the expansion of international tourism (ATAG, 2021). According to data from the World Tourism Organization (WTO, 2021), international tourist arrivals doubled from 1.08 billion in the period spanning 1995–2019 to 2.4 billion. This rapid increase in tourism activity has placed a significant strain on transportation infrastructure, particularly air travel, resulting in a substantial environmental footprint (Kongbuamai et al., 2020). The link between tourism and aviation emissions highlights the urgent need for sustainable solutions to mitigate the environmental impact of the tourism industry. While tourism contributes to economic growth and cultural exchange, unchecked growth in air travel poses risks to the environment, public health, and global climate stability (Işik et al., 2017). Addressing the challenge of escalating emissions requires a multi-faceted approach that combines technological innovation, policy intervention, and behavioral change. Advancements in designs, propulsion systems, and alternative fuels hold promise for reducing the carbon intensity of aviation. Initiatives to improve fuel efficiency, optimize traffic routes, and enhance traffic management can help minimize emissions and mitigate environmental impact. However, such practices require careful analysis of the present situation, legacy challenges, and future technologies to attain sustainable development. Fig. 12.1 visually illustrates the stark disparity between the growth of absolute emissions and the desired growth rate necessary to effectively limit carbon emissions, underscoring the urgency of adopting sustainable practices within the tourism industry.

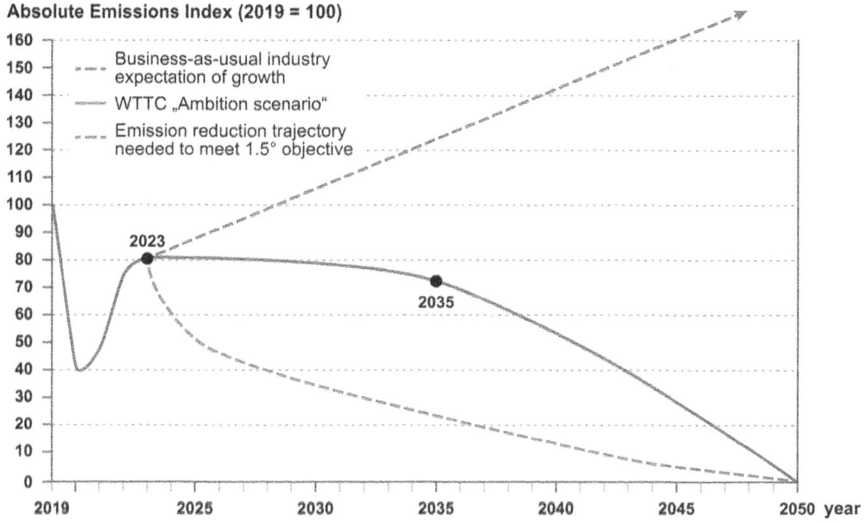

Fig. 12.1. Absolute Carbon Emissions. *Source*: Adapted from WTTC–UNEP–UNFCCC (2021).

Fig. 12.1 illustrates the mitigation challenge confronting the tourism industry in mitigating carbon emissions. It delineates the expected emission growth line (red line), the ambition curve (green line), and a trajectory toward net-zero emissions aligned with the 1.5°C objective (blue line). The figure provides the uncertainty surrounding how the gap between these trajectories will be bridged (Gossling et al., 2023). The required annual carbon emission reduction rate of 5% contrasts with the expected annual carbon growth rates of 3% (aviation) and 5% (other tourism).

Achieving the climate objective necessitates a decarbonization rate of approximately 10%. However, despite a global emission reduction rate of 6% observed during the COVID-induced shutdown, as noted by Friedlingstein et al. (2022), concerns persist regarding the mechanisms for carbon reduction in tourism and the adoption of sustainable tourism practices. This suggests the need for proactive measures within the tourism industry to address carbon emissions and advance sustainable practices.

The complex relationship between tourism and climate change can be characterized by a two-way mechanism that shapes and is shaped by environmental dynamics (Leal Filho et al., 2023). At the core of this intricate relationship lies the contribution of tourism activities to the emission of greenhouse gases (GHGs), prominent of them being carbon dioxide. The emissions originating from travel and accommodation, integral components of the tourism industry, constitute a significant driver of climate change. As carbon dioxide levels rise, the delicate balance of Earth's climate system is disrupted, precipitating a cascade of environmental consequences. Extreme weather events, ranging from devastating floods to unpredictable cloudbursts, emerge as stark manifestations of climatic instability exacerbated by tourism-related emissions. Further, the repercussions of heightened carbon emissions extend beyond episodic weather events, permeating the fabric of environmental equilibrium. Reduced water availability, a direct consequence of altered precipitation patterns and climatic conditions, poses a formidable challenge to ecosystems and human communities alike. The scarcity of water resources, compounded by tourism-induced climate change, amplifies vulnerabilities and strains the resilience of regions reliant on freshwater for sustenance and livelihoods. The impacts of tourism-related climate change reverberate far beyond the confines of environmental spheres, impinging upon societal dynamics and cultural landscapes. Alterations in climatic conditions can trigger migratory patterns, prompting population movements and demographic shifts. Displaced communities, grappling with the upheaval wrought by environmental degradation, navigate uncertain futures marked by resource scarcity, social upheaval, and economic dislocation.

Fig. 12.2 illustrates the mechanism delineating the relationship between climate change and tourism, shedding light on the multifaceted dynamics that underpin this complex interplay. It suggests the feedback loops and interdependencies that define the reciprocal relationship between climate dynamics and

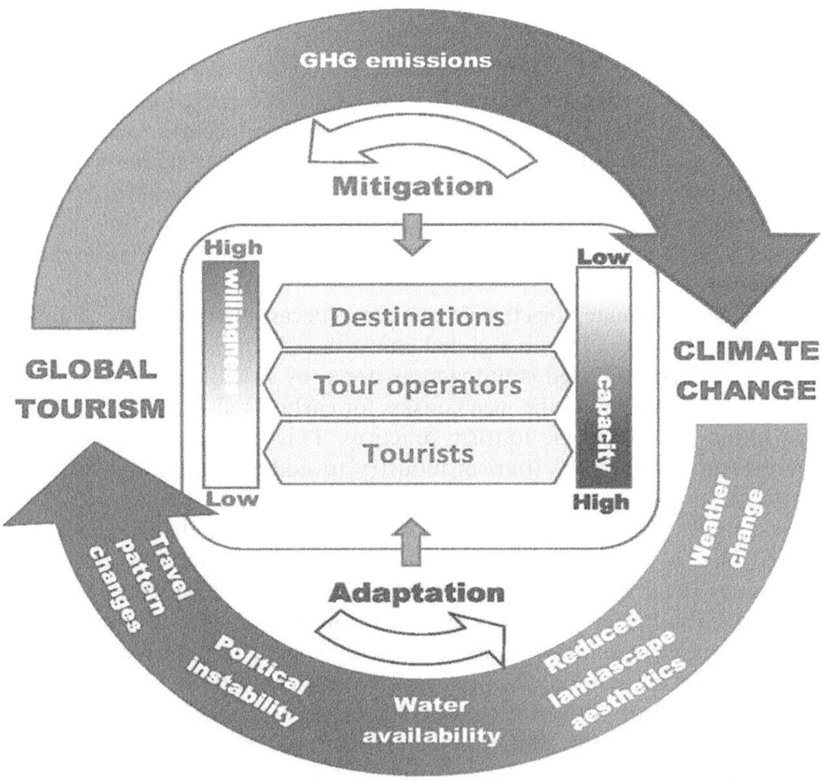

Fig. 12.2. Mechanism Between Climate Change and Tourism. *Source*: Adapted from Leal Filho et al. (2023).

tourism activities, offering insights into the pathways through which environmental changes influence and are influenced by the tourism sector. Central to this mechanism is the cyclical nature of interactions between climate change drivers and tourism-related variables. Climate change drivers, encompassing a spectrum of environmental stressors such as greenhouse gas emissions, deforestation, and ocean acidification, serve as catalysts for transformative shifts in climatic patterns and ecosystem stability. These changes, in turn, exert cascading effects on tourism destinations and activities, reshaping the landscapes and experiences that define the tourism industry. Within this framework, climate change manifests through a myriad of manifestations, ranging from shifts in temperature and precipitation patterns to the intensification of extreme weather events and the degradation of natural habitats. These environmental shifts disrupt the equilibrium of tourism destinations, altering the attractiveness, accessibility, and sustainability of travel experiences. Conversely, tourism activities, characterized by transportation, accommodation, and recreation, contribute to the exacerbation of climate change drivers through the emission of greenhouse gases, land-use changes, and resource depletion. The carbon footprint associated with travel, fueled by fossil fuel consumption and energy-intensive infrastructure, amplifies the anthropogenic footprint on the environment, perpetuating a cycle of environmental degradation and climate instability.

At the intersection of climate change and tourism lie intricate feedback mechanisms and adaptive responses that shape the resilience and vulnerability of both natural and human systems. Tourism destinations, faced with the dual challenges of environmental change and economic sustainability, navigate complex trade-offs between development aspirations and ecological integrity. Moreover, the mechanism depicted in Fig. 12.2 underscores the interconnectedness of global and local dynamics, highlighting the transboundary nature of climate change impacts and the need for collaborative, multilateral approaches to address shared environmental challenges. As tourism emerges as a key driver of economic growth and cultural exchange, the imperative to reconcile tourism development with environmental sustainability grows ever more urgent. Mitigation efforts, aimed at reducing the drivers of climate change and limiting its adverse impacts, play a crucial role in shaping the future trajectory of tourism sustainability. These efforts encompass a spectrum of interventions, including the adoption of renewable energy sources, improvements in energy efficiency, and the implementation of carbon offsetting measures within the tourism industry. By mitigating greenhouse gas emissions and minimizing environmental footprints, tourism destinations can lessen their contribution to climate change while fostering long-term resilience. Adaptation, on the other hand, involves strategies to enhance the capacity of tourism systems to withstand and respond to the impacts of climate change. This may involve investing in climate-resilient infrastructure, diversifying tourism offerings to minimize reliance on climate-sensitive activities, and integrating ecosystem-based approaches to enhance natural resilience. By adapting to changing environmental conditions, tourism destinations can better safeguard their ecosystems, cultural heritage, and socio-economic livelihoods against the vagaries of climate change.

The relationship between tourism, economic development, and environmental impact is profoundly intricate and multifaceted, as underscored by a plethora of research findings. Studies consistently reveal that the growth of tourism activities is closely intertwined with increases in carbon dioxide emissions, thereby amplifying environmental concerns on a global scale. Notably, empirical evidence suggests that for every 1% surge in tourist expenditure, there is a corresponding uptick in carbon emissions, with rates varying across different income brackets. Specifically, low-income countries experience a more substantial impact, with carbon emissions rising by 0.084% per unit increase in tourist spending, followed by middle-income countries at 0.072%, and high-income countries at 0.069% (Voumik et al., 2023). The differential environmental impact of tourism across various economic contexts is a significant focal point of investigation. Research indicates that the causal relationship between tourism and emissions tends to be more pronounced in lower-income nations, where the ecological footprint of tourism activities is disproportionately high compared to their economic contribution. Conversely, higher-income countries exhibit a stronger causality between tourism expenditure and carbon emissions, reflecting the intricate dynamics between economic prosperity, consumption patterns, and environmental consequences.

Empirical studies, such as those conducted in Greece, offer compelling insights into the environmental ramifications of burgeoning tourism sectors. Evidence from Greece elucidates a direct correlation between the expansion of tourism activities and environmental degradation, particularly in terms of increased carbon emissions. The observed positive correlation coefficient of 0.132 between tourism and environmental degradation underscores the urgency of adopting alternative tourism models that prioritize sustainability and mitigate adverse ecological impacts. These findings collectively emphasize the imperative for proactive measures aimed at mitigating the environmental footprint of tourism while simultaneously fostering sustainable economic development in regions heavily reliant on tourism revenues. Implementing comprehensive strategies that integrate environmental conservation, community engagement, and economic diversification is paramount to achieving long-term sustainability in tourism-dependent areas. Such initiatives may include promoting ecotourism ventures, investing in renewable energy infrastructure, enforcing stringent environmental regulations, and fostering community-based tourism initiatives that empower local stakeholders.

Ravinthirakumaran and Ravinthirakumaran (2022) conducted analysis with cross-sectional dependence and heterogeneous panel estimation techniques to explore the intricate relationship between economic growth, energy consumption, tourism, trade openness, and CO_2 emissions spanning the period from 1995 to 2017. Their study affirmed the stationarity of the variables at their first differences, signifying the stability of the data over time. The application of the Westerlund panel cointegration test corroborated their findings, unveiling a one-sided causality between tourism activities and CO_2 emissions. In a parallel study by Sherafatian-Jahromi et al. (2017), the focus shifted toward the rapid expansion of the tourist industry in Southeast Asia and its implications for carbon emissions. Employing pooled mean group techniques and panel cointegration methodologies spanning the period from 1995 to 2013, their research uncovered an intriguing inverted U-shaped relationship between tourism growth and carbon intensity. This finding suggests a non-linear correlation wherein the initial stages of tourism development may witness a surge in carbon intensity, followed by a subsequent decline as sustainability measures and environmental regulations are implemented. The integration of the Environmental Kuznets Curve (EKC) framework provides valuable insights into the complex interplay between economic development, environmental quality, and tourism expansion. The application of this framework allowed for a nuanced understanding of the relationship between economic growth and environmental degradation, offering policymakers and stakeholders strategic guidance in balancing economic prosperity with ecological sustainability (Bilen et al., 2017).

Rahman et al. (2022) explored the determinants of carbon emissions, including economic growth, population, and energy intensity. Using the AutoRegressive Distributed Lag (ARDL) approach, they found an insignificant impact of electricity and natural gas on carbon emissions but significant impacts of tourist arrivals, coal usage, and oil usage on carbon emissions. They recommended the adoption of cleaner energy sources and the imposition of carbon taxes. Wang and Wu (2021) analyzed the factors influencing the relationship between tourism and carbon emissions in countries associated with the Regional Comprehensive Economic Partnership (RCEP) framework. Their study, using various tests and models over the period 1995–2018, identified urbanization as a Granger cause of both tourism and carbon emissions. The study highlighted the complex dynamics between urbanization, tourism, and environmental impact. Le and Nguyen (2021) investigated tourism development and carbon emissions across 95 countries from 1998 to 2014 using the STIRPAT model and EKC framework. They found that tourism increased CO_2 emissions from the transportation sector and that a rise in tourist arrivals correlated with an increase in CO_2 emissions per capita.

The study emphasized the importance of prioritizing green transportation infrastructure and practices within the tourism industry to mitigate environmental impact. The existing literature sheds light on the relationship between tourism growth and carbon emissions, focusing on economic development, energy consumption, and population growth. However, a critical gap remains in assessing the specific role and efficacy of sustainable tourism practices in mitigating carbon emissions. The limited emphasis on sustainable tourist practices suggests the need for empirical examination of how eco-friendly initiatives within the tourism industry may contribute to environmental sustainability. Moreover, the insufficient integration of control variables indicates a limited examination of potential confounding factors. Further limitations are observed in comparing the global perspective with regional variances in the assessment of tourism and carbon emissions. The primary objective is to empirically investigate whether the adoption of sustainable tourism practices serves as an effective mechanism for mitigating carbon emissions. To achieve this goal, the study incorporates a robust statistical approach, leveraging multiple iterations to enhance the reliability and validity of the findings.

Control variables are strategically introduced to account for external factors that might influence the relationship between sustainable tourism and carbon emissions. These variables include economic indicators, population density, and energy consumption patterns, which are applied iteratively to account for control and moderation effects. Through multiple regression analyses, the study aims to quantify the extent to which sustainable tourism practices contribute to carbon emission reduction while examining the relationship between sustainable tourism and carbon emissions. The findings from this research may have implications for policymakers, industry stakeholders, and environmental advocates seeking evidence-based strategies to promote eco-friendly tourism practices and address the challenges posed by carbon emissions in the tourism sector. Fig. 12.3 depicts the conceptual framework for conducting an analysis of the relationship between sustainable tourist practices and carbon emissions concerning conventional tourist practices. The conceptual framework builds up a trade-off between tourism intensity and carbon emissions. As tourism intensity increases, carbon emissions also increase. It provides that there are ways to mitigate the increase in carbon emissions, such as by implementing sustainable tourism practices.

The research aimed to assess sustainable tourism intensity based on the sustainable tourism initiative of G20 countries. Data from these 20 countries spanning from 1995 to 2020 was gathered from World Development Indicators, World Bank, and the United Nations Statistical Yearbook, with carbon emissions (measured by CO_2 emissions) serving as the dependent variable (C). STIRPAT variations were modified (Destek & Aydın, 2022). Independent variables included tourist arrivals (TA), energy consumption measured by oil consumption (ENC), urbanization percentage (U), and GDP 2005 levels (GDP), all of which varied the construct C (Balsalobre-Lorente et al., 2020). To analyze the relationship, we applied the STIRPAT model, which involved converting the data into logarithmic series where applicable. SIRPAT is a statistical model used in environmental studies to analyze and predict the impact of human activities on the environment.

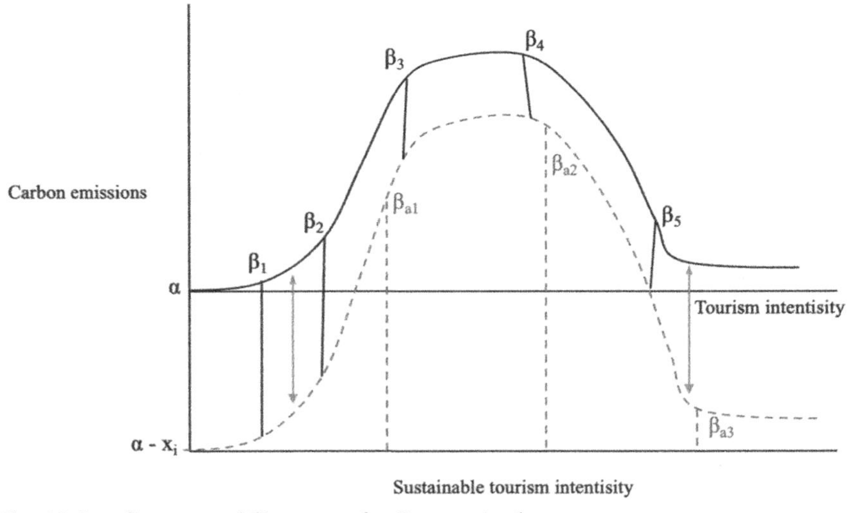

Fig. 12.3. Conceptual Framework. *Source*: Authors.

The STIRPAT model identifies environmental impact (I) as a function of population (P), affluence (A), and technology (T) by incorporating stochastic (random) elements into the analysis, allowing for more sophisticated and nuanced assessments of environmental impacts. The inclusion of stochastic elements acknowledges the inherent uncertainty and variability in human-environment interactions. This allows accounting for unpredictable factors and random fluctuations that may influence environmental outcomes. The STIRPAT model typically employs regression analysis to quantitatively examine the relationships between human activities and environmental impacts. Causality relationships were assessed using the pooled mean group estimator and cointegration test methodologies. These statistical techniques allowed for a comprehensive examination of the interactions between sustainable tourism initiatives and carbon emissions across the G20 countries, providing insights into the dynamics and potential drivers of sustainable tourism practices and their environmental impact. In applying the STIRPAT model, the data underwent a logarithmic transformation to ensure compatibility with the model's assumptions and requirements. This transformation facilitated the examination of relationships between the dependent and independent variables while accounting for potential nonlinearities and scale effects. The pooled mean group estimator was employed to account for both cross-country heterogeneity and time-series dynamics within the dataset. This estimator allowed for robust estimation of coefficients while considering potential differences in the impact of independent variables across countries and over time. Cointegration tests were conducted to explore long-term relationships among the variables, providing insights into their interdependencies and potential causal mechanisms. By assessing whether the variables shared common stochastic trends, the cointegration tests helped ascertain the presence of stable, long-term relationships that could inform policy and decision-making processes related to sustainable tourism and carbon emissions reduction. The following equation represents the model applied:

$$St_{!"} = \beta_{\#}Sd_{\$\%} + \beta_{\&}TA_{\$\%} + \beta,C_{\$\%} + (SAI_{\$\%} + \beta)NC_{\$\%} + \beta_{*}U_{\$\%} + e_{\$\%} \tag{1}$$

St_{AI} represents the level of sustainable tourism with respect to AI integration. Sd_{it}, TA_{it}, C_{it}, SAI_{it}, ENC_{it}, and U_{it} represent various factors influencing sustainable tourism. The coefficients quantify the impact of each variable on the level of sustainable tourism with AI integration. The analysis using the STIRPAT model demonstrated a significant positive correlation between tourist arrivals (TA) and carbon emissions (C). This finding suggests that higher levels of tourism activity are associated with increased carbon emissions. Energy consumption (ENC), as measured by oil consumption, showed a notable positive relationship with carbon emissions. This implies that higher levels of energy consumption, particularly from oil sources, contribute to elevated carbon emissions within the tourism sector. Additionally, urbanization percentage (U) exhibited a significant positive association with carbon emissions, indicating that greater urbanization rates corresponded to higher levels of carbon emissions from tourism-related activities. GDP levels (GDP) from 2005 showed a positive relationship with carbon emissions, suggesting that higher economic development levels are linked to increased carbon emissions within the tourism sector. The findings from the pooled mean group estimator and cointegration tests supported the robustness of these relationships, indicating consistent patterns across countries and over time. These findings bear important implications for various stakeholders. Sustainable tourism practices can be integrated into policy frameworks to mitigate carbon emissions and promote environmental stewardship. Industry stakeholders should prioritize sustainability initiatives to minimize the carbon footprint of tourism operations and align with consumer preferences for eco-friendly travel options.

The results emphasize the importance of international cooperation and investment in sustainable tourism infrastructure and innovation to address environmental challenges on a global scale. Artificial Intelligence (AI) can assist in the implementation of sustainable tourism practices and combating carbon emissions within the industry. Through advanced data analytics, AI algorithms can parse vast datasets related to carbon emissions, energy consumption, and tourist behavior, extracting valuable insights to inform strategic decision-making. Predictive modeling, powered by AI, enables stakeholders to forecast future trends in tourism demand and emissions, facilitating proactive planning and resource allocation (Jain & Jain, 2020). By harnessing smart energy management systems driven by AI, tourism facilities can optimize energy usage in real-time, minimizing waste and reducing their carbon footprint. AI-driven optimization algorithms can revolutionize transportation systems, optimizing routes and schedules to minimize fuel consumption and emissions. These smart transportation solutions analyze traffic patterns, passenger demand, and environmental factors to streamline operations and promote sustainability.

Through personalized recommendation systems, AI can empower travelers with tailored suggestions for eco-friendly accommodations, low-carbon transportation options, and sustainable activities, fostering responsible travel behaviors. AI facilitates real-time tracking and monitoring of carbon emissions, enabling tourism stakeholders to assess their environmental impact accurately.

By integrating data from various sources, including transportation, accommodations, and tourist activities, artificial intelligence (AI) algorithms offer valuable insights for devising effective strategies to reduce emissions in the tourism sector. Through sophisticated data analysis techniques, AI can identify patterns and trends that inform targeted interventions aimed at mitigating carbon footprints and promoting sustainable practices throughout the industry (Majid et al., 2023). This analytical capability extends to understanding consumer behavior and preferences, thereby enabling the design of tailored interventions and marketing campaigns that encourage the widespread adoption of sustainable tourism practices. The principal domains where AI has the potential to make substantial contributions to the sustainability of tourism include:

(i) *Environmental Monitoring and Conservation*: AI-enabled monitoring systems play a crucial role in tracking environmental indicators such as air and water quality, biodiversity, and ecosystem health in tourist destinations. Through the integration of sensor networks, satellite imagery, and other data sources, AI technologies facilitate the collection, analysis, and interpretation of environmental data in real time. These monitoring systems enable stakeholders, including environmental agencies, conservation organizations, and tourism operators, to gain insights into the health and integrity of natural habitats and ecosystems within tourist destinations. By detecting changes in environmental parameters and identifying areas of concern, AI helps prioritize conservation efforts and allocate resources effectively to address pressing environmental challenges. AI algorithms can analyze patterns in air and water quality data to detect pollution hotspots or identify signs of ecological stress. By alerting authorities and stakeholders to emerging environmental risks, AI empowers proactive interventions to prevent further degradation and mitigate potential impacts on local ecosystems and public health. By generating actionable insights from complex environmental data sets, AI helps policymakers and practitioners develop targeted strategies and interventions to address specific environmental issues, such as habitat loss, pollution, and climate change adaptation.

(ii) *Visitor Management*: AI-based solutions manage visitor flows and crowd control in popular tourist attractions to prevent overcrowding and minimize negative impacts on local ecosystems and communities. By analyzing visitor behavior and traffic patterns, AI helps optimize visitor experiences while preserving the integrity of natural and cultural sites. One of the primary functions of AI-based visitor management systems is to prevent overcrowding and mitigate the negative impacts associated with mass tourism. By monitoring visitor numbers in real-time and predicting future trends, AI algorithms can provide insights into peak visitation periods and identify areas prone to congestion. This enables destination managers and tourism operators to implement proactive measures, such as capacity limits, timed entry slots, and alternative itineraries, to distribute visitor flows more evenly and reduce pressure on sensitive ecosystems and infrastructure. AI-powered visitor management solutions facilitate personalized experiences for tourists while minimizing environmental footprint and cultural disruption. By analyzing visitor preferences, interests, and feedback, AI algorithms can tailor recommendations and itineraries to match individual preferences, thereby enhancing visitor satisfaction and promoting responsible tourism practices. For instance, AI-driven recommendation systems can suggest off-the-beaten-path attractions, eco-friendly accommodations, and sustainable transportation options based on user profiles and environmental criteria. AI-powered visitor management solutions facilitate personalized experiences for tourists while minimizing environmental footprint and cultural disruption. By analyzing visitor preferences, interests, and feedback, AI algorithms can tailor recommendations and itineraries to match individual preferences, thereby enhancing visitor satisfaction and promoting responsible tourism practices. For instance, AI-driven recommendation systems can suggest off-the-beaten-path attractions, eco-friendly accommodations, and sustainable transportation options based on user profiles and environmental criteria.

(iii) *Sustainable Transportation*: AI algorithms optimize transportation networks and routes to reduce carbon emissions and minimize environmental impact.

By analyzing traffic patterns, optimizing vehicle routing, and promoting the use of eco-friendly transportation options such as electric vehicles and public transit, AI contributes to the development of sustainable transportation systems in tourism destinations. At the core of AI-driven sustainable transportation solutions lies the ability to analyze vast amounts of data related to traffic flow, congestion patterns, and environmental considerations. By leveraging real-time data from sensors, GPS devices, and traffic monitoring systems, AI algorithms can identify inefficiencies within transportation networks and propose optimal routing strategies that minimize fuel consumption, reduce emissions, and alleviate congestion. One of the primary advantages of AI in sustainable transportation is its capacity to optimize vehicle routing and scheduling. Through predictive analytics and machine learning algorithms, AI systems can forecast demand patterns, anticipate traffic fluctuations, and dynamically adjust transportation routes in response to changing conditions. This enables more efficient use of resources, reduces idle time, and enhances overall system performance while lowering environmental impact. AI-powered transportation solutions facilitate the integration of eco-friendly modes of transportation, such as electric vehicles (EVs) and public transit, into the tourism ecosystem. By analyzing factors such as charging station availability, range limitations, and user preferences, AI algorithms can identify opportunities to promote the adoption of EVs among tourists and local residents alike. AI-driven route optimization algorithms can prioritize the use of public transit and shared mobility services, thereby reducing reliance on single-occupancy vehicles and promoting sustainable travel behavior.

(iv) *Crisis Management and Resilience*: AI technologies assist in crisis management and disaster response efforts in tourism destinations. By analyzing real-time data and social media feeds, AI can detect early warning signs of natural disasters or emergencies, enabling timely evacuation and resource allocation to ensure the safety and well-being of tourists and local residents (Wu et al., 2020). One of the primary functions of AI in crisis management is its ability to analyze vast amounts of data from diverse sources, including weather sensors, satellite imagery, and social media feeds. By leveraging machine learning algorithms, AI systems can detect patterns, trends, and anomalies that may indicate potential threats, such as severe weather events, natural disasters, or civil unrest. This proactive approach enables authorities and emergency responders to anticipate crises, assess risks, and implement preventive measures to mitigate potential impacts on tourism infrastructure and public safety. AI-powered crisis management solutions enable real-time monitoring and situational awareness during emergencies. By aggregating and analyzing data streams from various sources, including surveillance cameras, IoT devices, and citizen reports, AI algorithms provide decision-makers with up-to-date information on evolving threats, incidents, and response efforts. This situational awareness enables authorities to make informed decisions, coordinate resources effectively, and communicate timely alerts and instructions to tourists and residents in affected areas. In addition to crisis response, AI-powered solutions contribute to the resilience and recovery of tourism destinations in the aftermath of disasters. By analyzing post-event data, assessing damage assessments, and identifying recovery needs, AI systems support decision-makers in prioritizing reconstruction efforts, allocating resources, and implementing measures to restore tourism infrastructure and community livelihoods. Moreover, AI-driven sentiment analysis of social media feeds enables authorities to gauge public perceptions, address concerns, and foster community resilience in the face of adversity.

(v) *Data-Driven Decision-Making*: AI-driven analytics generate actionable insights from large volumes of data, enabling tourism stakeholders to make informed decisions that promote sustainability and environmental responsibility. By analyzing trends, predicting future demand, and identifying areas for improvement, AI empowers tourism businesses, policymakers, and destination managers to develop strategies that balance economic growth with environmental conservation and social equity. AI has the capacity to analyze diverse data sources, including customer feedback, market trends, environmental indicators, and socioeconomic data. By leveraging advanced machine learning algorithms, AI systems can identify patterns, correlations, and predictive trends within these datasets, thereby providing stakeholders with valuable insights into consumer behavior, market dynamics, and sustainability challenges. AI-driven analytics enable tourism stakeholders to anticipate and respond to changing market conditions and consumer preferences in real time. By analyzing social media feeds, online reviews, and booking patterns,

AI algorithms can identify emerging trends, assess customer sentiment, and tailor marketing strategies and product offerings to meet evolving demand. This agile approach to decision-making allows businesses to stay competitive, enhance customer satisfaction, and adapt to changing market dynamics while aligning with sustainability goals. AI-powered predictive modeling enables tourism stakeholders to forecast future demand, optimize resource allocation, and mitigate risks associated with over-tourism, environmental degradation, and climate change. By integrating historical data, market trends, and environmental indicators, AI algorithms can generate forecasts and scenario analyses that inform strategic planning, investment decisions, and policy formulation. This proactive approach enables stakeholders to anticipate challenges, identify opportunities, and develop resilient strategies that balance economic growth with environmental stewardship. In addition to strategic planning and forecasting, AI-driven analytics support continuous improvement and innovation within the tourism industry. By analyzing performance metrics, customer feedback, and operational data, AI systems help identify areas for optimization, process efficiency, and sustainability improvements. This data-driven approach enables businesses to streamline operations, reduce waste, and enhance resource efficiency while delivering high-quality experiences to travelers. AI-powered analytics facilitate collaboration and knowledge-sharing among diverse stakeholders within the tourism ecosystem. By providing accessible and interpretable insights, AI systems foster transparency, accountability, and dialogue among policymakers, businesses, communities, and tourists, thereby promoting collective action toward shared sustainability goals.

The integration of artificial intelligence (AI) with environmental considerations represents a critical frontier in contemporary discourse, yet this intersection has received insufficient scholarly attention. As AI continues to permeate various sectors, from healthcare to finance, its potential to address environmental challenges remains largely untapped. While there is recognition of AI's transformative capabilities, the incorporation of environmental considerations into AI applications has not been adequately explored. One reason for the limited scholarly attention may stem from the complexity of the issues involved. Environmental challenges, such as climate change, biodiversity loss, and pollution, are multifaceted and interconnected. Addressing these issues requires a comprehensive understanding of ecological systems, coupled with innovative technological solutions. However, the integration of AI into environmental initiatives necessitates interdisciplinary collaboration between environmental scientists, computer engineers, policymakers, and other stakeholders. Another factor contributing to the lack of scholarly attention is the perceived complexity and uncertainty surrounding AI's role in environmental management. While AI offers promising capabilities, its implementation requires careful consideration of ethical, legal, and societal implications.

Concerns about data privacy, algorithmic bias, and unintended consequences of AI-driven decisions have prompted cautious approaches to its deployment in environmental contexts (Li et al., 2022). However, despite these challenges, there is a growing recognition of the urgent need to harness AI for environmental sustainability. Researchers, policymakers, and practitioners are increasingly exploring ways to leverage AI technologies to address pressing environmental issues. One key area where AI demonstrates significant potential is in enhancing our understanding of complex environmental systems. By analyzing vast amounts of data from various sources, including satellite imagery, sensor networks, and scientific studies, AI algorithms can identify patterns, correlations, and causal relationships within environmental data sets. This enables more accurate modeling and prediction of environmental phenomena, such as climate change, deforestation, and biodiversity loss. By gaining deeper insights into the dynamics of ecological systems, policymakers and conservationists can develop more effective strategies for mitigating environmental degradation and promoting sustainability (Nosheen et al., 2021). AI-powered monitoring systems play a crucial role in facilitating real-time tracking of environmental indicators, such as air and water quality.

By deploying sensor networks and IoT devices equipped with AI algorithms, environmental agencies, and organizations can collect and analyze data on pollutant levels, temperature fluctuations, and ecosystem health in real time. This enables proactive interventions to prevent pollution, mitigate environmental risks, and protect public health. For instance, AI-driven predictive analytics can forecast the spread of pollutants in water bodies or identify areas at high risk of air pollution, allowing authorities to implement targeted measures to reduce emissions and improve environmental quality.

Fig. 12.4 illustrates the applicable roles wherein we can develop policies for integrating AI to support sustainable tourism. Through its multifaceted capabilities,

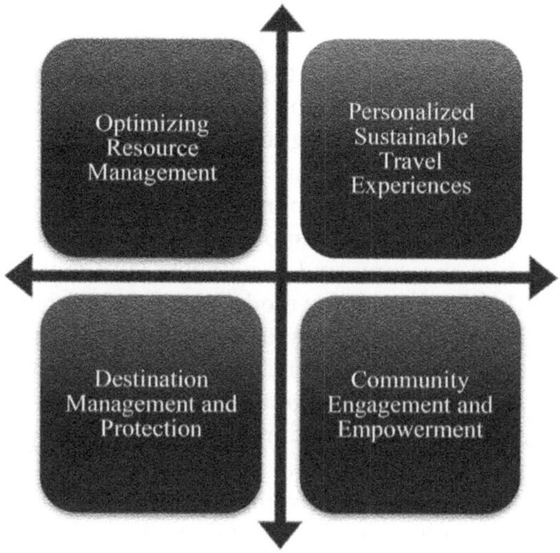

Fig. 12.4. Applicable Role of AI for a Sustainable Tourism Future.
Source: Authors.

AI contributes significantly to the advancement and enhancement of sustainable practices within the tourism industry. The key roles have been explained as follows:

(i) *Optimizing Resource Management*: AI plays a pivotal role in optimizing resource management within the context of sustainable tourism. By leveraging data analytics and predictive modeling, AI systems help tourism operators and destination managers to better understand and anticipate resource demands. Through the analysis of historical data and real-time inputs, AI can optimize the use of resources such as water, energy, and waste management systems. For instance, AI-driven algorithms can adjust lighting and HVAC systems in hotels based on occupancy rates, thereby reducing energy consumption. Similarly, AI-powered waste management systems can optimize collection routes and schedules, minimizing environmental impact and operational costs.

(ii) *Personalized Sustainable Travel Experiences*: AI enables the delivery of personalized and sustainable travel experiences tailored to the preferences and values of individual travelers. By analyzing vast amounts of data, including past travel behavior, demographic information, and environmental preferences, AI algorithms can generate personalized recommendations for accommodations, activities, and transportation options that align with sustainable practices. For example, AI-driven travel platforms can suggest eco-friendly accommodations, carbon-neutral transportation options, and responsible tourism activities based on user profiles and sustainability criteria. By offering personalized recommendations, AI enhances visitor satisfaction while promoting sustainable tourism practices and minimizing negative environmental impacts.

(iii) *Destination Management and Protection*: AI supports destination management and protection efforts by providing real-time monitoring, analysis, and decision support tools for destination managers and conservation organizations. Through the integration of remote sensing data, geospatial analysis, and environmental monitoring technologies, AI systems can assess the health and integrity of natural and cultural assets within tourist destinations. For instance, AI-powered monitoring systems can detect changes in land use patterns, biodiversity hotspots, and ecosystem health, enabling proactive conservation measures and sustainable land management practices. Additionally, AI-driven predictive modeling can help forecast the impacts of tourism activities on sensitive ecosystems and cultural heritage sites, informing sustainable tourism planning and management strategies.

(iv) *Community Engagement and Empowerment*: AI fosters community engagement and empowerment by facilitating inclusive and participatory decision-making processes within local communities and indigenous groups. Through community-based monitoring programs, AI tools enable residents and stakeholders to collect, analyze, and share data on environmental conditions, cultural heritage, and socio-economic impacts of tourism. By empowering communities with access to information and decision-making tools, AI promotes greater transparency, accountability, and social equity in tourism development initiatives. Furthermore, AI-driven communication platforms and social media analytics can facilitate dialogue, collaboration, and knowledge exchange among diverse stakeholders, fostering a sense of ownership and stewardship toward sustainable tourism development goals.

Despite the potential benefits of AI in environmental sustainability, several challenges and considerations must be addressed to ensure the responsible and equitable deployment of AI technologies. First and foremost, concerns about data privacy and security loom large, especially considering the sensitive nature of environmental data and the potential for misuse or unauthorized access. Robust data governance frameworks, encryption protocols, and privacy-enhancing technologies are essential to safeguarding the confidentiality and integrity of environmental data. These measures not only protect sensitive information but also promote transparency and accountability in AI-driven decision-making processes, fostering trust among stakeholders and ensuring that data is used ethically and responsibly. Addressing algorithmic bias is another critical concern in the deployment of AI for environmental sustainability. Biases inherent in data sets or algorithmic design can lead to discriminatory outcomes and inequitable access to environmental resources and benefits. To mitigate bias, efforts to diversify data sets, incorporate diverse perspectives, and engage stakeholders from different backgrounds are essential. Additionally, ongoing monitoring and evaluation of AI models are necessary to identify and correct biases that may emerge during development and deployment. Ensuring fairness in AI models requires a multidisciplinary approach that involves collaboration among scientists, technologists, policymakers, and communities. By fostering interdisciplinary collaboration and knowledge-sharing, stakeholders can leverage diverse perspectives and expertise to develop AI solutions that address complex environmental challenges while promoting social equity and inclusion. This collaborative approach also facilitates the co-creation of solutions that reflect the needs and priorities of local communities, ensuring that AI technologies are responsive to the diverse contexts and concerns of stakeholders. Promoting ethical AI deployment requires careful consideration of the potential social and environmental impacts of AI technologies. Stakeholders must assess the risks and benefits of AI applications, considering factors such as data privacy, environmental justice, and community well-being. Transparent and participatory decision-making processes are essential to ensure that the deployment of AI technologies aligns with sustainability goals and respects the rights and dignity of affected communities.

The study is in confirmation with the findings of Li et al. (2022) who explored emphasized the integration of smart tourism principles with big data artificial intelligence. Technologies such as the Internet of Things, cloud computing, and intelligent data mining are suggested to enhance tourism experiences, industrial development, and administrative management. The role of artificial intelligence algorithms emphasizes their capacity to simulate and extend human intelligence activities. In conclusion, this comprehensive study has illuminated the multifaceted relationship between sustainable tourism practices, carbon emissions, and the transformative potential of Artificial Intelligence (AI). Through analysis and inquiry, we have unraveled the intricate dynamics shaping the tourism industry's environmental footprint and the role AI can play in fostering sustainability. Our findings highlight the need for concerted action to address the environmental challenges confronting the tourism sector. AI is a powerful tool capable of catalyzing change and driving innovation. Through advanced data analytics, predictive modeling, and smart decision-making, AI empowers stakeholders to navigate complex environmental issues with precision and foresight. However, AI innovations derive their value more when they work toward establishing sustainability (Brown, 2021). From optimizing energy consumption in tourism facilities to revolutionizing transportation systems and empowering travelers with sustainable choices, AI offers a pathway to a greener, more sustainable future for tourism. This study serves as a clarion call – a call to action for all stakeholders to work toward sustainable tourism unite in pursuit of a shared vision: a vision of tourism that not only enriches the lives of travelers but also safeguards the planet we call home. Together, let us embark on this journey toward a brighter, more sustainable future for tourism – a future where AI serves as a catalyst for change, and where sustainability is not just a goal, but a way of life.

References

ATAG. (2021). Balancing growth in connectivity with a comprehensive global air transport response to the climate emergency: A vision of net-zero aviation. *Waypoint 2050* (2nd ed.). ATAG.

Balsalobre-Lorente, D., Driha, O. M., Shahbaz, M., & Sinha, A. (2020). The effects of tourism and globalization over environmental degradation in developed countries. *Environmental Science and Pollution Research, 27*(7), 7130–7144.

Bilen, M., Yilanci, V., & Eryüzlü, H. (2017). Tourism development and economic growth: A panel Granger causality analysis in the frequency domain. *Current Issues in Tourism*, *20*(1), 27–32.

Brown, A. (2021). *How AI and a strong need for sustainability is affecting the travel industry's evolution: A discussion with Jason Fong*. Forbes. https://www.forbes.com/sites/anniebrown/2021/10/11/how-ai-and-a-strong-need-forsustainability-is-affecting-the-travel-industrys-evolution-by-jason-fong/?sh=7bce8a9913af

Destek, M. A., & Aydın, S. (2022). An empirical note on the tourism and sustainable development nexus. *Environmental Science and Pollution Research*, *29*, 34515–34527.

Friedlingstein, P., Jones, M. W., O'Sullivan, M., Andrew, R. M., Bakker, D. C. E., Hauck, J., Le Quéré, C., Peters, G. P., Peters, W., Pongratz, J., Sitch, S., Canadell, J. G., Ciais, P., Jackson, R. B., & Simone, R., … Zeng, J. (2022). Global carbon budget 2021. *Earth System Science Data*, *14*(4), 1917–2005.

Gossling, S., Balas, M., Mayer, M., & Sun, Y-Y. (2023). A review of tourism and climate change mitigation: The scales, scopes, stakeholders and strategies of carbon management. *Tourism Management*, *95*, 1–18.

Işik, C., Kasımatı, E., & Ongan, S. (2017). Analysing the causalities between economic growth, financial development, international trade, tourism expenditure and/on the CO2 emissions in Greece. *Energy Sources, Part B: Economics, Planning, and Policy*, *12*(7), 665–673.

Jain, P., & Jain, P. (2020). Understanding the concept of HR analytics. *International Journal on Emerging Technologies*, *11*(2), 644–652.

Kongbuamai, N., Zafar, M. W., Zaidi, S. A. H., & Liu, Y. (2020). Determinants of the ecological footprint in Thailand: The influences of tourism, trade openness, and population density. *Environmental Science and Pollution Research*, *27*(32), 40171–40186.

Le, T. H., & Nguyen, C. P. (2021). The impact of tourism on carbon dioxide emissions: Insights from 95. *Applied Economics*, *53*(2), 235–261.

Leal Filho, W., Ng, A.W., Sharifi, A., Janová, J., Özuyar, P. G., Hemani, C., Heyes, G., Njau, D., & Rampasso, I. (2023). Global tourism, climate change and energy sustainability: Assessing carbon reduction mitigating measures from the aviation industry. *Sustainability Science*, *18*, 983–996. https://doi.org/10.1007/s11625-022-01207-x

Lee, D. S., Fahey, D. W., Skowron, A., Allen, M. R., Burkhardt, U., Chen, Q., Doherty, S. J., Freeman, S. Forster, P. M., Fuglestvedt, J., Gettelman, A., De León, R. R., Lim, L. L., Lund, M. T., Millar, R. J., Owen, B., Penner, J. E., & Pitar, G., …, Wilcox, L. J. (2021). The contribution of global aviation to anthropogenic climate forcing for 2000 to 2018. *Atmospheric Environment*, *244*, 117834.

Lenzen, M., Sun, Y. Y., Faturay, F., Ting, Y. P., Geschke, A., & Malik, A. (2018). The carbon footprint of global tourism. *Nature Climate Change*, *8*(6), 522–528.

Li, D., Du, P., & He, H. (2022). Artificial intelligence-based sustainable development of smart heritage tourism. *Wireless Communications and Mobile Computing*, 5441170, 1–13. https://doi.org/10.1155/2022/5441170

Majid, G. M., Tussyadiah, I., Kim, Y. R., & Pal, A. (2023). Intelligent automation for sustainable tourism: A systematic review. *Journal of Sustainable Tourism*, *31*(11), 2421–2440. https://doi.org/10.1080/09669582.2023.2246681

Nosheen, M., Iqbal, J., & Khan, H. U. (2021). Analysing the linkage among CO2 emissions, economic growth, tourism, and energy consumption in the Asian economies. *Environmental Science and Pollution Research*, *28*(13), 16707–16719.

Rahman, A. R. A., Shaari, M. S., Masnan, F., & Esquivias, M. A. (2022). The impacts of energy use, tourism and foreign workers on CO2 emissions in Malaysia. *Sustainability*, *14*(4), 2461.

Ravinthirakumaran, K., & Ravinthirakumaran, N. (2022). Examining the relationship between tourism and CO_2 emissions: Evidence from APEC region. *Anatolia*, *34*(3), 306–320.

Sherafatian-Jahromi, R., Othman, M. S., Law, S. H., & Ismail, N. W. (2017). Tourism and CO_2 emissions nexus in Southeast Asia: New evidence from panel estimation. *Environment, Development and Sustainability, 19*(4), 1407–1423. https://doi.org/10.1007/s10668-016-9811-x

Shi, H., Li, X., Zhang, H., Liu, X., Li, T., & Zhong, Z. (2020). Global difference in the relationships between tourism, economic growth, CO_2 emissions, and primary energy consumption. *Current Issues in Tourism, 23*(9), 1122–1137.

UNWTO. (2022). *Transforming tourism for climate action.* https://www.unwto.org/sustainable-development/climate-action.

Voumik, L.C., Islam, M.A., & Nafi, S.M. (2023). Does tourism have an impact on carbon emissions in Asia? An application of fresh panel methodology. *Environment, Development and Sustainability*. https://doi.org/10.1007/s10668-023-03104-4.

Wang, C. M., & Wu, T. P. (2021). Does tourism promote or reduce environmental pollution? Evidence from major tourist arrival countries. *Environment, Development and Sustainability*. https://doi.org/10.1007/s10668-021-01568-w

WTTC–UNEP–UNFCCC. (2021). Driving climate action: A net zero roadmap for travel & tourism. https://wttc.org/Portals/0/Documents/Reports/2021/WTTC_Net_Zero_Roadmap.pdf

Wu, W., Su, Q., Li, C., Yan, C., & Gozgor, G. (2020). Urbanization, disasters, and tourism development: Evidence from RCEP countries. *Sustainability, 12*(3), 1–14.

Zhang, S., & Liu, X. (2019). The roles of international tourism and renewable energy in environment: New evidence from Asian countries. *Renewable Energy, 139*, 385–394.

Chapter 13

Sustainable Tourism: A Decision Support System Proposal for Policymaking

Raffaele Boffardi[a], *Carlotta D'Alessandro*[b] *and Luisa De Simone*[a]

[a]*Department of Human and Social Science, University of Naples L'Orientale, Italy*
[b]*Department of Economics, University of Messina, Italy*

Abstract

Over the last decades, tourism has become a fundamental engine of both developed and developing economies. A wide set of negative social, economic and environmental implications of tourism have been recognised along with the recognised positive effects. In this vein, sustainable tourism principles have been developed and widely accepted as a means to guarantee the long-term sustained and sustainable growth of local economies relying on tourism. The incorporation of these principles into the policymaking process requires the simultaneous consideration of a broad set of interests and goals, often conflicting among them. Thus, promoting sustainable tourism requires governments to adopt a strategic approach to planning, through goal-oriented, participatory and comprehensive governance mechanisms. Based on these, the following chapter is aimed at presenting a novel decision support system for policymakers, in which sustainable tourism principles and measures are accounted for, together with other purely economic and strategic issues. This strategic approach is operationalised through a multi-objective optimisation model. The main results show that the inclusion of sustainable tourism principles into the planning phase of public investments allows to boost the potentially positive effects of tourism on local economies, while not hampering the achievement of other policy goals, such as expenditure efficiency and consistency of investments.

Keywords: Sustainability; multicriteria techniques; public investments; tourism; planning

The Need for Sustainable Tourism in an Era of Global Climate Change: Pathway to a Greener Future, 269–295

doi:10.1108/978-1-83608-668-020241037

Introduction

During the last five decades, tourism has become a crucial component of the economies of several developing and developed countries, with reported tourist arrivals equal to 1.5 billion in 2019 (UNWTO, 2020). Before the pandemic crisis, the travel and tourism sector accounted for about 10% of the global GDP and 10% of all jobs (WTC, 2022). Notwithstanding the devastating implications for the sector deriving from the limitations imposed during the pandemic, in 2022, the sector still accounted for 7.6% of global GDP (WTC, 2022).

Indeed, tourism is widely considered an essential engine of economic growth, engaging broad groups of stakeholders in multidisciplinary activities (Chiosova, 2015). The benefits deriving from tourism involve an increase in employment of local communities due to its inherent nature of labour-intensive activity, a flywheel to private entrepreneurship and the incentive for public investments in facilities and infrastructures for tourists (Yanes et al., 2019; Zapata et al., 2013). Moreover, it enhances local economies thanks to purchases by tourists, thus being an important component of tourist destinations aggregate demand (Archer et al., 2012). The impacts of tourism-related activities spread far beyond these direct ones, with potential indirect gains in terms of opportunities generation for local communities, women and minorities (Medina-Muñoz et al., 2016). From a social and cultural point of view, tourism destinations enjoy a higher degree of both historic building heritage preservation and industrial areas and derelict land renovation (García et al., 2015). Moreover, it stimulates the improvement of local communities' skills through the undertaking of education and training, together with the cultivation of local traditions (Medina-Muñoz et al., 2016). Along with positive impacts, literature widely recognises the existence of potentially severe adverse economic, social and environmental effects of tourism-related activities (Frent, 2016; Kozak, 2017; Öztürk et al., 2015). The main negative economic implications of tourism activities involve local inflation effects deriving from both increasing demands and higher purchasing power of tourists, distortion of resources from local communities, and over-dependency of local economies on tourism activities (World Economic Forum (WEF), 2022). Similarly, intense tourism activities may generate negative environmental impacts, due to increased noise and air pollution, garbage generation, overcrowding and destruction of the natural landscape of tourist areas, among others (Ghulam Rabbany et al., 2013). Finally, improperly planned tourism activities may result in harsh conflicts between tourists and local communities, if the latter perceive excess social costs deriving from tourist presence or asymmetries in resource allocations (WEF, 2022). Other risks derive from seasonality and, mainly, from the potential loss of cultural identities because of the rapid massification of tourism in specific areas (González-García et al., 2022).

Therefore, as recognised by the Rio+20 outcome document *The Future We Want*, tourism activities have the potential to affect all three pillars of sustainable development both positively and negatively. This potential has led academics and policymakers to develop a novel approach towards tourism promotion: Sustainable Tourism (Yadav et al., 2018). Thus, tourism in the field of sustainable development is envisaged as a model in which tourism develops 'in such a manner and at such a scale that it remains viable over an indefinite period and does not degrade or alter the environment (human and physical) in which it exists to such a degree that it prohibits the successful development and wellbeing of other activities and processes' (Butler, 1993, p. 29). In this setting, tourism, environment, society and economy are unified and well-coordinated to achieve resource management and preservation, human well-being, environment and economic development (Dowling, 2020).

In this vein, the achievement of sustainable tourism requires coordination and balancing among different goals and different agents, who are all equally involved in the development of tourist activities. Moreover, a collaborative approach based on consultation and involvement of the actors, directly and indirectly interested in tourism activities – e.g., interest groups, institutions and private actors – is required to grant a proper balancing of the concerned interests (Ballantyne et al., 2009; Petrosillo et al., 2009).

Based on this acknowledgement, self-regulation and voluntariness in the tourism sector are not sufficient approaches (Bramwell, 1998). Proper planning, management and control practices are, thus, required to ensure that tourism acts as a catalyst for both the growth of local economies and the protection of social and natural capital (Sun et al., 2020). In this vein, decision-making becomes a central moment in the whole planning phase of tourism policies. Indeed, this process relies on the definition of a road set of actions to be evaluated based on variables that are related to other aspects than purely economic ones – e.g., social, environmental, and cultural (Fraiz et al., 2020).

Based on these premises, the present chapter aims to show that a goal-oriented, participatory and comprehensive approach to planning of tourism activities might allow them to maximise their contribution to the sustainable development of local areas. More in detail, it aims to present a decision support system for supporting policymakers in the inclusion of sustainable tourism criteria in the process of policy formulation, by informing them of the potential impacts of the planned tourism activities on the different aspects of sustainability. This strategic approach is operationalised through a multi-objective optimisation model. It supports the selection of investments to be implemented in tourist areas by identifying 'best compromise solutions' in which economic, social and environmental benefits of tourism investments are maximised. In doing so, it does not only consider the interests of public administrations (mainly, budget constraints), but it balances them with the goals and interests of the communities involved (Yasarata et al., 2010). Therefore, it shows different alternative solutions in which the different features of sustainable development are maximised, thus showing the advantages for the concerned stakeholders implementing specific types of policy actions.

The model presents some operative advantages. First, it enhances the active involvement of the public agents, who are required to set the constraints to be included in the model and constant interaction with both analysts and other stakeholders involved in tourism policymaking. Second, it is an iterative instrument, allowing to generation of different scenarios according to the novel constraints included time by time. In doing so, it avoids setting ex-ante priorities among the different goals underlying the policymaking process, which are often difficult or not practicable.

The Background of Sustainable Tourism: Definitions, Principles and Planning

Following the definition of Sustainable Development on behalf of the 1987 Brundtland Report, both literature and policymakers have started including the principles of SD in tourism activities definition (Higgins-Desbiolles, 2018).

The first definitions of the topic were provided by Bramwell and Lane (1993), who are considered pioneers of the inclusion of sustainability concepts in the tourism industry, as well as in the tourism policymaking process (Zolfani et al., 2015). In their formulation, sustainable tourism could be the means to deal with the main impacts on the environment, local cultures and traditions deriving from tourism activities. Subsequently, Butler (1999) distinguished between 'sustainable tourism' and the implementation of tourism activities in the framework of sustainable development policies. The former was defined as a type of 'tourism which is in a form which can maintain its viability in an area for an indefinite period of time' (Butler, 1999, p. 36). The latter engages in a different behaviour, which can be defined as tourism developed in a way that does not over-exploit or alter natural, human and physical resources and to such an extent that it is not impossible to successfully develop other natural, human and physical processes (Butler, 1999). According to Zolfani et al. (2015), a progressive development of the topic lead literature to identify sustainable tourism as a source of positive benefits to environment, society and economy, rather than as a solution to the negative externalities of mass tourism.

From a policy perspective, the first definition of sustainable tourism was introduced in 1996 by the United Nations World Tourism Organization (UNWTO) claiming that it is 'tourism which leads to management of all areas, in such a way, that the economic, social and environmental needs are being fulfilled with the cultural integration, ecological processes, biodiversity and supporting the development of societies'. In 2001, UNWTO added that sustainable tourism allows to satisfy the necessities of tourists and local communities, while simultaneously creating, keeping and boosting opportunities for future tourists and locals. As it results from the above definitions, sustainable tourism involves policy frameworks that deal not only with specific tourist activities but also with the creation of management schemes for all typologies of tourism (Streimikiene et al., 2021).

According to Janusz and Bajdor (2013), based on the definitions provided internationally, sustainable planning should account for different – often conflicting – goals, responding to the necessity of granting environmental, social and social sustainability of tourism activities.

First, on the side of environmental sustainability, resource exploitation should be minimised and optimised, and environmental management schemes and biodiversity protection activities should be implemented by both private and public actors involved. These actions allow for the optimisation of the use of natural resources while keeping intact the fundamental processes characterising the local natural heritage and biodiversity. Second, tourism activities should enable lasting and effective economic growth processes, enhancing the well-being of communities in terms of income generation and decent work conditions creation. Moreover, these benefits should be fairly distributed across the concerned communities, thus contributing to the general goal of poverty reduction. Third, all activities should be implemented with due regard for the cultural and traditional heritage of local communities and to foster the empowerment of women and minorities (Medina-Muñoz et al., 2016).

Finally, all these priorities need to be balanced with each other, thus granting the achievement of the highest social, environmental and economic returns for all actors involved and affected by tourism activities (Buckley, 2009). This requires a cooperative relationship among the various stakeholders in the process of sustainable tourism development, based on a constant balancing of their goals and interests in view of emphasising the fairness of tourism-related opportunities (Guo et al., 2019). This framework is summarised in Fig. 13.1.

Recently, the WEF has introduced a set of guidelines to guide a broad group of stakeholders – businesses, governments, destinations, tourists and local communities – in the definition of sustainable tourism practices (Table 13.1; WEF, 2022). The guidelines are articulated along ten principles, strictly interconnecting

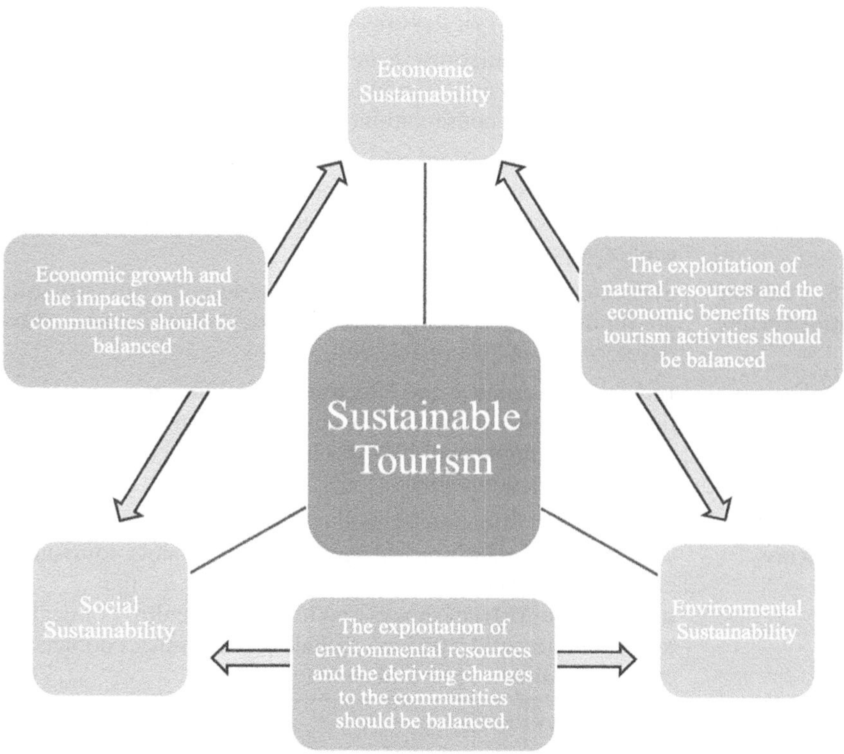

Fig. 13.1. Key Aspects of Sustainable Tourism. *Source*: Authors elaboration on Janusz and Bajdor (2013).

Table 13.1. WEF's Principles of Sustainable Tourism.

Principles	Definition
P.1: Certify and monitor	Sustainable destinations implement standards, certifications and a balanced set of evidence-based metrics to enable the transition to a more sustainable, inclusive and resilient sector.
P.2: Cultivate the workforce	Sustainable destinations nurture and retain a motivated and trained workforce to deliver services at the quality and efficiency required while ensuring workers share the values and purpose associated with pursuing sustainable and resilient destinations.
P.3: Prioritise communities	Sustainable destinations prioritise and engage communities in planning and stewarding a tourism destination.
P.4: Align visitors	Sustainable destinations adequately manage visitors in line with their carrying capacity and enable visitors to choose and act sustainably through marketing, education, management and monitoring.
P.5: Protect heritage	Sustainable destinations protect the culture and heritage of tourism destinations in a thoughtful way that ensures the preservation of their historic, cultural and natural heritage.
P.6: Protect nature	Sustainable destinations recognise that nature underpins healthy societies and economies and encourage stakeholders to prioritise nature conservation, management and restoration.
P.7: Produce and consume responsibly	Sustainable destinations design and implement strategies to promote responsible consumption and production by all stakeholders and establish mechanisms to promote circular economy pathways in tourism.
P.8: Calibrate infrastructure	Strategic planning and investment in sustainable infrastructure are fundamental enabling factors for the balanced and inclusive development of sustainable destinations.
P.9: Govern effectively	Inclusive governance and effective regulations are essential factors to guide, enable and incentivise tourism sector players and users to implement business practices and behaviours promoting sustainability.
P.10: Embed resilience	Sustainable destinations embed resilience as a critical component of their tourism agenda. Governments, tourism enterprises and communities do so to prepare for the inevitable short-term shocks and long-term health, political and environmental stressors, including pandemics, conflicts, extreme weather events and climate change.

Source: Authors elaboration on WEF (2022).

with the UN Sustainable Development Goals (Streimikiene & Korneeva, 2020). They aim to provide a demonstration that (i) adequate management and governance schemes of destinations, (ii) redesign of policy interventions in a sustainable manner, (iii) creation of a common ground for the participation of stakeholders, and (iv) implementation of proper evaluation schemes are necessary to transform sustainable tourism into a flywheel for sustained and sustainable development and resilience.

As can be noticed from the above discussion, achieving sustainability in tourism policymaking requires a simultaneous consideration – and balancing – of both the interests and goals of several stakeholders, which must be accounted for in all steps of sustainable tourism planning (García-Melón et al., 2012). This not only includes the different levels of government and the industries directly involved in tourist activities (Yasarata et al., 2010). Embedding social, environmental and economic aspects into tourism policymaking with cultural and political aspects can only be achieved by fostering the participation of all parties whose lives and activities are – directly or indirectly – influenced by tourism (Hall, 1994; Torres-Delgado & Palomeque, 2012). This implies a growing consideration of the territorial dimension, in which all stakeholders must consider and promote the dissemination of sustainability through their actions on a territorial scale (Blancas et al., 2010; Camagni, 2017; Landforf, 2009). Subsequently, promoting sustainable tourism policies requires the creation of proper governance mechanisms, through either multidisciplinary or multi-object planning (Arbolino et al., 2020; Castellani & Sala, 2010).

Methodological Approach to Sustainable Tourism Planning

Tourism policy planning is a very complex process, especially when economic, social and environmental dimensions of sustainability are considered as policy drivers. Several issues might influence and reduce the effectiveness of both planning and implementation of a policy promoting sustainable tourism: (i) the number of actors involved, (ii) the presence of conflicting goals, and (iii) the need for resource rationalisation (Önder et al., 2013).

Literature has elaborated several methodological approaches to deal with the complexity of the policymaking process. Among others, multicriteria techniques present several operational arrangements allowing to solve some inherent problems in the planning phase, such as the possibility of (i) including different targets in the process of policy alternative selections – which can be evaluated by resorting to their original units of measurement, (ii) easing the comparison among various alternative, thus (iii) presenting different alternatives to the policymakers who can make their own decision according to their judgements (Soltani et al., 2015). Different multicriteria techniques have been employed for tourism policy planning, such as the Analytic Hierarchic Process (AHP), the Analytic Network Process (ANP; García-Melón et al., 2012), the Technique for Order of Preference by Similarity to Ideal Solution (TOPSIS; Önder et al., 2013; Ulkhaq et al., 2016) or the ELECTRE-the techniques (Michailidou et al., 2016).

However, following Soltani et al. (2015) and Arbolino et al. (2020), these techniques present several weaknesses that might hamper both the effectiveness and robustness of sustainable tourism planning. First, these techniques simplify the decision problems, thus risking neglecting some relevant information (i.e., oversimplification of the problem). Second, most of these techniques require decision-makers to state weights ex-ante, with a subsequent risk of biases related to subjectivity. Finally, they assume the technical feasibility of each alternative and, accordingly, rank them: they do not allow to present the 'best' solution.

In this research, we present a different approach, based on objective optimisation techniques, integrated with a novel set of metrics to evaluate sustainable tourism, which allows us to overcome several shortcomings related to the most well-known multicriteria techniques.

In order to face the multidimensionality of sustainable tourism, we propose a methodology that is useful when the transformation of a multi-objective decision problem into a single objective one is required. We implement this transformation through the identification of an aggregate objective which – following the statement of various constraints – has to be maximised or minimised. The main advantage of this approach compared to other techniques can be identified in the possibility of elaborating different scenarios, computed by looking at the potential impacts on the various objectives stated.

Specifically, by applying an optimisation technique, we elaborate different scenarios respecting a list of constraints applied to the different targets. By exploiting a specific calculation system, the different scenarios – showing the impacts on each objective stated by the policymaker – can be compared. In doing so, the decisionmaker can easily use them to provide its own 'value judgments'.

In this vein, this technique ensures both an interaction and an iteration between analysts and decision-makers during the evaluation and selection process. Indeed, the decision-maker identifies its targets; then the analyst computes the impacts of each alternative by applying the model. The final decision is based on the results allowing to achieve the stated priorities. The process can lead to (i) a Pareto-optimal solution – in which specific objective functions are maximised or minimised; (ii) a non-satisfactory solution, for which other iterations are required by including new constraints. Thus, each new iteration generates new scenarios requiring further testing.

The calculation of each solution is based on the simplex algorithm (Dantzig & Thapa, 2003), which applies an integer number programming with the association of each decision variable with a binary variable. This 'existence' variable takes value one if the alternative is selected and zero, otherwise. Thus, a set of projects – namely those maximising the objective function and simultaneously meeting all the constraints – is identified. Subsequently, by including further constraints the same problem can be solved in different ways allowing the identification of a different solution (i.e., a new list of projects to be funded). This iterative approach supports the subsequent inclusion of several targets into the evaluation and selection process of activities to be funded and implemented.

Different steps are required to achieve the best compromise solution:

1. Definition of the decision problem.
2. Definition of significant decision criteria and indicators.
3. Construction of the model.
4. Identification of the alternatives, evaluation and selection.

The Decision Problem

The first step is defining the decision problem. In our case, the main issue is the selection of tourism initiatives to be implemented that are able to both maximise the utility of stakeholders involved in (both public and private actors) and minimise the negative impacts on both territory and community (to ensure high levels of sustainability). In addition, the selected projects have to comply with a set of exogenous constraints, such as budget constraints.

So, we assume that the decision-maker uses a call for tender to collect a set of design alternatives, which will be evaluated through a set of selection criteria, divided into specific objectives, and measured through specific indicators.

Definition of Significant Decision Criteria and Indicators

Each alternative has to be analysed according to a set of criteria, whose selection needs to be consistent with both the goals stated by the policymakers and the definitions of sustainable tourism. For doing so, we identified eight groups of criteria, further grouped into four dimensions.

The first dimension includes criteria evaluating project consistency and applicant entity eligibility (Table 13.2). Together with the adherence of the proposal to the requirements of the concerned administration, this set of criteria evaluates the degree of public stakeholder participation in the planning process and the level of cooperation between public and private entities expected. Co-participation of different typologies of stakeholders is a fundamental driver of the selection process, thus complying with the goal of promoting an inclusive governance system, as stated by WEF (2022) and, overall, by the Sustainable tourism literature.

For the evaluation of sustainability, we followed the partition into three blocks (economic, environmental and social dimensions) and, subsequently, selected the most appropriate indicators from the lists presented in the European Tourism Indicators System for Sustainable Destination Management (ETIS; European Commission, 2016) and by the UNWTO (2004).

The second dimension refers to the environmental dimension of sustainability. As previously stated, to be sustainable, tourism initiatives should be implemented by minimising the impacts on the environment through the reduction of natural resource use and waste prevention and/or recycling (Obersteiner et al., 2021). In terms of natural resource exploitation, the model includes indicators for assessing the actions implemented for granting proper energy management, water availability and conservation and the implementation of waste management practices in the tourism projects involved. Finally, the provision of environmental

Table 13.2. Consistency Criteria.

Dimension	Criteria	Objective	Weight	Level/Indicator	Score
Consistency of proposal and of applicant entity Max 30 score	Project consistency Max 20	Consistency with priorities and objectives stated by programming documents	10	High (fully consistent)	1
				Medium (partially consistent)	0.7
				Low (consistent with one objective)	0.3
		Coherence	5	High	1
				Low	0.5
		Integration with (realised and none) public initiatives for promoting local tourism and communicating	5	Integration with more than one initiative	1
				Integration with only one	0.5
				No integration	0
	Features of applicant entity Max 10	Investments for promoting tourism in the last 3 years	3	> € 300.000,00	1
				<= € 300.000,00	0.5
		N. of privately funded projects for promoting tourism implemented in the last three years	2	> 5	1
				<= 5	0.5

	Degree of public stakeholder participation in the planning process	3	Involvement of more than four institutions	1
			Between two and three institutions	0.7
			Involvement of one institution	0.3
			No institutions involved	0
	Public–private cooperation	2	Involvement of more than four public and private bodies	1
			Between 2 and 3 entities involved	0.7
			One entity involved	0.3
			No involvement of other public or private entity	0

Source: Author's elaboration.

administrative units and the adoption of management certification on behalf of the involved stakeholders are also evaluated features.

The selection of indicators has been performed accordingly (Table 13.3). They assess if the proposer aims to:

- Reduce the consumption of energy – e.g., by applying energy-saving techniques – and enhance renewable energy sources exploitation.
- Minimise water exploitation and its reuse.
- Promote campaigns aimed at increasing awareness, reducing waste consumption and incentivising separate collection.
- Create environmental management units and adopt quality certification.

The third dimension refers to the economic sustainability of the proposed initiatives. Indeed, sustainable tourism should aim at enhancing local economies through the fair distribution of resources, the creation of stable and decent jobs and the mitigation of risks derived from seasonality (González-García et al., 2022). Based on these criteria, we selected indicators that allow to both control the impact of proposed tourism-related projects on the degree of competitiveness of the area where the initiatives should be implemented and to evaluate the economic impacts generated (Table 13.4). Moreover, in the optimisation model, we included a further criterion based on whom at least 50% of total financial support available should be granted to areas where tourism activities are less developed. This criterion should support the development of tourism initiatives in those areas with high potential (e.g., due to cultural or natural heritage availability), thus responding to the objective of a fair territorial distribution of tourism-related resources and advantages. Therefore, the selected indicators aim to assess the following:

- the potential generation of tourism demand on the territory;
- the level of seasonality, whose reduction is a main target;
- the direct and indirect impacts of the proposed tourism activities (i.e., beds occupancy rate and employment).

The fourth dimension is addressed to measure the social dimension of sustainability, in terms of impacts on local communities. Thus, the propensity towards the engagement of locals in tourism activities is assessed as well as the existence of intermodal transportation systems to make destination accessible. The provision of actions paying special attention to people with disabilities is also considered. Indeed, literature and policymakers have widely acknowledged that the creation of equal opportunities and the consideration of human and social aspects is fundamental for granting the ethical aspects of tourism, as well as its social sustainability (Happ & Bolla, 2022). Therefore, in the model, the implementation of actions supporting the involvement of people who are usually marginalised is awarded as a means for creating equal opportunities, i.e., accessible tourist practices (Biddulph & Scheyvens, 2018; Qiao et al., 2022). Other criteria refer to communication skills, through which reach as many people as possible and involve them in the project. Table 13.5 presents the considered indicators.

Table 13.3. Environmental Sustainability Criteria.

Dimension	Criteria	Objective	ETIS/WTO Issue and/or Indicator	Weight	Level/Indicator	Score
Environmental sustainability max 30 scores	Energy management max 7 scores	Reduction of energy consumption, application of energy-saving techniques	Businesses participating in energy conservation programs, or applying energy-saving policies and techniques (percentage)	3	High (adoption of more than one energy-efficient equipment)	1
					Low (adoption of one energy-efficient equipment)	0.5
		Enhancement of renewable energy sources exploitation	Energy consumption from renewable resources (at destinations, establishments) (% of total energy consumption)	4	High (more than 50% of energy exploited is expected to be generated from RES)	1
					Medium (between 50% and 20%)	0.7
					Low (less than 20% from RES)	0.3
	Water availability and conservation max 5 scores	Efficient water consumption	Water use (total volume consumed and litres per tourist per day)	2	High (more than one equipment for reducing water consumption)	1
					Low (one or no equipment for reducing water consumption)	0.5
		Possibility of reuse of water	Water saving (% reduced, recaptured or recycled)	3	Yes	1
					No	0

(*Continued*)

Table 13.3. (*Continued*)

Dimension	Criteria	Objective	ETIS/WTO Issue and/or Indicator	Weight	Level/Indicator	Score
	Solid waste management max 10 scores	Reduction of waste production	Waste reduction	3	High (initiatives for promoting waste reduction)	1
					Low (no initiatives)	0
		Potential percentage of waste recycled	Volume of waste recycled (m3) / Total volume of waste (m3)	4	High (more than 50% of waste expected to be recycled)	1
					Medium (between 20% and 50% of waste expected to be recycled)	0.7
					Low (less than 20% of waste expected to be recycled)	0.3
		Installation of domestic composting facilities	Number of recycling bins for plastic, aluminium, paper products, organic waste	3	Yes	1
					No	0
	Environmental management max 8 scores	Creation of an environmental management unit	Tourism managers with environmental training	4	Yes	1
					No	0
		Quality certification	Presence of a certified Environmental Management System	4	Yes	1
					No	0

Source: Authors elaboration.

Table 13.4. Economic Sustainability Criteria.

Dimension	Criteria	Objective	ETIS/WTO issue and/or indicator	Weight	Level/Indicator	Score
Economic sustainability max 20 scores	Economic benefits max 20 scores	Total Potential tourist demand	Expected number of turist arrivals	6	More than 3,000 visitors	1
					Less than 3,000 visitors	0.5
		Level of seasonality	Ratio of number of tourists in peak month to lowest month	4	High (the activity takes place in peak months or season)	0
					Low (the activity takes place in off-peak months or season)	1
		Occupancy rates in accommodation establishments	Tourism revenue: Occupancy rates in accommodation establishments	4	More than 50%	1
					Less than 50%	0.6
					0%	0
		Potential (direct and indirect) employment	Expected employment	6	High (more than 100 employees)	1
					Medium (between 30 and 100 employees)	0.7
					Low (less than 30 employees)	0.3

Source: Authors elaboration.

Table 13.5. Social Sustainability Criteria.

Dimension	Criteria	Objective	ETIS/WTO issue and/or indicator	Weight	Level/Indicator	Score
Social sustainability max 20 scores	Effects on local community max 20 scores	Existence of engagement activities for local community	Degree of involvement of local citizens and business	4	Yes	1
					No	0
		Use of avenues/channels to promote sustainable tourism (e.g., audiovisual and printed media, events, Internet)	Use of avenues/channels for promoting sustainable tourism	4	Yes	1
					No	0
		Number of information points in the destination	Number of places in the destination where information is available	4	High (>2)	1
					Low (<=2)	0
		Existence of disabled-friendly policy	Existence of disabled-friendly policy	4	Yes	1
					No	0
		Existence of integrated transportation mode	Existence of multi-modal platforms (yes / no)	4	Yes	1
					No	0

Source: Authors elaboration.

The Model

For testing our approach, twenty projects aimed at implementing tourism-related activities are considered as alternatives, among which the decision-maker should make their selection. Each alternative is labelled as A_j where j = 1,2,...,20. For each criterion determined, the scores obtained by each alternative is computed and reported in the input matrix, which is partially reported in Table 13.6.

These scores are computed based on the following:

$A_j = \Sigma_i w_{ij} * r_{ij}$

where:

- Aj is the j-th alternative, j = 1, 2, ... n;
- w_{ij} is the weight assigned to each objective (as in Tables 13.2–13.5);
- r_{ij} is the value assigned to each indicator assessing the objective to which it refers.

Moreover, in the matrix (Table 13.6), the financial data of each proposal are showed, such as: the investment cost and the public contribution to it (both in absolute terms and in the percentage of the total amount – max. 80%). This latter data – public aid – is described according to the territorial destination: whether it is an area in the stage of tourist congestion or an area with potential tourism growth.

Based on the matrix data, through the optimisation model, it is possible to compute and compare various scenarios, with different formulations and related solutions.

Scenario 0

Scenario 0 is based on the selection procedure following the traditional 'weighted sum', but also considering the budgetary constraint, represented by the total financial amount available to the decision-maker (1 million euro). Thus, the alternatives are chosen considering both their high scores and the scarcity of financial resources, which requires their efficient use. In this vein, the multi-objective problem is treated as a mono-objective one, by maximising the aggregate score, under a list of constraints – exogenously determined. The function to be maximised is given by:

$$\max f(x) = \sum_j TS_j * x_j$$

in which:

- x is the vector of decision variables;
- TS is the total score achieved by the j-th alternative;
- x_j is a binary variable of existence [0,1]: if the alternative is excluded from the 'optimum', x_j takes value 0 and the product is zero. Subsequently, it is excluded from the computation of the Objective function.

Table 13.6. Input Matrix (Partial).

Objectives			A1	A2	Aj	A19	A20
C1.1	*max*	Consistency with priorities and objectives stated by programming documents	10	3	...	7	7
C1.2	*max*	Coherence	3	3	...	5	5
C1.3	*max*	Integration with (realised and none) public initiatives for promoting local tourism and communicating	5	0	...	3	5
C2.1	*max*	Investments for promoting tourism in the last 3 years	3	2	...	3	3
C2.2	*max*	No of privately funded projects for promoting tourism implemented in the last three years	2	1	...	1	2
C2.3	*max*	Degree of public stakeholder participation in the planning process	2	1	...	0	2
C2.4	*max*	Level of cooperation with public and/or private entities	1	0	...	1	1
C3.1	*max*	Reduction of Energy consumption, application of energy-saving techniques	3	2	...	3	2
C3.2	*max*	Enhancement of renewable energy sources exploitation	3	1	...	1	1
C4.1	*max*	Efficient water consumption	1	1	...	1	1
C4.2	*max*	Possibility to reuse water	0	0	...	0	0
C5.1	*max*	Reduction of Waste production	0	3	...	3	3
C5.2	*max*	Potential percentage of waste recycled	3	4	...	2	3
C5.3	*max*	Installation of domestic composting facilities	0	0	...	0	0
C6.1	*max*	Creation of an environmental management unit	0	4	...	0	0
C6.2	*max*	Quality certification	4	4	...	0	4

C7.1	*max*	Existence of engagement activities for local community	4	4	…	4	0
C7.2	*max*	Use of avenues/channels to promote sustainable tourism (e.g., audio-visual and printed media, events, Internet)	4	4	…	4	4
C7.3	*max*	Number of places in the destination where information is available	0	0	…	0	0
C7.4	*max*	Existence of disabled-friendly policy	4	0	…	4	4
C7.5	*max*	Existence of integrated transportation mode	0	4	…	0	0
C8.1	*max*	Potential tourist demand	6	3	…	3	3
C8.2	*max*	Level of seasonality	0	0	…	0	0
C8.3	*max*	Occupancy rates in accommodation establishments	2	2	…	4	0
C8.4	*max*	Potential (direct and indirect) employment	2	4	…	2	6
C9	*max*	Total Score	62	49	…	50	56
C10	*max*	Environmental sustainability	14	19	…	10	14
C11	*max*	Social sustainability	12	12	…	12	8
C12	*max*	Economic sustainability	10	10	…	9	9
C13	*min*	Investment costs (€)	65.200	46.890	…	105.000	23.650
C14	*min*	Public aid (€)	48.900	37.512	…	78.750	18.920
C15	*min*	Public aid %	75%	80%	…	75%	80%
C16	*min*	Public aid – developed tourism destination (€)	48.900,0	37.512,0	…		
C17	*max*	Public aid – in growth tourism destination (€)	-	-	…	78.750,0	18.920,0

Source: Authors elaboration.

Subject at:
$\Sigma\, PA_j * x_j \leq €1,000,000.00$
where:

- PAj means the requested public financial aid for each j-th tourist project alternatives;
- x_j is a binary variable of existence [0,1].

Scenario 1

We constructed Scenario 1 after maximising the score referring to the dimension 'Consistency of proposal and of applicant entity' and, at the same time, meeting the constraint that the total allocation of resources on behalf of policymakers have to not exceed total available funds (i.e., budgetary constraint). For building this scenario, another constraint has been introduced in the model: at least to 50% of total available budget has to be allocated to areas with a high tourist potential that is already unexploited, the so called 'in-growth tourism destination'. This constraint follows one of the main stated policy objectives: encouraging the development in those areas where the present artistic and cultural heritage could generate significant socioeconomic benefits for local community. So, the model has been formulated as follows:

$$\max f(x) = \sum_j CS_j * x_j$$

Subject at:
$\Sigma\, PA_j * x_j \leq €1,000,000.00$
$\Sigma\, InGrPA_j * x_j \geq 50\%$ PA
where:

- CS is the 'Consistency of proposal and of applicant entity' score achieved by the j-th alternative;
- PAj means the requested public financial aid for each j-th tourist project alternatives;
- *InGrPA* means the requested public financial aid for each j-th tourist project alternatives in-growth tourism destination.

Scenario 2

If we consider sustainability, the alternative selection process should focus on all sustainability issues (environmental, social and economic aspects), so that only tourist activity generating low negative externalities could be selected. Moreover, for the elaboration of Scenario 2, we maintain the financial constraint that should ensure that resources are partially directed to those areas with an unexploited cultural heritage. So, the 'Environmental Sustainability Score' is chosen as Objective

Function to be maximised, conditioned to the financial and policy constraints already defined. Thus, the applied model is the following:

$$\max f(x) = \sum_j EnvS_j * x_j$$

Subject at:
$\sum PA_j * x_j \leq$ €1,000,000.00
$\sum InGrPA_j * x_j \geq 50\%$ PA
where:

- *EnvS* is the Environmental Sustainability score achieved by the j-th alternative;
- PAj means the requested public financial aid for each j-th tourist project alternatives;
- *IngrPA* means the requested public financial aid for each j-th tourist project alternatives in the growth tourism destination.

Tourism Project Selection Through the Optimisation Model

For the current analysis, three scenarios have been built and compared with each other, in order to identify the best solution:

- Scenario 0 – in which the objective function to be maximised considered both criteria related to consistency and the total sustainability score (from an economic, social and environmental point of view). Therefore, it aims at maximising the total score (comprehending both consistency and sustainability scores).
- Scenario 1 – in which only aspects related to the consistency of the project to the objectives of the call to the priorities set by the policymakers (i.e., tourism development) and the features of the applicant entity are assessed. Indeed, Scenario 1 is built by considering a policymaking process in which sustainability issues are not considered relevant by the decision-makers.
- Scenario 2 – in which the impacts of tourism activities on the environment are considered as the priority of policymakers. In doing so, the maximisation of the sustainability score is the main target.

In Scenario 0, once the Objective Function has been described, it is possible to identify the 'best compromise' solution, balancing between financial, socioeconomic and environmental sustainability and consistency constraints (Table 13.7). This scenario generates a first hypothetical allocation of the available resources: 17 out of 20 project proposals are selected resulting in the expenditure of 97.58% of the available resources.

Total Score – whose maximisation was the aim underlying the construction of this scenario – is equal to 966.5, while environmental sustainability, social sustainability and economic sustainability scores are, on average, equal to 14.5, 11.3 and 11.7, respectively. Based on this scenario, around 51% of resources are allocated

Table 13.7. Scenarios from Optimisation Model.

Scenario:	Scenario 0			Scenario 1			Scenario 2		
Objective Function	**Max Total Score**			**Max Consistency Score**			**Max Sustainability Score**		
		Constraints	**Impact**		**Constraints**	**Impact**		**Constraints**	**Impact**
Total score			966.5			945.6			890.5
Environmental sustainability score (average)			14.47			14.68			16.57
Social sustainability score (average)			11.29			11.25			11.73
Economic sustainability score (average)			11.73			12.16			12.17
N. Projects			17			16			15
N. Projects – developed tourism destination			9			7			7
N. Projects – in growth tourism destination			8			9			8
Public Budget (€)	<=	1,000,000.00 €	975,803.30 €	<=	1,000,000,00 €	987,760.90 €	<=	1,000,000,00 €	995,491.10 €
Public Budget – developed tourism destination (€)			466,806.90 €			359,596.50 €			456,333.10 €
Public Budget – in growth tourism destination (€)			508,996.40 €	>=	500,000.00 €	628,164.40 €	>=	500,000.00 €	539,158.00 €
% Public Budget			97.58%			98.78%			99.55%
% Public Budget – developed tourism destination			46.68%			35.96%			45.63%
% Public Budget – in growth tourism destination			50.90%			62.82%			53.92%

Source: Authors elaboration.

to underdeveloped areas (so-called 'in-growth tourism destinations'), thus contributing to the goal of granting a fair territorial distribution of resources.

In Scenario 1 – whose objective function to be maximised is the Consistency Score – a lower total score is achieved, compared to Scenario 0 (945.6 versus 966.5). Overall, more financial resources are used – 98.78% in comparison with 97.58% – ensuring greater financial efficiency, even if fewer projects are selected (16 projects versus 17 of Scenario 0). However, the set of selected projects achieves a social sustainability score lower than in the previous Scenario, while the levels of environmental and economic dimensions of sustainability are almost equal. Overall, a sensibly larger share of resources is devoted to in-growth destinations (62.8%), rather than in both Scenarios 0 and 2.

Finally, Scenario 2 shows improved results in comparison with the previous scenario, and, obviously with Scenario 0. Firstly, a greater financial amount is allocated to fund tourism initiatives – 99.55% of the available budget, more than in Scenarios 0 and 1. Moreover, a higher share of resources (53.9%) is allocated to projects to be implemented in in-growth tourist areas rather than in already developed ones (45.63%). Notwithstanding the reduction of the share of resources allocated in underdeveloped areas in comparison to Scenario 1, the proposed allocation still contributes to the goal of sustaining the fair distribution of resources across the concerned territories. Therefore, the alternatives selected (even if fewer than the previous scenarios) can ensure both a more efficient resource management and a limitation of the effects on local communities. Compared to the previous scenarios analysed, higher scores are achieved in terms of the scores evaluating the social, economic and environmental sustainability of the selected set of projects. First, in this scenario, the average environmental score value is substantially higher, thus implying a clear reduction of the impacts on the natural environment of the areas where tourism-related initiatives are going to be implemented. Similarly, slightly higher social and economic sustainability scores are achieved. This result implies that a tourism policy must necessarily encompass the sustainable dimension as it allows to incentivise cultural and tourist promotion by reducing the pressure on the physical environment (Kurt Konakoglu et al., 2019).

Discussion and Conclusion

Over the last decades, tourism has become a relevant social, cultural, economic and environmental phenomenon (Muresan et al., 2016). Despite the unprecedented negative effects of the Coronavirus pandemic on the activities related to the tourism sector – mainly due to the limitations imposed to movement and social contacts – in 2022, it still generated around 7.6% of global GDP (WTC, 2022). As a matter of fact, tourism has become a non-negligible component of economic growth and national aggregate demands (Chiosova, 2015).

However, together with a wide set of potential positive implications, tourism is deemed to generate relevant negative consequences on the economic, social and environmental sides. Among other, mass tourism and mismanaged tourist flows are likely to disrupt economic activities of tourism destination, by generating inflation, resource distortion, overdependency and seasonality. Similarly, from a social point of view, the uncontrolled massification of tourism threatens local communities, with the risk of local cultural and traditional heritage. Finally, huge negative environmental implications of tourism have been globally reported, including increased noise and air pollution, garbage generation, overcrowding and destruction of the natural landscape of tourist areas, among others. A potential solution to these issues is presented by the implementation of the principles of Sustainable Tourism, implying the embedment of sustainable development principles in both design and implementation of tourism-related activities.

In this framework, the role of all actors involved in the planning phase of tourism is fundamental, to include sustainability in their tourism strategies and action plans (Budeanu et al., 2016). However, three main issues risk damaging the achievement of sustainable tourism goals: (i) the high number of stakeholders involved, (ii) their conflicting goals and (iii) the necessity of rationalising resource expenditure. Involving a plurality of subjects in the sustainable tourism programming phases complicates the decision-making process, often leading to a non-optimal allocation of resources (Önder et al., 2013). The existence of these criticisms might be easily overcome by adopting decision-support instruments for sustainable tourism policymaking, based on multicriteria techniques.

Therefore, the main objective of this chapter has been to construct a decision support system for tourism policymakers, allowing them to include as many as possible sustainable tourism criteria in the process of policy formulation and informing them of the potential impacts of the planned tourism activities. It relies on a multi-objective optimisation model, which operationalised a goal-oriented, participatory and comprehensive approach to tourism activities planning, responding to the principles of Sustainable tourism presented by the main international organisations.

The model was tested on a hypothetical call for tender presented by a public administration willing to promote tourism in a well-defined territorial area. Three different scenarios have been presented: the former has been elaborated by maximising the total score achieved by a set of projects proposed by different entities; the second one only considers the consistency of the project to the goals of the administration; the third one includes all the principles of economic, social and environmental sustainability analysed in the previous sections and aims at maximising them. The main evidence provided by our analysis is that the maximisation of sustainability scores – subject to a set of exogenous constraints – still allows policymakers to achieve other desirable goals, such as the efficiency of public expenditure and the development of in-growth areas. In spite of the achievement of reduced total scores and the funding of a smaller number of projects, the concerned administration would be able to reduce a wider set of negative implications on the economy, society and the environment deriving from tourism policies.

In this chapter, we exemplified a possible application of the presented methodology. Indeed, this planning instrument might be applied to different contexts (i.e., different policies, different targets and constraints) following a simple modification of the list of variables considered and a review of the constraints to be applied.

Apart from the achievement of a preferable solution in terms of the concerned example, the proposed approach for incorporating sustainable tourism principles into policy planning presents several advantages. First, the model relies on an iterative approach, in which the results provided by each iteration allow the generation of further scenarios, thus increasing the knowledge and awareness of the policymaker on the decision problem. Moreover, it implies an interactive approach, in which the decision-makers constantly collaborate with a wide set of actors, especially concerning (i) local communities – who support their own interests and goals that the policymaker can decide to include in the model, thus requiring new iterations; and (ii) businesses and actors involved in tourism activities, who are required to implement higher standards of sustainability in their activities. Moreover, it spurs a constant interaction between policymakers and the analysts, who are in charge of elaborating the model and including the judgements presented by the policymakers.

References

Arbolino, R., Boffardi, R., De Simone, L., & Ioppolo, G. (2020). The evaluation of sustainable tourism policymaking: A comparison between multicriteria and multi-objective optimisation techniques. *Journal of Sustainable Tourism*, *29*(6), 1000–1019.

Archer, B., Cooper, C., & Ruhanen, L. (2012). The positive and negative impacts of tourism. In W. F. Theobald 3rd ed., *Global tourism* (pp. 79–102). Routledge. eISBN 9780080478043

Ballantyne, R., Packer, J., & Hughes, K. (2009). Tourists' support for conservation messages and sustainable management practices in wildlife tourism experiences. *Tourism Management*, *30*(5), 658–664.

Biddulph, R., & Scheyvens, R. (2018). Introducing inclusive tourism. *Tourism Geographies*, *20*(4), 583–588.

Blancas, F. J., González, M., Lozano-Oyola, M., & Perez, F. (2010). The assessment of sustainable tourism: Application to Spanish coastal destinations. *Ecological Indicators*, *10*(2), 484–492.

Bramwell, B. (1998). User satisfaction and product development in urban tourism. *Tourism Management*, *19*(1), 35–47.

Bramwell, B., & Lane, B. (1993). Interpretation and sustainable tourism: The potential and the pitfalls. *Journal of Sustainable Tourism*, *1*(2), 71–80.

Budeanu, A., Miller, G., Moscardo, G., & Ooi, C. S. (2016). Sustainable tourism, progress, challenges and opportunities: An introduction. *Journal of Cleaner Production*, *111*, 285–294.

Butler, R. W. (1999). Sustainable tourism: A state-of-the-art review. *Tourism Geographies*, *1*(1), 7–25.

Butler, R.W. (1993). Tourism: An evolutionary perspective. In J. G. Nelson, R. W. Butler, & G. Wall (Eds.), *Tourism and sustainable development: Monitoring, planning, managing* (pp. 27–44). University of Waterloo.

Camagni, R. (2017). On the concept of territorial competitiveness: Sound or misleading? In R. Capello (Ed.), *Seminal studies in regional and urban economics: Contributions from an impressive mind* (pp. 93–113). Springer

Castellani, V., & Sala, S. (2010). Sustainable performance index for tourism policy development. *Tourism Management*, *31*(6), 871–880.

Chiosova, J. (2015). Transformation of approaches to the definition of tourism in the context of socio-economic importance. *Socio-Economic Research Bulletin*, *4*, 26–31.

Dantzig, G. B., & Thapa, M. N. (2003). Two-stage stochastic programs. In G. B. Dantzing & M. N. Thapa (Eds.), *Linear programming: 2: Theory and extensions* (pp. 335–366). Springer. https://doi.org/10.1007/b97283

Dowling, R. (2020). Sustainable tourism: Principles, contexts and practices, *Journal of Ecotourism, 19*(3), 1–490. https://doi.org/10.1080/14724049.2020.1763009

European Commission. (2016). *European tourism indicators system for sustainable destination management*. https://ec.europa.eu/docsroom/documents/21749

Fraiz, J. A., de Carlos, P., & Araújo, N. (2020). Disclosing homogeneity within heterogeneity: A segmentation of Spanish active tourism based on motivational pull factors. *Journal of Outdoor Recreation and Tourism, 30*, 100294.

Frent, C. (2016). An overview on the negative impacts of tourism. *Revista de turism-studii si cercetari in turism, 22*, 1–4.

García-Melón, M., Gómez-Navarro, T., & Acuña-Dutra, S. (2012). A combined ANP-delphi approach to evaluate sustainable tourism. *Environmental Impact Assessment Review, 34*, 41–50.

García, F. A., Vázquez, A. B., & Macías, R. C. (2015). Resident's attitudes towards the impacts of tourism. *Tourism Management Perspectives, 13*, 33–40.

Ghulam Rabbany, M., Afrin, S., Rahman, A., Islam, F., & Hoque, F. (2013). Environmental effects of tourism. *American Journal of Environment, Energy and Power Research, 1*(7), 117–130.

González-García, R. J., Mártínez-Rico, G., Bañuls-Lapuerta, F., & Calabuig, F. (2022). Residents' perception of the impact of sports tourism on sustainable social development. *Sustainability, 14*(3), 1232.

Guo, Y., Jiang, J., & Li, S. (2019). A sustainable tourism policy research review. *Sustainability, 11*(11), 3187.

Hall, C. M. (1994). *Tourism and politics: Policy, power and place*. John Wiley & Sons. ISBN (Hardback): 978-0-471-94919-0

Happ, É., & Bolla, V. (2022). A theoretical model for the implementation of social sustainability in the synthesis of tourism, disability studies, and special-needs education. *Sustainability, 14*(3), 1700.

Higgins-Desbiolles, F. (2018). Sustainable tourism: Sustaining tourism or something more? *Tourism Management Perspectives, 25*, 157–160.

Janusz, G. K., & Bajdor, P. (2013). Towards to sustainable tourism–framework, activities and dimensions. *Procedia Economics and Finance, 6*, 523–529.

Kozak, M. W. (2017). Innovations in tourism policy: The case of Poland. *Almatourism Journal of Tourism, Culture and Territorial Development, 8*(7), 160–177.

Kurt Konakoglu, S. S., Hełdak, M., Kurdoglu, B. C., & Wysmułek, J. (2019). Evaluation of sustainable development of tourism in selected cities in Turkey and Poland. *Sustainability, 11*(9), 2552.

Landorf, C. (2009). Managing for sustainable tourism: A review of six cultural world heritage sites. *Journal of Sustainable Tourism, 17*(1), 53–70.

Medina-Muñoz, D. R., Medina-Muñoz, R. D., & Gutiérrez-Pérez, F. J. (2016). The impacts of tourism on poverty alleviation: An integrated research framework. *Journal of Sustainable Tourism, 24*(2), 270–298.

Michailidou, A. V., Vlachokostas, C., & Moussiopoulos, N. (2016). Interactions between climate change and the tourism sector: Multiple-criteria decision analysis to assess mitigation and adaptation options in tourism areas. *Tourism Management, 55*, 1–12.

Muresan, I. C., Oroian, C. F., Harun, R., Arion, F. H., Porutiu, A., Chiciudean, G. O., Todea, A., & Lile, R. (2016). Local residents' attitude toward sustainable rural tourism development. *Sustainability, 8*(1), 100.

Obersteiner, G., Gollnow, S., & Eriksson, M. (2021). Carbon footprint reduction potential of waste management strategies in tourism. *Environmental Development, 39*, 100617.

Önder, E., Yıldırım, B. F., & Ozdemir, M. (2013). Multi criteria decision making approach for evaluating tourism destinations in Turkey. *Academic Journal of Tourism and Management Researches*, *1*(1), 1–15.

Öztürk, A. B., Ozer, O., & Çaliskan, U. (2015). The relationship between local residents' perceptions of tourism and their happiness: A case of Kusadasi, Turkey. *Tourism Review*, *70*(3), 232–242.

Petrosillo, I., Valente, D., Zaccarelli, N., & Zurlini, G. (2009). Managing tourist harbors: Are managers aware of the real environmental risks? *Marine Pollution Bulletin*, *58*(10), 1454–1461.

Qiao, G., Ding, L., Zhang, L., & Yan, H. (2022). Accessible tourism: A bibliometric review (2008–2020). *Tourism Review*, *77*(3), 713–730.

Soltani, N., Mousavi, M. N., & Sharifi, B. (2015). Strategic planning for tourism development in Urmia. *Journal of Urban Social Geography*, *2*(1), 79–98.

Streimikiene, D., & Korneeva, E. (2020). Economic impacts of innovations in tourism marketing. *Terra Economicus*, *18*(3), 182–193.

Streimikiene, D., Svagzdiene, B., Jasinskas, E., & Simanavicius, A. (2021). Sustainable tourism development and competitiveness: The systematic literature review. *Sustainable Development*, *29*(1), 259–271.

Sun, Y. Y., Lin, P. C., & Higham, J. (2020). Managing tourism emissions through optimizing the tourism demand mix: Concept and analysis. *Tourism Management*, *81*, 104161.

Torres-Delgado, A., & Palomeque, F. L. (2012). The growth and spread of the concept of sustainable tourism: The contribution of institutional initiatives to tourism policy. *Tourism Management Perspectives*, *4*, 1–10.

Ulkhaq, M. M., Akshinta, P. Y., Nartadhi, R. L., & Nugroho, W. S. (2016). Assessing sustainable rural community tourism using the AHP and TOPSIS approaches under fuzzy environment. In *MATEC Web of Conferences* (Vol. 68, p. 09003). EDP Sciences.

UNWTO (United National World Tourism Organisation). (2004). *Indicators of sustainable development for tourism destinations: A guidebook* (p. 507). eISBN 978-92-844-0726-2. https://www.e-unwto.org/doi/epdf/10.18111/9789284407262?role=tab

UNWTO (United National World Tourism Organisation). (2020, January 5). *UNWTO world tourism barometer and statistical annex*. https://www.e-unwto.org/doi/abs/10.18111/wtobarometereng.2024.22.1.1

WEF (World Economic Forum). (2022, September). *Ten principles for sustainable destinations: Charting a new path forward for travel and tourism* [White paper]. https://www3.weforum.org/docs/WEF_Ten_Principles_for_Sustainable_Destinations_2022.pdf

WTC (World Travel & Tourism Council). (2022). *Travel & tourism: Economic impact 2022 global trends*. https://assets-global.website-files.com/6329bc97af73223b575983ac/647df24b7c4bf560880560f9_EIR2023-APEC.pdf

Yadav, N., Sahu, N. C., Sahoo, D., & Yadav, D. K. (2018). Analysis of barriers to sustainable tourism management in a protected area: A case from India. *Benchmarking: An International Journal*, *25*(6), 1956–1976.

Yanes, A., Zielinski, S., Diaz Cano, M., & Kim, S. I. (2019). Community-based tourism in developing countries: A framework for policy evaluation. *Sustainability*, *11*(9), 2506.

Yasarata, M., Altinay, L., Burns, P., & Okumus, F. (2010). Politics and sustainable tourism development: Can they co-exist? Voices from North Cyprus. *Tourism Management*, *31*(3), 345–356.

Zapata, M. J., Hall, C. M., Lindo, P., & Vanderschaeghe, M. (2013). Can community-based tourism contribute to development and poverty alleviation? Lessons from Nicaragua. In J. Saarinen, C. M. Rogerson, & H. Manwa (Eds.), *Tourism and the millennium development goals* (pp. 98–122). Routledge. eISBN 9781315868530

Zolfani, S. H., Sedaghat, M., Maknoon, R., & Zavadskas, E. K. (2015). Sustainable tourism: A comprehensive literature review on frameworks and applications. *Economic Research Ekonomska istraživanja*, *28*(1), 1–30.

Chapter 14

Effects and Response Strategies of the Tourism Industry to Tropical Storms: Case of a Climate-vulnerable Island

Adjnu Damar Ladkoo

Department of Management, University of Mauritius, Mauritius

Abstract

Cyclones affect several aspects of a destination and its communities. The tourism and hospitality sector, which is one of the most thriving yet vulnerable industries, is mostly affected by cyclones and specifically those in Small Island Developing States. However, such effects have been scantily documented in the academic literature. Using Mauritius, a Small Island Developing State also known to be a climate-vulnerable island located in the Western Indian Ocean, this study investigated on the consequences of cyclones on tourism businesses and their response strategies. The study adopted a qualitative approach by conducting semi-structured in-depth interviews with diverse stakeholders of businesses in the tourism arena. Their views on the impacts of storms on their businesses were sought as well as their strategies to cope with the latter. As compared to earlier-done researches, this study showed that discussion of impacts is more on the secondary bearings rather than primary ones and this could be associated with the fact that there is a prevalent joint determination from the authorities and key stakeholders of the sector to strengthen the island's resiliency to cyclones and its repercussions. The study contributed to portraying the recently felt impacts of storms on the tourism businesses and also provided additional knowledge that could aid in fortifying existent response strategies. Future researchers can excavate deeper into each of the impacts and proposed response strategies so that detailed plans for overcoming cyclones by tourism businesses in the context of small islands, can be developed.

Keywords: Strategies; tourism; storms; climate; island

The Need for Sustainable Tourism in an Era of Global Climate Change: Pathway to a Greener Future, 297–317

doi:10.1108/978-1-83608-668-020241040

1. Introduction

The climatic conditions around the planet have drastically evolved over the last few years. There is noticeably a cacophony in the predictability of the intensity or frequency and types of climatic calamities. Severe hurricanes and cyclones increase in blizzards, extreme heat waves, and flash floods are among these natural catastrophes. As per The Times of India (2023), studies put forward that a heating climate may bring more devastations and the storms may make the elevated sea levels more fatal. Least alterations in weather are hugely impacting on the economies of various countries; specially those who are more prone and vulnerable to the climate crisis. While developed countries are also not spared from the latter, Small Island Developing States (SIDS) are the most affected. Mauritius, Seychelles, Haiti, Vanuatu, and Maldives are some of the concerned SIDS. The economic activities of the latter get halted because important economic zones such as airports and ports cannot operate in harsh climatic conditions which eventually lead to losses despite the increased involvement of insurance firms. Thus, more than ever, proactively planning the response strategies to such natural calamities is imperative. This study focuses on Mauritius, which is a Small Island Developing States in the Indian Ocean and which is situated in an active cyclone zone; making it experience an accelerated climate crisis and hence, much more vulnerable. The island thrive on its diverse industries such as textile, food, tourism and finance amongst others. This research emphasises particularly on the tourism sector of the island given that it is among the most exposed one to climate change and its impacts.

Mauritius offers mostly coastal tourism and with the upsurge of storms, diverse of its coastal hotels and beauties are getting destroyed. Some of these devastations include coastal beach erosions, coral bleaching, rise in the sea level and change of seawater temperatures which entails the destruction of diverse marine fauna and flora and destruction in river paths and hence, loss of freshwater bodies downwards to the coastal regions. It is to be pointed out that since Mauritius is in a tropical region of the planet, the term cyclone is frequently accompanied by the word Tropical. Thus, known as tropical cyclone for this region of the globe. The tourism and hospitality sector of Mauritius is highly reliant on the coast and its surrounding areas but storms' new intensities, unpredictability and frequencies are restraining the coastal tourism to flourish and expand due to fears of physical destruction of the regions, death and cognitive dissonance in the minds of customers who want but cannot or may not visit this island due to its heightened vulnerability. Yet, stakeholders of the tourism industries are showing their endeavours to rebuild a positive image of its tourism industry and to some extent while they are successful, they may have to gear up to tackle new characteristics of cyclones and other associated impacts such as flash floods and heavy winds. This chapter has as objectives: (1) to understand the impacts of cyclones on the tourism industry of Mauritius, and (2) to recognise the response strategies adopted by tourism and hospitality businesses to tactfully combat the severe climatic conditions associated with cyclones.

The chapter starts with an explanation about cyclones in Mauritius followed by an elucidation of cyclones in the tourism industry and eventually impacts of cyclones on businesses and their response strategies have been presented. Following the above, the research methodology adopted for this research has been presented. As per Leoni and Boto-García (2023), natural catastrophes are vital impediments for the tourism sector's supply – and demand-related activities. So, this chapter eventually particularises on the impacts of cyclones on the supply and demand of tourism businesses while themes generated from this study portray as well the response strategies of the latter. The chapter ends with a concluding paragraph.

2. Cyclones in Mauritius

Different terminologies are associated with natural calamities such as cyclones, hurricanes or typhoons. It is, therefore, vital to understand what a tropical cyclone is because each natural calamity might be different from each other. Meteorologists use a broad term to define a tropical cyclone which for them is a spinning, arranged structure of clouds and thunderstorms that initiates over tropical or sub-tropical waters and has locked low-level movement and once the latter touches maximum constant airstreams of 74 miles per hour or higher, it is categorised as a hurricane, typhoon or cyclone reliant on where it originates in the sphere (National Ocean Service, 2018). As per Mauritius-Holidays-Discovery. Com (2018), tropical storms mostly form in the Inter Tropical Convergence Zone (ITCZ), a place which is near the equator where breezes from the northern hemisphere come across those from the south and small waves on the ITCZ grow into vortexes that occasionally build up into tropical cyclones (see Fig. 14.1).

Since several years now, Mauritius has been facing this natural calamity at various intervals and various intensities. As per Mauritius-Holidays-Discovery. Com (2018), once a cyclonic structure is perceived, usually, nearby the Diego Garcia region positioned south of the equator and around 1,000 miles southwest of India, it is baptised and followed by the local climatological station. A cyclone guard will be generated. Mauritius is likely to be knocked out by tropical storms in the summer period between November and April. The hazards are greater in the more dynamic period, from mid-December to March when the sea temperature becomes warmer. The cyclonic term formally starts on 15 November and culminates on 15 May. The general summer period for Mauritius falls from November to around April. Storms mostly form during these moist and hot months.

Imperatively, both responsive and proactive actions have to be taken by businesses because while the commerce will be highest in summer, it is also the moment when the island is inflicted by storms. This shows that firms should be prepared more than ever to tackle the problems related to the occurrence of cyclones. Nevertheless, embracing the correct response policies is the real defy because the high unpredictability about the evolution of the storms might diffuse all care and actions taken. Storms that evolve with higher forces might necessitate dissimilar response strategies as for one with lower power because the impacts would equally be unalike and sterner. Different countries have varied warning systems – while some use quantitative means such as numbers others use qualitative means like colour indicators to aid their authorities and public to understand the strength of the storms. As per Walshe (2022), Mauritius possibly founded the globe's initial telegraph storm early warning system. Table 14.1 illustrates the

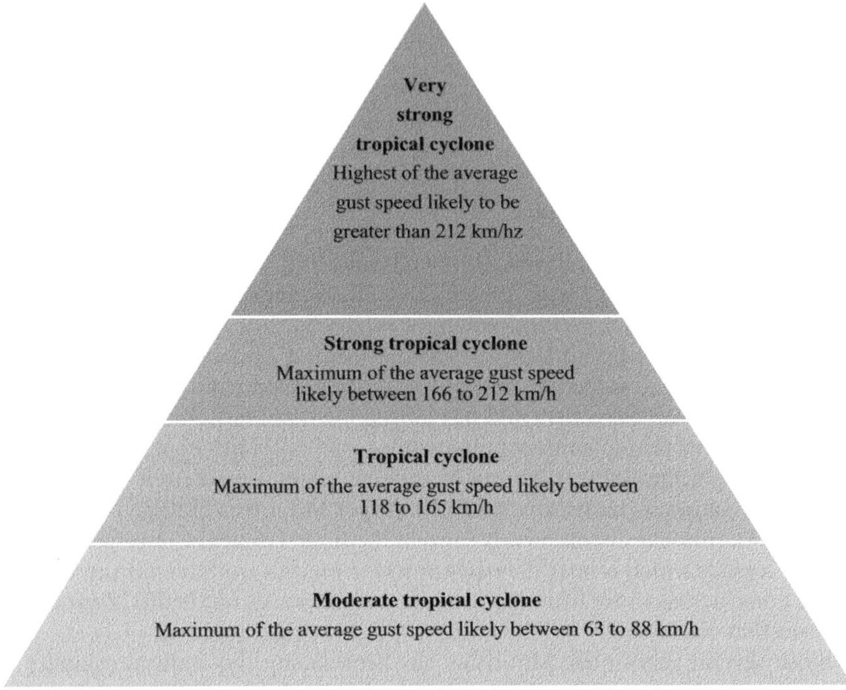

Fig. 14.1. A Pictorial Depiction of Some of the Definitions of Tropical Cyclones with Different Intensities. *Source*: Adapted from Mauritius Meteorological Services (n.d.).

different storm warning plans in place for Mauritius and Rodrigues, which is another island part of the Republic of Mauritius.

Mauritius has a prevalent National Disaster Risk Reduction and Management Act 2016 which crafts a charter for risks and disaster management that also englobes formation and mounting of required and appropriate disaster management committees. Impacts of cyclones on Mauritius can be minimal to severe, depending on various factors, amongst which are the available infrastructures, the architectural designs of buildings and houses, the level of reactiveness of a country's citizens to warning systems and to safety measures prescribed by authorities before, during and after cyclones, the accuracy of predicting the intensity of the cyclone by concerned authorities, dispositions taken by governmental and private bodies such as rapidity in implementing measures by disaster management committees.

2.1. Tourism Industry of Mauritius and Cyclones

Out of all industries which are helping Mauritius to thrive despite being a SIDS is that of the tourism industry. This is one of the pillars of the Mauritian economy. Yet, this flourishing sector is not spared by natural calamities like storms, flash

Table 14.1. Cyclone Warning Systems for Mauritius and Rodrigues.

Class	Issue Timing & Likely Gusts' Speed
I	36 to 48 hours before the country is stuck by gusts reaching 120 km/h.
II	To the extent possible, 12 hours of daylight before the manifestation of gusts of 120 km/h.
III	To the extent possible, 6 hours of daylight in advance of the occurrence of winds of 120 km/h.
IV	Issued when winds of 120 km/h have been noted in some places and are likely to last.
A safety bulletin	Issued for the purpose of removing the storm warning class III or IV, as the situation may be; and notifying the public of the presence of any severe weather settings related to the storm and other environmental peril, contingent on the nature and degree of the destruction happened during the passage of the storm.
A termination bulletin	Issued when there is no longer any danger of gusts beyond 120 km/h.

Source: Adapted from Mauritius Meteorological Services (n.d.).

floods, heavy rains or even climate crises. This sector was exposed to the devastating power of the COVID-19 pandemic and with proper strategies emanating from diverse stakeholders of the industry, the latter is right on its feet and thriving. But, the danger of chaos that can be caused by cyclones is ever-present given the tropical nature of the island and its exposure to deepened impacts of natural calamities as a result of climate change. Undoubtedly when mankind seems to be firmly gripping to meaningful advancement, nature seems to be putting further challenges to the later, for instance with the increase in the number of cyclones having greater power to destroy regions. Occurrences of cyclones are not new to Mauritius, yet the adaptation to the latter seems to be sometimes inadequate. One of the most recent incidents in the year 2024 relates to heavy flooding that emerged from heavy rains of a cyclone. Streets of the capital were flooded which lead to the loss of life as well as amenities and infrastructures. Such happenings deepen the perceptions of vulnerability of the island, especially of those regions which are nearer to water bodies and the sea. The tourism industry has to therefore face a raft of impacts such as a terrified or lowered number of visitors and a decrease in business activities such as in hotels and places of attractions among others. These may lead to a drop in the economic performance of this industry. As such apt measures have to be proactively planned and deployed by the tourism industry's stakeholders to cope with the new dynamism of cyclones in the region. The island is highly susceptible to cyclones and not being prepared to tackle the latter and underestimating the direct and indirect dangers will imply purposeful action to erase a great tourist destination from the business arena. Commendably,

Mauritius is amongst the rare SIDS which has portrayed great resilience and ability to resume its economic activities post-stormy periods despite the recent incidents. So, this study reinforces the need to continue to study and draw lessons on the impacts of cyclones on the tourism industry so that proper response strategies are brought forward proactively. The readiness, dedication and commitment of the stakeholders are some of the key factors to consider for ensuring rapid and efficient recovery posts cyclones which can help Mauritius to continue to thrive as one of the best island destinations. As such, this study elucidates the impacts and response strategies of businesses in the tourism arena of Mauritius.

2.2. Impacts of Cyclones and Responses of Tourism Businesses

Storms have diverse impacts on an island and its tourism sector. Amongst the impacts are economic, social, political and environmental. Many businesses in the tourism and hospitality industry such as restaurants, travel agencies, airports, ports, places of interest and hotels have to face the consequences of cyclones both straight-forwardly and indirectly. Firstly, there is a decreased possibility of potential tourists to opt for Mauritius as either a business or holiday destination. Many tourists want to be psychologically positive and in a peaceful mind-state when visiting a particular destination. Media coverage of happenings during cyclones may hinder the desire of choosing the island as an apt vacation place. Additionally, while most tourists appear to be happy with Mauritius whenever they have visited the island, storms may distort this feeling and lead to dissatisfaction or even create cognitive dissonance. Dissatisfaction may, for example, occur when the tourists planned some visits to places of attractions and due to harsh cyclonic conditions or closure, cannot fulfil this wish. Cognitive dissonance which is a state of confusion may arise when they have partly enjoyed themselves and have as well dealt with the impacts of the storms which may have hampered their planned visits. Another consequence may be higher uncertainty as regards confirmation of bookings which leads to tourists doing late bookings rather than early ones; even if they may have to pay a higher price. They also look for deals that extend to them refundable and cancellable options. One more consequence of cyclones on the tourism businesses may be related to the areas and departments that have to prepare for the welcome of tourists who might be affected by approaching cyclones.

Flight schedules might change and in some cases, there are flight cancellations which is a problem for firms who have to either drop or fetch tourists from airports as it involves putting at risk the lives of employees involved in such tasks. Changes in flight schedules might also cause annoyance in tourists who imperatively had to depart or arrive at Mauritius since other forthcoming activities that they had planned would have to be modified and consequences might be harsh in some cases whereby high cancellation fees have to be paid. For example, if a tourist already booked a safari for his family and due to cyclones, it had to be cancelled, the policies and conditions of the organisation might be that the tourist will have to pay a cancellation fee during prevailing natural calamities. Cyclones can also hugely affect both the front office and back office tasks of tourism businesses such as in restaurants or hotels. Supply Chain Managers as well as Food and Beverage Managers of hotels might be in real dilemmas since suppliers might not have adequate products to deliver. Alongside there might be delays in raw material delivery hence altering the time of lunch or dinner preparation for visitors of the island. Storms undoubtedly can stir a surge in the price of the required raw materials because of damages caused by cyclones to the plantations and crops of the island. Additionally, basic routine tasks such as housekeeping can be affected since employees might not have the possibility to attend work as a result of flooding, family concerns, and warnings emitted by authorities to avoid the roads and other areas prone to dangers like landslides.

All of the above can eventually cause tourists to get frustrated because it might be the first time that they are experiencing a cyclone on an island while also having to face many changes in their holiday or business schedules. The reputation of hotels and even that of the destination are at risk. It is, therefore, noted that consequences can be both physical and immaterial and when some are assessable, others cannot. The above makes the tourism and hospitality sector even more challenging to be managed whereby the right skills, training and manpower would be in high need to be able to deal considerately with all the glitches related to storms. As per Glaesser (2003), generally, there are three constitutive characteristics of tourism which are change of location, temporary stay and the existence of a motive. As a result of cyclones, visibly all three get affected because tourists prefer to opt for safer locations, less prone to cyclones, because even if the stay is temporary they are more certain that there will be positive valence which leads them to choose cyclone-free zones globally.

Response strategies adopted by tourism firms can be much dependent on the warning emitted by the National Meteorological Station of the island. It is imperative to understand the disposition taken by these businesses when cyclones are approaching, hitting or leaving the island. Documenting these response strategies and creating a brochure or guideline for potential and existing tourists can greatly minimise the damage caused to the image of both firms and the destination. Tourists being already aware of actions to be taken by authorities, can more easily welcome the safety measures and smoothly opt for a shift to in-hotel vacation and relaxation option rather than those which they would have chosen outside the hotel such as visits to places of interests or swimming in the sea. Moreover, predicting dangers of 'not following' the response strategies can greatly sensitise the tourists such that they become more alert and vigilant. The tourists can also, proactively understand the need to cater for an additional budget that might be needed in case of cancellation policies being applied on airlines and other services. Today we live in a world where information is available everywhere, making it possible to predict the approach of cyclones. Yet, the dark side is the unknown damage and impacts that it can cause to an island if the right response strategies are not taken. Bringing clarity to response strategies adopted by tourism businesses is thus a must. This can eventually make tourists better accept as well as tolerate natural calamities such as cyclones and the associated torrential rains or flooding.

3. Research Methodology

As per Faulkner (2001), tourism destinations in every corner of the globe face the virtual certainty of experiencing a disaster of one form or another at some point in their history and despite this, few destinations have properly developed disaster management plans in place to help them cope with such eventualities and among the reasons for this, is the limited amount of systematic research that has been carried out in the field. As such, this chapter attempted to shed light on the effects as well as on the response strategies of diverse businesses of the tourism sector in Mauritius during cyclones. It is against this background that a qualitative study was undertaken whereby semi-structured interviews were conducted with diverse categories of business people operating in the tourism sector of Mauritius such as either port or airport officers, owners of restaurants and shops selling products to tourists as well as employees of hotels and places of attractions. The below broad questions were considered for this research: (1) what are the impacts of cyclones on your tourism business, and (2) how is your business responding to the impacts of storms?

Thematic analysis as an independent qualitative descriptive approach is mainly described as 'a method for identifying, analysing and reporting patterns (themes) within data' (Braun & Clarke, 2006). Outputs of the thematic analysis of the textual material generated from the interview transcripts of 10 of these stakeholders of the tourism arena were themes. Be it qualitative or quantitative research, certain ethical principles have to be considered such as anonymity, confidentiality and even no pressurisation for answers. Given that this study adopted a semi-structured interview, prompting was included where deemed appropriate. Interestingly, being a resident of a small island developing state, the author found it valuable to amalgamate the discoveries from the interviews with personal observations of the latter regarding the impacts and responses of the tourism industry to cyclones. Alongside, for critical exposure of the data, references were made to some relevant books, websites and Google Scholar journal articles. These three procedures enhanced the soundness and trustworthiness of the data. Themes generated from this research were explained and importantly, categorised under the demand and supply sides of businesses in the tourism industry. The next section is dedicated to the findings from this study.

4. Findings on Impacts and Responses

The responses of the interviews were presented in an anonymous manner to preserve the identity of the stakeholder and hence presented as respondent one (R1), respondent two (R2), respondent three (R3) and so on. While some of the presented themes emerged from only one respondent, others were generated as a common theme from a diversity of respondents. Table 14.2 is an excerpt of the analysis of responses from two of the respondents along with the themes emerging from them, respectively.

Coping with cyclones is today a prerequisite for many tourism businesses, especially those situated near or around more vulnerable areas such as water bodies

Table 14.2. Excerpt of Responses and Themes from Two Respondents.

Respondent ID	Reply on Impact/s of Cyclones	Theme/s	Reply on Response Strategy/ies	Theme/s
R3	*The routine of work changes because of concerns from clients and our own colleagues. Safety comes first after all. E.g., visitors also do not to have their money wasted on stays and attractions they have booked already so they ask lots of questions.*	Desire to amend hotels' and places of attractions' bookings.	*A diversity of activities can be conducted indoors during cyclones. They have to be discovered together with the clients.*	Liaison with airports for updated flight status and changes.
		Request information on cancellation fees for hotels and attractions.	*We also have to contact the airports or places of attraction operation for updates regularly and then extend this info to our clients. Sometimes, we propose new visits.*	Alter location for visits to attractions.
R4	*We have to deal with visitors' anger and sometimes fear. They want to recover their money for lost bookings.*	Request information on cancellation fees for hotels and attractions.	*We need to answer to their concerns and in the long term with the support of other people like from the Government and other organisations, we need to find new ways to diversify to make the business survive. May be thinking of more indoor activities during storms.*	Substitute outdoor activities with indoor activities. Attend to queries of tourists.

Source: Author-generated.

and the ocean. Amelung et al. (2007) revealed that with climate change, when some locations are likely to experience substantial increases in attractiveness due to improvements in their weather conditions, others may become significantly less appealing to tourists, leading to shifts in the temporal patterns of visitation and/or actual declines in the number of visits. For Gök and Aylan (2024), COVID-19 and natural cataclysms such as floods, storms, forestry fires, droughts and earth tremors that affected either local or international contexts disturbed the evolution and progress of their tourism sectors. In the present study as well, it was noted that cyclones impacted on businesses in the tourism arena in diverse ways such as affecting visitors or tourists, employees as well as owners of certain businesses. Fig. 14.2 shows the number of themes that could be captured about impacts and response strategies.

Below are excerpts of two transcripts that illustrate answers about the impacts of storms on the stakeholders' businesses:

Respondent 2 (R2):

> We have to stop operating and prioritise our own and that of our employees' or clients' safety.

Respondent 6 (R6):

> My employees want to go home for their health and safety as well as concern for their families. They stay glued to their radios or phones for more cyclone-related information or they ask for it.

As compared to earlier-done research, this study showed that respondents' discussion of impacts was more on the secondary bearings also and not only on the primary impacts such as just economic and physical impacts as may have been felt in former years. This can be associated with the fact that with time, the

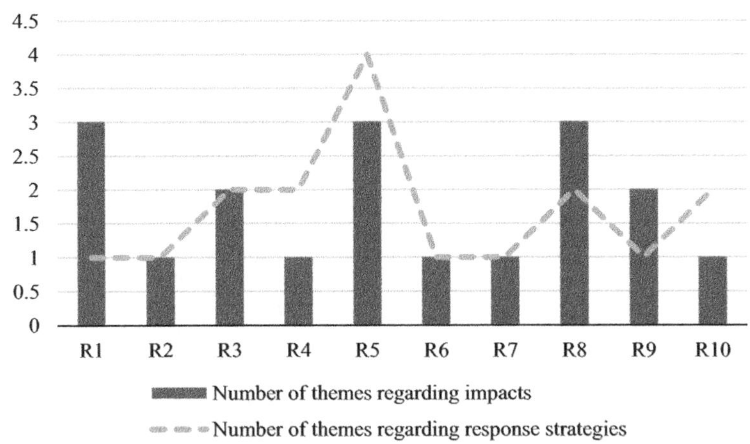

Fig. 14.2. Number of Themes Captured Per Respondent Regarding Impacts and Response Strategies. *Source*: Author-generated.

island has become more resilient to welcome and overcome storms notwithstanding that some unknown aftermaths of cyclones such as heavy rain and winds are still deeply affecting certain regions. As per RE/MAX 24 Mauritius (2024), the joint determinations of numerous sectors in guaranteeing a rapid return to normalcy post the cyclone not only diminished the impact on the island's economy but also solidified its standing as a reliable and beautiful target for worldwide tourists. Themes about the impacts of cyclones that emerged from the demand side showed that there is a perceived mix of economic, psychological, social and environmental impacts of cyclones on visitors or tourists.

Below are excerpts of two transcripts that illustrate answers about the response strategies of tourism businesses to cyclones:

Respondent 7 (R7):

> We plan an anticipated bad weather-stock. We also extend cyclone information to our visitors as they ask for it.

Respondent 10 (R10):

> So we have to amend the bookings and provide alternative timings. We need to extend necessary help to our employees too.

During the interview, the researcher's first felt impression was that tourism businesses did not seem to be in alarmed states given that Mauritius has been facing cyclones since a long time and to some extent, vigilance about storms has gained momentum. Being a resident of the context being studied, the latter confirmed that this could be also visibly noted during the approach phases of cyclones, whereby people appeared less panicked and more organised. More so, authorities are already working towards better readiness for help to diverse stakeholders of the sector for disaster risk reduction. As per Capacity for Disaster Reduction Initiative (2020) over the last 5 years, Mauritius has made substantial improvements in firming its strategy and institutional outline investment in cyclone readiness and its community-based disaster response players. The felt bearings of cyclones on businesses – from the supply side – that were captured from the interview transcripts were similar to those of the demand side with the addition of also the operational impacts of cyclones. Hence, Table 14.3 which emanated from this study, illustrates the two sides – demand and supply – on which cyclones have their impressions. Additionally, themes emanating allowed the researcher to identify key response strategies used by the respondents. Such findings could help tourism organisations to assess and fine-tune their strategies to cope with storms in tropical SIDS.

In a nutshell, it was found that impressions of cyclones on tropical SIDS like Mauritius could be dismissed by refining several components such as warning systems, heightening of weather insurance policies as well as enhanced health and safety measures along with better management of bookings of flights, leisure activities and supply chain. As such, the forthcoming section discusses major recommendations for aptly responding to cyclones and hence, help alleviate the above-presented impacts on a diversity of tourism stakeholders. Additional excerpts of interview transcripts have been presented along with the proposed recommendations.

Table 14.3. Impressions of Cyclones on the Demand and Supply Sides of Tourism Businesses.

Side	Impacts and Responses	Themes
Demand side (emanating from visitors'/tourists')	Perceived mix of impacts: psychological, social, economic and environmental	Request information on cyclone warning systems and evolution.
		Request information on changes in leisure activities and visits to places of attractions.
		Desire to amend hotels' and places of attractions' bookings.
		Request information on cancellation fees for hotels and attractions.
		Request information on flight schedules' changes or cancellations.
		Desire to understand security measures taken by the hotel.
		Desire to personally meet hotel management for assurance and clarifications.
		Request for availability of medical assistance for any urgency.
Supply-side (emanating from employees/ owners of tourism businesses)	Impacts and responses: operational, economic, social and environmental	Increased regular weather reporting with tourists.
		Liaison with airports for updated flight status and changes.
		Attend to queries of tourists.
		Alter the timing of leisure activities.
		Alter location for visits to attractions.
		Substitute outdoor activities with indoor activities.
		Propose weather insurance for hotels and clients.
		Consider revenue diversification (focus on new concepts).
		Suppliers reluctant to deliver goods.
		Provision of health and safety supports.
		Support to employees.
		Request support from tourism organisations/ government.

Source: Author-generated.

5. Recommendations for Curtailing the Impacts of Cyclones

5.1. Recommendation 1: Cyclone Warning Systems and Evolutions

Excerpt of the transcript of Respondent two (R2):

> Though there is an established cyclone warning system, it has been noted that there are some inconsistencies among those stakeholders who advise on weather evolutions. So this has to be remedied and also have a more inclusive approach.

As mentioned earlier, cyclones vary in intensity and different countries have varied warning systems. Yet, because of the ever-changing nature of cyclones, businesses have to be on their alert irrespective of the cyclone warning because lives – tourists or employees – are under their responsibilities. For Leroux et al. (2024) especially SIDS in the southwest region of the Indian Ocean are sensitive and are experiencing the turmoil associated with climate changeability and risky weather situations like tropical storms and thus, personalised climate services which take into consideration the exactitudes of the needs and contexts, are required for making apt adaptation measures. Businesses have to familiarise the tourists as well as employees about facts behind any approaching tropical cyclone and the businesses' emergency warning systems and signals have to be activated. When some firms can consider using sounds as alert systems, others can use visuals. However, to be a socially responsible and inclusive organisation means to caterfor all categories of people; meaning that novel warning systems for disabled people will have to be also brought forward as a proactive means to coping with cyclones. Moreover, hotels can put at the disposition of tourists or employees small radio transistors so that even if internet is not available, they can still stay tuned to local radio stations for weather news. Alternatively, a common radio can be put for several people in a common lounge and if internet is available, for instance, tourists can be guided to resourceful weather websites or social media pages.

5.2. Recommendation 2: Leisure Activities and Visits

Excerpt of transcript of Respondent two (R3):

> A diversity of activities can be conducted indoors during cyclones. They have to be discovered together with the clients.

Excerpt of transcript of Respondent two (R1):

> Anything can be done today with technology, so businesses need to utilise the potentials that the latter is offering. Like, I find it interesting if one could use VR headsets so as to give tourists a different scenery instead of focusing on the bad weather.

The impacts of cyclones are felt in diverse manners and one of them is how it tends to play with the satisfaction levels of tourists by disturbing their leisure activities and visits to places of attractions. As per Becken (2010), weather conditions are not always ideal and there are different behaviours a tourist can display that indicates how comfortable they are and to what extent they are able or willing to adapt. De Freitas (2003) categorised on-site behaviour by tourists as follows:

- Avoid areas of unfavourable weather- or climate-determined conditions (e.g., move from sun to shade).
- Change activity to suit weather conditions (e.g., swim more/less).
- Use structural or mechanical aids (e.g., umbrellas or windbreaks).
- Adjust thermal insulation of body (clothing).
- Adopt passive acceptance.

However, there can be cases where tourists simply feel that there is no possibility for adaptation; especially when there are extreme changes in weather conditions and such changes are noticeable when cyclone warnings are emitted whereby hotels' or restaurants' safety policies have to be followed and abided by at all costs. This can lead to frustration of tourists with eventually dissatisfaction since the perception of value for what they have paid drastically decreases. To avoid the above, tourism businesses are today, during cyclones, in dire need of solutions to diversify leisure, activity times and outdoor visits. Hotels that are financially strong can bring new concepts such as 'feel the outdoor while being inside'; whereby covered and protected areas – resistant to wind and water can be developed. More so, ballrooms or common lobbies accommodated with entertainment materials such as movies and toys for children & teenagers. Furthermore, indoor football, indoor cricket, covered water ponds, artificial beaches, covered swimming pools and the use of virtual realities are some of the indoor activities that can be considered to overcome the negative effect of cyclonic conditions on planned outdoor activities.

5.3. Recommendation 3: Flight Status and Changes

One of the most affected areas of the tourism and hospitality sector due to cyclones is that of the aviation sector. Undeniably, flights are cancelled or rescheduled which can directly impact on the airports' service delivery levels, facilities and amenities. Moreover, the tourists themselves will have to modify their plans for departures or arrivals; without having to mention the stress and pain that tourists have to undergo because of long queues or long waiting times in airports or hotels. As such, airports and hotels should envisage the possibility to cater for such tourists who are waiting and undergoing this stressful situation. Some airlines appeal to customers to themselves, postpone or cancel non-vital flights because of more flight cancellations, delays, busy air traffic and passenger backlog with the approach of a cyclone and intensification in cyclone warnings. Thus, while some tourism businesses closely work in collaboration with airports to facilitate these changes for their customers others, have yet to consider the same.

As such, airlines apply the 3Rs as a measure to cope with cyclonic conditions – rebooking for an alternative date, rerouting flights to other countries with transits and refunding passengers for cancellation of flights and bookings. Simultaneously, customers can be advised about such possibilities and thus, be encouraged to pay a small additional sum of money while booking flights to countries prone to be affected by climatic conditions. This will be like a warranty given to them in worst-case scenarios. This additional fee, however, can be much dependent on the type of airlines, services they offer and seat type. People having loyalty cards or frequent flyer loyalty cards can benefit from better tariffs. Tourists should equally be advised to provide the hotel, travel agencies and airlines with their exact contact details such that quick actions such as direct communications via a small messaging system or email or phone call can be done even in case of flight issues due to cyclones. Moreover, corporate communications or press releases or news bulletins should be escalated by the public relations departments of the concerned stakeholders so as to bring regular comfort to tourists who are undergoing the above stress due to cyclones.

A farfetched idea to deal with flight cancellations might be the possibility to swap flight costs with other alternatives of travel in case of calamities. Startlingly, some countries like China, India and Korea have developed underwater travel facilities. SIDS has not ventured into such transportation modes but, can be plausible means to deal with flight or cruise trip cancellations. As per NOAA Ocean Exploration (2024), scant studies have been conducted on the direct effects of storms on the deep sea during or after the latter because of its remote nature. Validation, regulations, safety, expert guidance, costs, acceptance and management are key issues to be considered for such changes.

5.4. Recommendation 4: Supply Management

Excerpt of transcript of Respondent two (R7):

> One big problem is supply of raw materials. Some supplier are reluctant to risk their lives to come and deliver products during bad weather so an anticipated bad weather-stock has to be there.

Communication with raw material suppliers and activation of replenishment protocols for food and beverages to ensure adequate food supply during cyclones should be also on the agenda of Managers in hotels or lodges and Airbnb accommodators. As such, the lists of suppliers and stakeholders who would be part of the disaster recovery services should be well established. Other important tasks also involve the following: verification of fuel supply, checking that generators are working, provision of glow sticks and provision of torch or flashlights to tourists and employees in case of power cut.

5.5. Recommendation 5: Hotel Booking & Attractions' Booking

Most hotels in hurricane-prone destinations such as the Caribbean, Mexico and Florida have hurricane policies in place, after all, they want to encourage travellers to visit during the off-season and are thus willing to offer assurance via the options of re-bookings, refunds, etc. (Reynolds, 2016). Three of the main attractions of Mauritius are the sea, sun and sand and cyclones do have drastic impacts on all of them and hotels or coastal restaurants cannot be that easily shifted to higher lands. As per Suarez (2011), even the best-planned strategies face challenges, as hotels have assets that cannot easily be relocated in the face of climate change and to account for this, future-thinking hotel companies with coastal properties might look for other business opportunities, such as investments in water-desalination technologies, or they may create policies to site new hotels well above the highest high tide line and seek alternative offerings for visitors that reduce dependency on sun, sand, and surf activities. It is essential to identify business continuity to remain profitable and not be affected by cyclones. As per Suarez (2011), business continuity will help companies prepare for long-term issues, even as they use disaster-response management to address more immediate issues. Hotels need to have robust and well-established contingency plans so as to cope with different types of calamities, including that of cyclones. In cases of cyclones, hotels should envisage, without additional costs, allow tourists to cancel and rebook within a specific time frame. Yet since cancellation and rebooking have an impact on all arenas of a hotel or attraction, tourists, as mentioned earlier, should be encouraged to opt for weather insurance packages with their travel agencies and hotels such that it is a win–win situation for both tourists and hotels and not a win–lose one.

5.6. Recommendation 6: Health and Safety Measures

Excerpt of transcript of Respondent two (R8):

> From a humanitarian perspective, I cannot put the lives of my employees or clients at risk. So, I need to ensure that I have all required medical products and facilities during harsh cyclonic conditions.

Tourism businesses always look forward to bringing customer satisfaction to their clients and one way of enhancing customer satisfaction would be to show genuine care and concern for the health and safety of the tourists. As such, there should be provision for emergency kits in each room of the hotels and if available already, tourists have to be reminded to keep them handy. Moreover, tourists should be advised of not to consume tap water and to wear clothes appropriate for their protection from potentially harsh cyclonic conditions. Additionally, tourists who have accompanying children have to be advised to neither venture in sea nor outside because there are potential dangers such as drowning in rough seas or getting hurt with branches falling off. Quick and rapid training sessions can be provided to tourists such that they know what to do in case buildings shake or break up and where to shelter.

Tourism businesses can consider inclusion of emergency buttons activated for their clients to use in case of problems during cyclones. The latter have to be made aware of the dangers revolving around the calm eye of a cyclone and to follow warnings and emergency protocols of the hotel. They have to be made aware that they need to stay alert for any activation of evacuation protocols – to act fast and immediately, wear the right shoes and clothes, avoid panicking and comfort children. Post cyclones, tourists have to be strictly warned of not exploring around and do sightseeing. Also, support has to be extended to disabled people with a planned allocation of a special caretaker during such crisis periods.

5.7. Recommendation 7: Weather Insurance

Excerpt of the transcript of Respondent two (R6):

> I think weather insurance is something on which we need to emphasise on as many other countries are already considering it. This can calm the stressful minds of for example tourists if they are visiting a place which is prone to getting storms.

Motivate tourists to secure insurance for such unpredictable conditions as heavy rains and flash floods during cyclones. Tourism businesses equally have to ensure that they are sufficiently covered and as per Wroten (2017) natural disasters can disrupt a hotel's regular business operations long after they have passed, and hoteliers need to ensure they have sufficient coverage in their insurance policies to carry them through until normal business levels return. When determining a property's losses, it requires looking at the actual physical damage to the property and whether that resulted in the complete or partial shutdown of the property, which means fewer guestrooms available to guests and some policies have coverage when employees, customers and/or suppliers cannot reach a hotel and allow it to operate normally (Wroten, 2017). The latter further added that if the roads are blocked or burned out, no one can reach the hotel and evacuation orders would also prevent people and suppliers from driving to hotels, so there are policies that cover that as well. In today's climate evolution, it has become very difficult to foresee disasters that can be caused by forthcoming sudden calamities and as such, if a tourism and hospitality organisation, like a hotel, does not have a business-interruption coverage, it would be high time to get one.

5.8. Recommendation 8: Collaboration and Support

Seetanah and Fauzel (2018) stated that climate change can negatively impact on tourism demand and thus lead to various negative effects on the economy of these countries mainly in terms of low growth rates and loss of employment. It, therefore, becomes mandatory for the government and tourism organisations to extend their support to tourism businesses that are prone to facing the harsh impacts of cyclones on them. A vulnerability index of impacts of cyclones on hotels can be brought forward which can aid to gauge in advance the potential impacts of cyclones on the latter and thus, accordingly plan and provide support. Support can be in the form of subsidies provided to manage disaster recovery services, provision of emergency medical caravans to tourists of hotels, and cleaning up facilities put at the disposition of firms, amongst others. For RE/MAX 24 Mauritius (2024) post a recent cyclone, the government establishments and service squads in this small island nation, were also enthusiastically engaged in clean-up operations across Mauritius, ensuring the island's attractiveness and safety were restored. As such, more such endeavours have to be reinforced across the island to extend the latter resilience and efficiency in dealing with the cyclones' aftermaths.

5.9. Recommendation 9: In-House Support to Tourism Employees

One immediate measure that should be implemented or improved would be the provision for year-round training to hoteliers or employees in the sector such that they can tactfully deal with natural calamities. One of the main concerns of employees while they are working during cyclonic conditions, would be the safety of their own families at home. To ensure that team members are performing diligently their work during cyclones without having such stress, for example, hotels can intervene by providing the right quantitative and qualitative support in the form of despatch of provision to families or proving morale support with regular intercom connection with them. Moreover, team members and families can be provided support by, for example, allowing them to stay in conference halls and ballrooms and provision for basic food and beverage items.

As per The Pharmacy Guild of Australia (2013), employee safety comes first so prepare, distribute and exercise your business emergency response plan and recovery and also establish a call-down procedure for warning and post-storm communications. Also, staff have to be called to return to work when safe to do so as per Mauritius' National Disaster Risk Reduction and Management Committee on the advice that it is safe to travel on roads. As such, the right resources such as transport facilities with accompanying ambulances, for emergency cases, can be considered.

5.10. Recommendation 10: Revenue Diversification

Tourism businesses in Mauritius should work and build better and necessary resilience towards cyclones. Hence, diversification strategies as a means to adapt to such climatic changes can be a potential solution. While tourists look for enjoying the sea, sand and sun, construction plans of hotels should be more sustainable; whereby the belief of 'being nearer to the sea is best' is diluted. Instead, new emerging hotels or tourism attractions or restaurants should consider moving inland so that rising tides and sand erosion do not impact on them during harsh cyclonic conditions. For Vousdoukas et al. (2023), consequent adaptation determinations will be necessary to lessen littoral flood impressions. At the same time, there will be a higher potential for these firms to allow the tourists to engage with the local communities and benefit from other inland attractions such as mountains and lakes. Necessary frameworks have to be established to avoid dilution of local culture with the mingling of tourists with the local communities. Today's buzzword is personalisation. When many tourists look for sea vacations, others visit Mauritius for its heritage, culture or food. As such, during the booking phase itself, tourists' needs can be better catered for by asking them what they truly desire to experience during their stays in Mauritius. As such, micro-marketing targeting strategies can be considered which can eventually lead to higher satisfaction levels of tourists. There should be less reliance of tourism businesses on the sea and sand and more integration of the locals; for example by promoting concepts such as 'the Mauritian style of living' or 'enjoy cyclones in the Mauritian style'. These concepts seem to be different and strange but can work amazingly well for tourists who truly want to enjoy vacations by following an anthropologist's pathway.

6. Conclusion

A tourism business' lifeline is truly a sensitive one because of such natural calamities like cyclones and the status is even worse for SIDS like Mauritius. Tourism businesses have to evaluate the efficacy of their warning systems and measures. Both should be addressing cyclonic conditions of two extremities – from the least dangerous to the worst-case scenarios. Taking onboard the social factor is key while working on the above as it involves living entities. Ghosh (2023) states that it is infrequent for someone to lose his life due to a storm in Mauritius and the cyclones do not stress the island's social setup as the country is finely organised for storms, and its peoples recognise clearly what to do when a fierce cyclone is upcoming. Yet, as per Ziegler et al. (2023), it is vital to add tourism to disaster risk management arrangements especially where menaces and the influx of tourists are elevated given that the latter is part of the temporary inhabitants who may not be aware of local settings and hence, not well prepared to deal with dangers. More so, several stakeholders put at risk their lives during cyclones. According to Ladkoo (2019), Mauritius is today confronting appalling concerns related to the climate crisis, making the habitual operations of many of its sectors unpredictable where one of them is the tourism business. Suppliers, customers and employees of tourism firms, who are the prime day-to-day drivers of the latter, have to be given due consideration through backup supply management systems, customer care and employee work comfort and safety. Also, there has to be the inclusion of novel indoor activities as a means to diversify the existent activities and not solely relying on outdoor ones so as to adapt to cyclonic conditions. Hotels, travel agencies and airports have to work in collaboration for better and enhanced ways of mediating flight cancellations, rebooking and rescheduling while encouraging customers to adopt weather insurance policies for their own betterment. The creation of a vulnerability index for gauging impacts of cyclones on tourism businesses can help to accordingly calibrate the support level needed per stakeholder category. Government intervention by way of subsidies can greatly help. While loss of lives has decreased in those islands and countries affected by cyclones, there has, however, been continuous loss of property. Tropical cyclones on SIDS have thus, both direct and indirect costs such as cost of damage, cost related to the warning systems, cost of planning and application of contingency plans, cost of comfort and safety given to several stakeholders and cost of business losses amongst others. Warning systems and measures to cope with cyclones should be in line with the physical capabilities of the people involved, the technological advancement of the society and financial–economic potential of the country and tourism businesses.

References

Amelung, B., Nicholls, S., & Viner, D. (2007). Implications of global climate change for tourism flows and seasonality. *Journal of Travel Research*, *45*(3), 285–296.

Becken, S. (2010). *The importance of climate and weather for tourism: Literature review*. Land Environment and People.

Braun, V., & Clarke, V. Using thematic analysis in psychology. *Qualitative Research in Psychology*, *3*, 77–101.

Capacity for Disaster Reduction Initiative. (2020, March 26). *Policy makers digest*. https://www.cadri.net/system/files/2021-06/Mauritius%20-%20Policy%20Makers%20Digest%20-%2005102020.pdf

De Freitas, C. R. (2003). Tourism climatology: Evaluating environmental information for decision making and business planning in the recreation and tourism sector. *International Journal of Biometeorology*, *48*, 45–54.

Faulkner, B., 2001. Towards a framework for tourism disaster management. *Tourism Management*, *22*(2), 135–147.

Ghosh, A. (2023). Reflections while in Mauritius. *Journal of Indentureship and Its Legacies*, *3*(1), 150–153.

Glaesser, D., 2003. *Crisis management in the tourism industry*. Elsevier.

Gök, H. S., & Aylan, F. K. (2024). The recovery of the tourism industry from pandemics and disasters. In *Dynamics of the tourism industry* (pp. 31–53). Apple Academic Press.

Ladkoo, A. D. (2019). Sensitization of tourists about climate change and its associated impacts on the tourism sector: The case of Mauritius. In *The Routledge handbook of tourism impacts* (pp. 327–337). Routledge.

Leoni, V., & Boto-García, D. (2023). The effect of natural disasters on hotel demand, supply and labour markets: Evidence from the La Palma Volcano eruption. *Environmental and Resource Economics*, *86*(4), 755–780.

Leroux, M. D., Bonnardot, F., Somot, S., Alias, A., Kotomangazafy, S., Ridhoine, A. O. S., Veerabadren, P., & Amélie, V. (2024). Advancing climate services for vulnerable islands in the Southwest Indian Ocean: A combined approach of statistical and dynamical Cmip6 downscaling. *Climate Services*, *34*, 1–26. https://doi.org/10.2139/ssrn.4728862

Mauritius Meteorological Services. (n.d., January 24). *Warning system*. http://metservice.intnet.mu/tropical-cyclone/warning-system.php

Mauritius-Holidays-Discovery.Com. (2018, January 30). *Mauritius weather: What is it like to go through a tropical cyclone?* http://www.mauritius-holidays-discovery.com/mauritius-weather.html

NOAA Ocean Exploration. (2024, March 22). *How do hurricanes impact the deep ocean?* https://oceanexplorer.noaa.gov/facts/hurricane-impact.html

National Ocean Service. (2018, January 10). *What is the difference between a hurricane and a typhoon?* https://oceanservice.noaa.gov/facts/cyclone.html

RE/MAX 24 Mauritius. (2024, March 10). *Now you can enjoy holiday in Mauritius: Tourism recovers from post cyclone Belal*. https://www.remax24.com/news/now-you-can-enjoy-holiday-in-mauritius-tourism-recovers-from-post-cyclone-belal/

Reynolds. (2016, May 28). *What you need to know about your hotel if a hurricane strikes.* https://www.oyster.com/articles/55336-what-you-need-to-know-about-your-hotel-if-a-hurricane-strikes/

Seetanah, B., & Fauzel, S. (2018). Investigating the impact of climate change on the tourism sector: Evidence from a sample of island economies. *Tourism Review*, *74*(2), 194–203.

Suarez, C. (2011, March 2). *How the tourism industry can prepare for climate change*. https://www.greenbiz.com/blog/2011/08/08/how-tourism-industry-can-prepare-climate-change

The Pharmacy Guild of Australia. (2013, May 10). *Responding to tropical cyclones*. https://www.guild.org.au/__data/assets/pdf_file/0014/5342/responding-to-tropical-cyclones-fact-sheet.pdf

The Times of India. (2023, March 22). *Tropical cyclones and storm surges: Why they are deadly?* https://timesofindia.indiatimes.com/world/rest-of-world/tropical-cyclones-and-storm-surges-why-they-are-deadly/articleshow/100209081.cms

Vousdoukas, M. I., Athanasiou, P., Giardino, A., Mentaschi, L., Stocchino, A., Kopp, R. E., Menéndez, P., Beck, M. W., Ranasinghe, R., & Feyen, L. (2023). Small island developing states under threat by rising seas even in a 1.5° C warming world. *Nature Sustainability*, *6*(12), 1552–1564.

Walshe, R. A. (2022). 'Who could have expected such a disaster?' How responses to the 1892 cyclone determined institutional trajectories of vulnerability in Mauritius. *Journal of Historical Geography*, *75*, 55–64.

Wroten, B. (2017, May 12). *Hotels navigate loss-of-business claims after disasters*. http://hotelnewsnow.com/Articles/254764/Hotels-navigate-loss-of-business-claims-after-disasters

Ziegler, A. D., Wasson, R. J., Sundriyal, Y., Srivastava, P., Sasges, G., Ramchunder, S. J., Ong, C. E., Nepal, S. K., McAdoo, B. G., Gillen, J., Bishwokarma, D., & Apollo, M. (2023). A call for reducing tourism risk to environmental hazards in the Himalaya. *Environmental Hazards*, *22*(1), 1–28.

Chapter 15

Understanding Impacts of Tourism for Destination Sustainability: A Residents' Perceptions Study of Leh, Ladakh

Sonam Angmo

Independent Researcher, Ladakh, India

Abstract

Resident perceptions of tourism impacts are a swiftly growing area in tourism research that has the potential to contribute to sustainability of destinations. Understanding and analyzing the perceptions of local communities for improving destination attractiveness is key to many destinations, the most popular being mountainous regions. Leh, Ladakh, in the northern part of the Trans-Himalayan region is a renowned Indian tourist place, which sees hordes of tourists in the short summer season. The influx of a large number of visitors in the short season has proved beneficial and also a problem for the destination as tourist numbers overpower the local population numbers leading to pressure on natural resources. Costs of mass tourism are becoming prominent each year and if not scrutinized can damage the likeability of the place. This research attempts to understand resident opinions of impacts of tourism in Leh through primary research. It also aims to find out the link between the demographic profile of residents and perceptions of tourism impacts using a Chi-square test. Results suggest that residents perceive many unfavorable impacts of tourism along with positive impacts. The Chi-square test reveals a significant association between demographic variables and tourism impacts.

Keywords: Resident perceptions; tourism in Ladakh; destination sustainability; impacts of tourism; perceptions of tourism impacts

The Need for Sustainable Tourism in an Era of Global Climate Change: Pathway to a Greener Future, 319–333

doi:10.1108/978-1-83608-668-020241042

Introduction

The tourism industry as one of the major economies of the world has undergone immense transformations and evolved over time (Michael Hall et al., 2021). From becoming a formalized sector to embracing digitization in sales and services to incorporating sustainability, the industry has been quick to adapt and continues to do so. Sustainable development as a theme emerged in 1987 from the Brundtland Commission Report which stated that development should look into the needs of both present as well as future generations (Brundtland Commission, 1987). The theme of sustainability is based on three pillars – economic, socio-cultural and environmental dimensions, also known as the triple bottom line (Haid et al., 2021). These three dimensions are the most affected by tourism development. Although tourism influences various interested parties, it is the local communities or residents who are affected directly by tourism (Sharpley, 2014). Their values, culture, behavior, etc., can be influenced which may or may not be acceptable to the larger population (Richards, 2007).[1]

In the last two decades, these influences have been observed carefully and the impacts of tourism have been studied by various tourism researchers. As residents are the most deeply influenced by tourism, it is vital to research the opinions of residents (Vareiro et al., 2013). Investigating these perceptions may help to maximize the welfare from tourism and lessen the adverse effects. If resident attitudes of tourism impacts are not examined, then governments or tourism management organizations will not be able to promote tourism in the correct manner at the destinations (Šegota et al., 2024). Their opinions with respect to tourism development projects and decision-making plans have to be considered as they interact with tourists regularly. Therefore, resident perceptions of tourism impacts have grown tremendously over several years with the aim of seeking to uncover the feelings and opinions of local communities (Johnson et al., 1994).

Local community attitudes are studied at all levels – international, national, regional and local levels (Lew et al., 2016). Different tourism destinations like islands, mountains, areas close to national parks, and wildlife sanctuaries may have communities with different perceptions toward tourism development. This is especially true of mountain destinations which have unique culture, lifestyle, climate and way of life (Johnston & Edwards, 1994). Leh, Ladakh, is one such mountainous destination that is famous among tourists world over for its natural beauty and adventure activities (Dar et al., 2022). Tourism in Leh, Ladakh, has been a prime contributor to its economic growth and prosperity (Pelliciardi, 2013). Be it infrastructure development, opportunities for business and employment, raising income and quality of life of people, tourism has no doubt benefited every resident of Leh. However, tourism is a double-edged sword (Martin, 2007), and therefore, has led to negative impacts as well at the destination. Accumulation of waste and plastic, air and water pollution, scarcity of water, adoption of Western culture, employment of outsiders and unplanned development are just a few of the negative impacts currently being witnessed by residents of

[1]A part of this paper has been published as a conference proceeding.

Leh (Chaudhary & Angmo, 2021). Moreover, these negative impacts are growing parallel to the benefits from tourism. Hence, it has become mandatory to study the impacts of tourism in Leh by understanding the perceptions of its residents.

The aim of this research is to examine the perceptions of residents of Leh toward the different impact dimensions through primary data. It seeks to understand specific impacts perceived by them – be it positive or negative. The study also attempts to understand if any association exists between demographic characteristics of locals and opinions of tourism impacts. Lastly, the study aims to contribute to the existing body of literature on tourism in Ladakh for sustainability of the place and to mitigate the unwanted impacts of tourism.

Review of Literature

Resident perceptions of tourism impacts are a broad field of study that is increasingly becoming important in today's times. Understanding the residents or the local community offers insights into the different aspects of tourism that a tourist or a tourism supplier might not perceive or feel (Sharma & Gursoy, 2015). It is the local people who stay at the destination and experience all the pros and cons of tourism development (Costa et al., 2020). It is the community who will engage with the tourists and make them feel comfortable and welcoming or unaccepted at the place. If locals are not happy with tourism, then the destination cannot prosper in the best way (Goodwin, 2002). Hence, it becomes compulsory to study the perceptions of residents toward the impacts of tourism and understand whether they support tourism or oppose tourism at their destination.

Studies on Resident Perceptions of Tourism Impacts

The majority of authors working on resident perceptions sought to discern the impacts of tourism and whether those residents favored tourism development or not. Milman and Pizam (1988) established that locals in Central Florida supported tourism and that residents found most economic impacts as positive but agreed that there were negative impacts as well. Gursoy et al. (2019) studied resident perceptions using Social Exchange Theory and suggested a strong link between impacts and support for tourism development but did not see improved results for environmental impacts. Mannaa and Abou-Shouk (2020) conducted a comparison study of residents and expatriates in the UAE and found both set of respondents to promote tourism at the destination. However, the resident group identified negative impacts on the economy and environment whereas the expatriate group identified more favorable economic impacts.

Studies on Tourism Impacts in Leh

Goering (1990) suggested issues like commercialization of culture, copying of tourist behavior and leakage of tourism revenues out of Leh. Fox et al. (1994) discussed tourism in Ladakh and learned about a shift in occupation of residents from traditional farming to a cash-based economy. Aengst (2014) discussed socio-cultural impacts in Ladakhi adolescents and found different practices emerging like dating, roaming, etc., happening among the youth of Ladakh. Khandekar (2017) highlighted water shortage issues, an increase in accommodation centers and waste generation near Pangong Lake, etc. Wani et al. (2018) stressed that waste problems in Leh exceeded the normal range of 1.2–1.6 kg per room/day and that there was a tremendous surge in waste accumulation in the peak summer season due to filling of large number of rooms.

Studies on Destination Sustainability: Impacts of Tourism

Economic Impacts of Tourism

The economic benefits of tourism are the most talked about aspect which helps a destination grow strongly. Tourism helps to bring foreign exchange, provides jobs for people, raises income levels of residents and improves the quality of life of its citizens (Marzuki, 2012). Public facilities are enhanced and infrastructure is developed. These economic aspects attract scores of people to invest in travel and tourism businesses (Boley et al., 2018). Literature on the impacts of tourism discusses different aspects and themes of economic impacts. Briedenhann and Wickens (2004) learned about economic impacts in rural areas and found that safety issues for tourists needed to be relooked and the government needed to assume responsibility for their safety. Kim and Chen (2006) discussed tourism development in Taiwan and stated that complete dependence on tourism industry should be reduced and other industries should be opened up. Lundberg (2011) conducted a research on two ice hotels in Sweden and stated that the hotel sector and transfers to the airport affected the destination economically. Dhakal (2017) established that tourism has enhanced economic opportunities and livestock production in Nepal but negatively affected the price of houses and goods.

Socio-Cultural Impacts of Tourism

Socio-cultural impacts are very important for a destination and its local communities. Tourism can bring with various changes to the socio-cultural fabric of the community which can have both good and bad effects (Gjerald, 2005). Positive changes can include preservation of traditions and rituals, pride in one's own culture and promotion of understanding between different cultures (Alhasanat & Hyasat, 2010). On the flip side, a destination may experience staged authenticity, adoption of Western culture and traditions, disrespect to monuments and attractions and so on (MacLeod, 2006). Andereck et al. (2005) revealed that locals who received benefits from tourism felt more positive impacts than those who incurred costs from tourism. Nyaupane et al. (2006) opined that tourism was affecting the schooling of children in tourist areas of Nepal and China but also admitted that it empowered women and helped in the preservation of culture. Afthanorhan et al. (2017) found that positive and negative perceptions have an influence on support for tourism development in which female respondents showed more enthusiasm. Pham et al. (2019) examined impacts in the United States of America and stated that a positive correlation was found between quality of life and support for tourism development.

Environmental Impacts of Tourism

Environmental impacts are the most evident at a destination as compared to social and economic impacts. Tourism can cause a whole lot of negative environmental impacts such as waste generation, pollution of water and air, disturbance to wild animals and birds, loss of biodiversity, soil erosion, etc. (Liu et al., 1987). Pickering et al. (2003) found that resort managers perceived negative impacts on the quality of air and water. Singh et al. (2009) suggested that the costs of tourism in Uttarakhand were deforestation, waste generation, etc., due to unplanned tourism. Zhong et al. (2011) based their study in China and found that the bulk of the impact studies were descriptive in nature and suggested investigation of environmental impacts and calculation of carrying capacity levels.

Association Between Demographic Characteristics of Residents and Perception of Impacts

A number of authors researched on the link between the demographic characteristics of residents and the impacts of tourism. Wu and Chen (2015) discussed resident perceptions in Macau and Singapore and discovered differences in relationships between social backgrounds and opinions toward gambling impacts. Wu et al. (2020) examined the attitudes of residents in Hainan, China, and affirmed that residents in varied professions can have different impacts on perceptions. Bhat and Mishra (2021) investigated demographic characteristics in Kashmir and again found a positive effect between demographics and opinions of tourism impacts. Almeida-García et al. (2016) conducted their research in Spain and affirmed that educational background, place of birth and length of residency showed a close connection with impact perceptions. Those who lived in the area for less than 5 years had more positive opinions of tourism impacts.

Thus, based on the above review, it can be deduced that

a. Residents are amongst the chief tourism stakeholders who experience all the positive and negative impacts of tourism.
b. Positive impact perception can lead to favoring of tourism in some cases.
c. Negative impact perception may not certainly lead to resistance toward tourism.
d. Tourism impact studies on Leh are scanty and focus on highlighting environmental issues like scarcity of water, waste generation, etc.
a. The association between demographic characteristics of residents and tourism impacts is significant but has given mixed results in different contexts.

As per the above findings, objectives were formed for the research:

Objectives of the Study

1. To understand resident perceptions toward economic, socio-cultural and environmental impacts of tourism in Leh.
2. To investigate the association between the demographic characteristics of residents and their opinions of tourism impacts.

Hypotheses of the Study

(i) Null Hypothesis $H0_1$: Residents majorly perceive negative economic, socio-cultural and environmental impacts of tourism.
(ii) Null Hypothesis HO_2: There is no significant association between the demographic characteristics of residents and their perceptions of tourism impacts.

Research Methodology

Research Area

Leh is one of the two main districts of the Ladakh region situated in the Trans-Himalayas (Müller et al., 2020). It was an important site on the historic Silk Route and is popular among adventure tourists around the world (Dey, 2018). Leh, Ladakh, remained unexplored to even most Indians till the 1960s due to harsh topography and high altitudes (Bera, 2015). Gradually, it started receiving tourists (mostly foreigners) in the 1970s and by the 2000s it was seeing a huge number of domestic as well as foreign tourists (Pelliciardi, 2010). The reason for its popularity was its natural landscape, Buddhist culture, and the warmth and hospitality of the locals which was popularized even more by the Bollywood movie "3 Idiots" (Vogel & Field, 2020). Its tourist attractions are simply incredible with pristine lakes like the Pangong Lake, ancient Buddhist monasteries like the Thiksey, Hemis, and Lamayuru and popular festivals like the Hemis festival, etc. (Sood, 2000).

The tourist attractions of Ladakh have attracted a large number of tourists and as a result, Leh district has rapidly evolved as a tourist destination. In 2018, tourist numbers reached three lakh mark for the first time (Bhatt, 2019). In 2023, the total number of tourist arrivals in Ladakh was 5,25,374 with 36,315 foreign tourists and 4,89,059 domestic tourists (Tourism Department Leh, 2023). The mass arrival of tourists over the years has contributed to economic growth and the development of infrastructure at the destination. The increased growth of tourism has also encouraged many residents to set up hotels, homestays, and travel agencies and buy taxis to accommodate the growing number of tourists (Gondhalekar et al., 2015). Changes have also been felt in the socio-cultural aspect wherein a large chunk of the society has embraced Western dressing and culture and few traditions have been modified. In spite of the sway of mainland culture and the effect of Westernization, the region has till now been able to preserve its unique culture to some extent (Lhaskyabs, 2018). Impacts of tourism-both positive and negative have been seen but the people of Ladakh have been welcoming toward tourists and have supported tourism development till now.

Data Collection

A questionnaire with a structured format was used to obtain primary data from the locals. The sample size for the research comprised of 280 residents. The questionnaire was split into three sections. The first part included 25 questions on the impacts of tourism. The second section comprised of demographic profile of the respondents. Responses were collected on a 5-point Likert scale with 1 being "completely agree," 2 being "agree," 3 being "neutral," 4 being "disagree" and 5 being "completely disagree" for negative statements and 5 being "completely agree," 4 being "agree," 3 being "neutral," 2 being "disagree" and 1 being "completely disagree" for the positive statements. Convenience sampling was employed to get responses across three summer seasons in 2018 and 2019. Respondents were surveyed at their homes, offices, cafes, marketplace, etc.

Research Findings and Analysis

Demographic Profile of Respondents

Before commencing with descriptive statistics of the data, it is important to have an idea about the demographic profile of respondents of the study. For this purpose, a frequency table was drawn and it is presented in Table 15.1.

Table 15.1. Demographic Profile of Residents.

Demographic Variable	Options	Number	Percentage
Gender	Male	128	45.7
	Female	152	54.3
Age	15–25 years	66	23.6
	26–36 years	98	35.0
	37–47 years	72	25.7
	48–58 years	38	13.6
	59 and above	6	2.1
Annual income	Less than 1 lakh	38	13.6
	1 lakh–2 lakh 99 thousand	102	36.4
	3 lakh–4 lakh 99 thousand	74	26.4
	5 lakh–6 lakh 99 thousand	34	12.2
	7 lakh and above	32	11.4
Educational Qualification	No schooling	52	18.6
	Upto 10th/12th class	82	29.3
	Graduate degree	80	28.6
	Master's degree	30	10.6
	Others	36	12.9
Marital Status	Single	130	46.4
	Married	140	50.0
	Widowed/divorced/separated	10	3.6

Source: Author's own research in SPSS.

The demographic profile informs that more females as compared to males participated in this study. Also, younger respondents constituted the majority and most had qualifications up to 10th standard. Most respondents were married and had an annual income in the band of 1–2.99 lakhs.

Impacts of Tourism as Perceived by Residents

Residents perceive different impacts of tourism – economic, socio-cultural and environmental. To get a better comprehension of these impacts, this research used means to analyze resident perceptions. The results are given in Table 15.2.

For economic impacts, two negative impacts stand out – "increased cost of goods and services" (1.64) and "occupation transformation by locals" (1.69). This signifies that locals feel that tourism has contributed to an increase in the cost of goods and services due to the increased influx of tourists and resultant increases in prices. Residents also feel that locals of Leh have changed their occupations and moved to tourism as they see greater income and profits from pursuing tourism businesses. With regards to positive economic impacts, two impacts show the highest means "Increased employment opportunities" (4.39) and "increased business opportunities" (4.40). This stands true because tourism has opened a lot of avenues in Leh and youngsters have the opportunity to get employed at hotels, cafes, restaurants and other tourism-related businesses like transportation, etc. It has also opened business opportunities as residents can set up their own travel and tourism enterprises.

For socio-cultural impacts, two negative impacts stand out – "conversion of monasteries into tourist attractions" (1.94) and "disturbance in family" (2.96) (see Table 15.3). Local residents opine that tourism has led to monasteries becoming

Table 15.2. Economic Impacts as Perceived by Residents.

Impact statements	Mean
Negative impacts*	
➢ Increased price of land and housing	1.64
➢ Occupation transformation by locals	1.69
➢ Increased cost of goods and services	1.71
➢ Increased employment of outsiders	1.86
➢ Increased cost of living	2.11
Positive impacts**	
➢ More investments in tourist infrastructure	2.54
➢ Improved public convenience	3.52
➢ Increased employment opportunities	4.39
➢ Increased business opportunities	4.40

Source: Adapted from Wu and Chen (2015).

*(1– completely agree, 2 – agree, 3 – neutral, 4 – disagree, 5 – completely disagree).

**(1 – completely disagree, 2 – disagree, 3 – neutral, 4 – agree, 5 – completely agree).

Table 15.3. Socio-cultural Impacts as Perceived by Residents.

Impact Statements	Mean
Negative impacts*	
➢ conversion of monasteries into tourist attractions	1.94
➢ disturbance in family values	2.18
➢ adoption of Western culture	2.23
➢ crime and violence	2.41
➢ disrespect and damage to monasteries	2.96
Positive Impacts*	
➢ Improved education system	3.33
➢ pride in culture among locals	3.64
➢ empowerment of women	3.69
➢ preservation of monasteries	4.00

Source: Adapted from Wu and Chen (2015).
*(1 – completely agree, 2– agree, 3 – neutral, 4 – disagree, 5 – completely disagree).
**(1– completely disagree, 2 – disagree, 3 – neutral, 4 –agree, 5 – completely agree).

attractions for tourists which is not desirable for a society and its cultural fabric. They also perceive family disturbances due to tourism as parents are engaged in the hotel and travel agency business and do not possess sufficient time for their children. However, residents seem to perceive two impacts very positively – "preservation of monasteries" and "empowerment of women." This signifies that locals of Leh feel that tourism has led to better preservation of monasteries as they are maintained on a regular basis to attract all sorts of visitors – be it domestic or international. Residents also feel that women are empowered in Leh because of tourism. This stands true as a large segment of women earn their livelihoods through working in hotels, cafes and restaurants. Many Ladakhi women have also set up their own tourism businesses rendering them financially independent and thus, empowered.

Based on Table 15.4, it can be inferred that two negative impacts have the highest means – "loss of land to hotels" (1.65) and "pollution and shortage of water" (1.69). This shows that residents of Leh perceive that tourism has induced the loss of land to make way for the construction of hotels and has also caused pollution and shortage of water in Leh. As far as positive impacts are concerned, awareness of environmental issues is high (3.60) among residents because of tourism and tourism-related initiatives focusing on the preservation of the environment.

Therefore, the first hypothesis – "Residents of Leh majorly perceive negative economic, socio-cultural and environmental impacts of tourism" stands rejected as residents perceive both positive and negative impacts of tourism in all three domains – i.e., economic, socio-cultural and environmental. This stands especially true for economic and social impacts in which an ample number of positive

Table 15.4. Environmental Impacts as Perceived by Residents.

Impact Statements	Mean
Negative impacts*	
➢ Loss of lands to hotels	1.65
➢ Pollution and shortage of water	1.69
➢ Loss of vegetation and wildlife	2.50
Positive impacts**	
➢ Love and respect for nature	3.39
➢ Awareness of environmental issues	3.60

Source: Adapted from Wu and Chen (2015).
*(1– completely agree, 2– gree, 3 – neutral, 4 – disagree, 5 – completely disagree).
**(1– completely disagree, 2 – disagree, 3 – neutral, 4 – agree, 5 – completely agree).

impacts are seen. Hence, residents of Leh feel that tourism has caused a mix of positive and negative impacts and it would be incorrect to say that only negative impacts are perceived by them.

Association Between Demographic Characteristics of Residents and Their Perceptions of Tourism Impacts

Literature on resident perceptions of tourism impacts has frequently mentioned the role and importance of demographic characteristics of residents wherein demographic variables like age, educational qualification, marital status, etc., have affected resident perceptions in different contexts (Sharma & Dyer, 2009). For this study as well, the link between demographic characteristics and perceptions of tourism impacts was calculated. The results are given in Table 15.5.

Results show that there is a strong association between the demographic characteristics of residents and their perceptions of tourism impacts. The demographic variable "age" depicts a significant association with both positive economic and negative economic impacts. Age, however, shows a significant relationship with negative socio-cultural and positive environmental impacts. This means that residents of different ages will have varied perceptions of tourism impacts. Aged or elder residents might perceive negative socio-cultural and environmental impacts from tourism due to their vast experience in life in comparison to younger residents who might view tourism in a positive light due to its quick economic benefits.

Marital status is related to both positive and negative economic impacts. It also shows a significant relationship with negative socio-cultural and negative environmental impacts. Married residents may have a different perception toward impacts in contrast to singles or divorced residents. Married men and women will have a positive perception of economic impacts because of the engagement of their spouses in tourism business. They may also perceive economic impacts negatively due to the same logic because being husband-wife means their perceptions

Table 15.5. Association Between Demographic Characteristics of Residents and Perceptions of Tourism Impacts.

Perception of Tourism Impacts	**Demographic Characteristics of Residents**				
	Age	**Marital Status**	**Size of Family**	**Educational Qualification**	**Annual Income**
	Chi-square value				
Positive economic impacts	64.795*	52.777*	32.819	74.844*	74.572
Negative economic impacts	100.365*	70.818*	87.761*	80.996	104.244*
Positive socio-cultural impacts	112.632	55.153	68.595*	111.718	108.105
Negative socio-cultural impacts	105.409*	90.011*	69.960*	123.775*	106.858
Positive environmental impacts	123.679*	17.897	58.021*	66.743	58.317
Negative environmental impacts	49.984	34.676*	51.677*	65.439*	56.865*

Source: Adapted from Wu and Chen (2015).
*$p<0.05$.

are shaped by their spouse's perception and his/her connection to the business. Married people may also perceive negative socio-cultural and negative environmental impacts as they will be more concerned about their kids and the negative impact of tourism on their daily lives. They may also be worried about the environmental impacts of tourism as they would like their kids to grow up in a clean and sustainable environment. Single residents may not be so concerned about the social fabric and the environment as much as married couples would.

Size of family is another important demographic variable that shows a robust link with almost all the tourism impacts except positive economic impacts. Residents with different family sizes will have varied perceptions toward tourism impacts. Those with large family sizes may have more negative perceptions toward economic, socio-cultural and environmental impacts due to greater interactions of family members with negative tourism experiences. Smaller size families may perceive more positive socio-cultural and environmental impacts due to lesser interactions with negative experiences and might view tourism more positively.

Educational qualification shows a positive association with economic impacts and a negative association with socio-cultural and environmental impacts. This could be due to the fact that highly educated people might perceive positive economic impacts because they have more information and awareness about the benefits of tourism and the tourism projects of the destinations. Similarly, they might also be more knowledgeable about the negative socio-cultural and environmental impacts emanating from tourism activities at the destination.

The last demographic variable is annual income which is negatively related to all three impacts – economic, socio-cultural and environmental. Residents drawing low annual incomes from tourism might have negative perceptions as they would not find tourism to be contributing much to their livelihoods or might feel that it is a seasonal source of occupation. For this same reason, they might blame tourism for its negative socio-cultural and environmental impacts which provides little income but leads to a lot of negative problems.

Thus, the second hypothesis – "There is no significant difference in the perceptions of residents toward tourism impacts with respect to their demographic profile" stands rejected as the study has pointed out significant differences in the perceptions of residents toward tourism impacts with respect to their demographic profile.

Conclusion

This study made an attempt to investigate the perceptions of residents toward the impacts of tourism in Leh through primary research for destination sustainability. The results of this research are consistent with research studies from the literature which state that residents of a destination perceive both positive and negative impacts but mainly positive impacts. As far as negative impacts are involved, residents believed that tourism has led to a rise in the price of goods and services, conversion of religious places into attractions for tourists and a rise in pollution and water scarcity. However, they also acknowledged that opportunities for employment and business have increased in Leh because of tourism. Further, women of Leh, Ladakh, have become empowered due to tourism. Locals also discovered that there is awareness among individuals toward environmental issues due to tourism. This signifies that residents of Leh are satisfied with tourism development and support it and in no way, do they seem to oppose tourism even after perceiving few of the negative impacts of tourism.

As far as the link between demographic characteristics and perception of tourism impacts was concerned, it was revealed that a significant relationship emerged for all the demographic variables be it age, educational qualification, marital status and annual income. Therefore, this paper shows that there is a relationship between demographic variables and tourism impacts for the residents of Leh.

Tourism departments and destination management organizations can utilize this research to understand the negative impacts and work on them to provide better facilities to the tourists as well as the residents. They can also conduct discussion sessions and seminars with the local residents and try to understand their problems related to tourists and tourism. Positive impacts such as business opportunities and preservation works can be further strengthened to promote tourism at the destination. Adverse impacts like pollution and commodification of religious places can be addressed so as to incorporate sustainability at the destination. Lastly, local communities and their suggestions should be utilized in tourism policies and plans of Leh, Ladakh, for a better, sustainable future of both Leh, Ladakh and its tourism.[2]

References

Aengst, J. (2014). Adolescent movements: Dating, elopements, and youth policing in Ladakh, India. *Ethnos*, *79*(5), 630–649.

Afthanorhan, A., Awang, Z., & Fazella, S. (2017). Perception of tourism impact and support tourism development in Terengganu, Malaysia. *Social Sciences*, *6*(3), 106.

Alhasanat, S., & Hyasat, A. S. (2010). Sociocultural impacts of tourism on the local community at Petra. Jordan. *European Journal of Scientific Research*, *44*(3), 374–386.

Almeida-García, F., Pelaez-Fernandez, M. A., Balbuena-Vazquez, A., & Cortés-Macias, R. (2016). Residents' perceptions of tourism development in Benalmádena (Spain). *Tourism management*, *54*, 259–274.

Andereck, K. L., Valentine, K. M., Knopf, R. C., & Vogt, C. A. (2005). Residents' perceptions of community tourism impacts. *Annals of Tourism Research*, *32*(4), 1056–1076.

Bera, T. R. (2015). *Ladakh: A glimpse of the roof of the world.* Partridge Publishing.

Bhatt, A. (2019, June 25). *Why are tourists staying away from Leh-Ladakh this season?* https://www.livemint.com/news/india/why-are-tourists-staying-away-from-leh-ladakh-this-season-1561433264997.html

Bhat, A. A., & Mishra, R. K. (2021). Demographic characteristics and residents' attitude towards tourism development: A case of Kashmir region. *Journal of Public Affairs*, *21*(2), e2179.

Boley, B. B., Strzelecka, M., & Woosnam, K. M. (2018). Resident perceptions of the economic benefits of tourism: Toward a common measure. *Journal of Hospitality & Tourism Research*, *42*(8), 1295–1314.

Briedenhann, J., & Wickens, E. (2004). Tourism routes as a tool for the economic development of rural areas—Vibrant hope or impossible dream? *Tourism Management*, *25*(1), 71–79.

Brundtland Commission (1987). *Our common future: Report of the world commission on environment and development*. Oxford University Press.

Brunt, P., & Courtney, P. (1999). Host perceptions of sociocultural impacts. *Annals of Tourism Research*, *26*(3), 493–515.

Chaudhary, M., & Angmo, S. (2021). Developing tourism impact index: A measure of tourism suppliers' perceptions in Leh, India. *International Journal of Tourism Policy*, *11*(3), 289–308.

Costa, T., Umbelino, J., de Lurdes Calisto, M., Nunes, S., & Afonso, V. A. (2020). Impacts of tourism and residents' perceptions: A study in the regional tourism area of Lisbon. *European Journal of Tourism, Hospitality and Recreation*, *10*(1), 28–40.

Dar, S. N., Shah, S. A., & Wani, M. A. (2022). Geospatial tourist information system for promoting tourism in trans-Himalayas: A study of Leh Ladakh India. *GeoJournal*, *87*(4), 3249–3263.

[2]A part of this paper has been published as a conference proceeding.

Dey, P. (2018, July 29). *Here is why you should visit Lamayuru, 'moonland' of Ladakh.* https://timesofindia.indiatimes.com/travel/destinations/here-is-why-you-should-visit-lamayuru-moonland-of-ladakh/as65193751.cms

Dhakal, B. (2017). Using factor analysis for residents' attitudes towards economic impact of tourism in Nepal. *International Journal of Statistics and Applications*, *7*, 250–257.

Fox, J. L., Nurbu, C., Bhatt, S., & Chandola, A. (1994). Wildlife conservation and land-use changes in the Trans-Himalayan region of Ladakh, India. *Mountain Research and Development*, 39–60.

Gjerald, O. (2005). Sociocultural impacts of tourism: A case study from Norway. *Journal of Tourism and Cultural Change*, *3*(1), 36–58.

Goering, P. G. (1990). The response to tourism in Ladakh. *Cultural Survival Quarterly*, *14*(1), 20–25.

Gondhalekar, D., Nussbaum, S., Akhtar, A., & Kebschull, J. (2015). Planning under uncertainty: Climate change, water scarcity and health issues in Leh town, Ladakh, India. *Sustainable Water Use and Management: Examples of New Approaches and Perspectives* (pp. 293–312).

Goodwin, H. (2002). Local community involvement in tourism around national parks: Opportunities and constraints. *Current Issues in Tourism*, *5*(3–4), 338–360.

Gursoy, D., Ouyang, Z., Nunkoo, R., & Wei, W. (2019). Residents' impact perceptions of and attitudes towards tourism development: A meta-analysis. *Journal of Hospitality Marketing & Management*, *28*(3), 306–333.

Haid, M., Albrecht, J. N., & Finkler, W. (2021). Sustainability implementation in destination management. *Journal of Cleaner Production*, *312*, 127718.

Johnson, J. D., Snepenger, D. J., & Akis, S. (1994). Residents' perceptions of tourism development. *Annals of Tourism Research*, *21*(3), 629–642.

Johnston, B. R., & Edwards, T. (1994). The commodification of mountaineering. *Annals of Tourism Research*, *21*(3), 459–478.

Khandekar, N. (2017). *Ladakh responds to tourism's demands on its water.* https://www.thethirdpole.net/en/climate/ladakh-water-tourism-demands-india/

Kim, H. J., & Chen, M. H. (2006). Tourism expansion and economic development: The case of Taiwan. *Tourism Management*, *27*(5), 925–933.

Lew, A., Ng, P.T., Ni, C., & Wu, T., 2016. Community sustainability and resilience: Similarities, differences and indicators. *Tourism Geography*, *18*(1), 18–27. https://doi.org/10.1080/14616688.2015.1122664.

Lhaskyabs, S. (2018, December 10). *Will all-weather connectivity and more tourists help Ladakh? The wire: Will all-weather connectivity and more tourists help Ladakh*?

Liu, J. C., Sheldon, P. J., & Var, T. (1987). Resident perception of the environmental impacts of tourism. *Annals of Tourism Research*, *14*(1), 17–37.

Lundberg, E. (2011). *Evaluating tourism impacts – A sustainable perspective* [Doctoral dissertation, University of Gothenburg]. https://gupea.ub.gu.se/bitstream/handle/2077/27968/gupea_2077_27968_1.pdf?sequence=1

MacLeod, N. (2006). Cultural tourism: Aspects of authenticity and commodification. *Cultural Tourism in a Changing World: Politics, Participation and (re) Presentation*, *7*, 177.

Mannaa, M. T., & Abou-Shouk, M. (2020). Students' perceptions towards working in the tourism and hospitality industry in the United Arab Emirates. *Al-Adab Journal*, *1*(135), 33–52.

Martin, C. B. (2007). *Tourism in the mountain south: A double-edged sword.* University of Tennessee Press.

Marzuki, A. (2012). Local residents' perceptions towards economic impacts of tourism development in Phuket. *Tourism: An International Interdisciplinary Journal*, *60*(2), 199–212.

Michael Hall, C., Scott, D., & Gössling, S. (2020). Pandemics, transformations and tourism: Be careful what you wish for. *Tourism Geographies*, *22*(3), 577–598. https://doi.org/10.1080/14616688.2020.1759131

Milman, A., & Pizam, A. (1988). Social impacts of tourism on central Florida. *Annals of Tourism Research*, *15*(2), 191–204.

Müller, J., Dame, J., & Nüsser, M. (2020). Urban mountain waterscapes: The transformation of hydro-social relations in the Trans-Himalayan town Leh, Ladakh, India. *Water*, *12*(6), 1698.

Nyaupane, G. P., Morais, D. B., & Dowler, L. (2006). The role of community involvement and number/type of visitors on tourism impacts: A controlled comparison of Annapurna, Nepal and Northwest Yunnan, China. *Tourism Management*, *27*(6), 1373–1385.

Pham, K., Andereck, K., & Vogt, C. (2019). Local residents' perceptions about tourism development. *Travel and Tourism Research Association: Advancing Tourism Research* Globally, 74. https://scholarworks.umass.edu/ttra/2019/research_papers/74

Pelliciardi, V. (2010). Tourism traffic volumes in Leh district: An overview. *Ladakh Studies*, *26*, 14–23.

Pelliciardi, V. (2013). Estimating total receipts for 2011 from tourism in Leh district. *Ladakh Studies*, *29*, 6–12.

Pickering, C. M., Harrington, J., & Worboys, G. (2003). Environmental impacts of tourism on the Australian Alps protected areas. *Mountain Research and Development*, *26*(3), 247–254. www.citeseerx.ist.psu.edu/viewdoc/download

Richards, G. (2007). *Cultural tourism: Global and local perspectives*. Psychology Press.

Šegota, T., Mihalič, T., & Perdue, R. R. (2024). Resident perceptions and responses to tourism: individual vs community level impacts. *Journal of Sustainable Tourism*, *32*(2), 340–363.

Sharpley, R. (2014). Host perceptions of tourism: A review of the research. *Tourism Management*, *42*, 37–49.

Singh, R. B., Mal, S., & Kala, C. P. (2009). Community responses to mountain tourism: A case in Bhyundar Valley, Indian Himalaya. *Journal of Mountain Science*, *6*(4), 394–404.

Sharma, B., & Dyer, P. (2009). An investigation of differences in residents' perceptions on the Sunshine Coast: Tourism impacts and demographic variables. *Tourism geographies*, *11*(2), 187–213.

Sharma, B., & Gursoy, D. (2015). An examination of changes in residents' perceptions of tourism impacts over time: The impact of residents' socio-demographic characteristics. *Asia Pacific Journal of Tourism Research*, *20*(12), 1332–1352.

Sood, V. K. (2000). Dynamics of tourism development and aspects of sustainability: A case study of Ladakh. *Tourism Recreation Research*, *25*(3), 115–127.

Tourism Department Leh. (2023). *Month Wise Tourist Arrival Statement for the year 2023*.

Vareiro, L. M. D. C., Remoaldo, P. C., & Cadima Ribeiro, J. A. (2013). Residents' perceptions of tourism impacts in Guimarães (Portugal): A cluster analysis. *Current Issues in Tourism*, *16*(6), 535–551.

Vogel, B., & Field, J. (2020). (Re) constructing borders through the governance of tourism and trade in Ladakh, India. *Political Geography*, *82*, 102226.

Wani, M. A., Dada, Z. A., & Shah, S. A. (2018). Sustainable integrated solid waste management in the trans-Himalayan accommodation sector. *African Journal of Hospitality, Tourism and Leisure*, *7*(2), 1–17.

Wu, H., Kim, S., & Wong, A. K. F. (2020). Residents' perceptions of desired and perceived tourism impact in Hainan Island. *Asia Pacific Journal of Tourism Research*, *25*(6), 583–601.

Wu, S. T., & Chen, Y. S. (2015). The social, economic, and environmental impacts of casino gambling on the residents of Macau and Singapore. *Tourism Management*, *48*, 285–298.

Zhong, L., Deng, J., Song, Z., & Ding, P. (2011). Research on environmental impacts of tourism in China: Progress and prospect. *Journal of Environmental Management*, *92*(11), 2972–2983.

Chapter 16

Interplay Between Global Tourism and Global Climate Change: Recent Trends

Nishi Pathak[a], Renu Ghosh[b], Lucky Gupta[a] and Jyoti Kasana[c]

[a] *Department of Management Studies, Raj Kumar Goel Institute of Technology, Ghaziabad, India*
[b] *Department of Management Studies, Netaji Subhas University of Technology, Delhi, India*
[c] *Department of Commerce, Kirori Mal College, University of Delhi, India*

Abstract

Tourist policies and businesses today are not far beyond the effect of climate change. Worldwide, tourist demand, and popular locations have been hit hard by an increase in the frequency and severity of extreme weather events such as floods, heat waves, droughts, storms, and wildfires. Significant tourist attractions are also impacted by climate change. Sea level rise and ocean warming endanger resources like beaches and coral reefs, while snowfall has become less predictable in many winter locations. The expense of travel is also expected to rise due to climate change. The tourism is contributing to the carbon emissions and increasing hazards to the environment. What these trends have in common is their potential to alter the face of tourism throughout the world. Research on the topic of climate change and tourism has been around for a while, but this paper lays out the field's evolution, outlines recent trends, and discusses the impact of climate change on tourism.

Keywords: Climate change; global risks; tourism; weather extremes; environment hazards; carbon emissions

1. Introduction

Because of its multi-sectoral nature and heavy reliance on environmental factors, the tourism industry is especially susceptible to the effects of climate change

The Need for Sustainable Tourism in an Era of Global Climate Change: Pathway to a Greener Future, 335–346

doi:10.1108/978-1-83608-668-020241044

(Scott & Gössling, 2022; UNWTO, 2008). Understanding and addressing the complex interplay between climate change and tourism requires a focus on geographical factors. These factors include the distribution of destinations, their exposure to changing weather patterns, the loss of significant tourism assets, and the subsequent effects on demand for tourism, the competitiveness of destinations, and the economic viability of the sector as a whole (Scott et al., 2012). Tourism locations possess distinctive natural landscapes and climatic conditions that contribute to their allure. The projected acceleration of climate change consequences is expected to bring about a significant modification in the geographical distribution of tourism (Gössling et al., 2012). While passengers have the freedom to choose their destinations freely, companies are restricted to certain locations and encounter difficulties such as asset loss, severe weather events, fluctuations in the cost of travel, and decreasing socio-economic stability. Recently, there has been a growing number of recorded consequences, with winter locations witnessing a decrease in snow cover and shorter winter seasons. Weather extremes, such as intense rains, storms, and heat waves, have caused disruptions to access and on-site conditions (Pereira et al., 2019). The occurrence of wildfires, which formerly had little impact on tourism, now increasingly affects several places and poses a danger to significant tourism zones, such as those in the Mediterranean, California, and Australia. Understanding the interconnected and sometimes sequential effects on destinations is still a significant area of knowledge that has to be filled (Scott & Gössling, 2022).

The geographical perspective also considers the economic aspects of tourism, including the expenses associated with travel and the sustainability of resources, which will be affected by different climate change scenarios. Severe weather phenomena have the potential to cause harm or destruction to transportation infrastructure, resulting in delays, cancelations, and higher expenses for upkeep. The increased occurrence and severity of these disasters need expenditures in adaptive measures, which puts further pressure on the financial viability of the sector. Climate change has a direct impact on agricultural output, causing unpredictable changes in the supply of food, which contributes to the increasing expenses associated with tourism. In light of this context, the study of tourism and climate change focuses on the geographic distribution of tourist sites according to their natural and climatic features, as well as their resources (such as beaches, biodiversity, and snow), socio-economic stability, and transport expenses. The area also examines the impact of tourism on climate change, as well as the economic, political, environmental, and social difficulties related to reducing carbon emissions.

According to UNWTO (2008), there are several ways of climate change like changes in mitigation policies, severe weather conditions affecting the footfall of tourists, impact on biodiversity, etc., which impact the tourism, but tourism also contributes to climate change through greenhouse gas emissions. Tourism and climate both are important for the growth of the economies worldwide. Tourism and climate change may be considered as two sides of the same coin; they cannot be separated and will continue to affect each other; but the salient aspect is the affect can be controlled by taking corrective measures for each. This study aims to study the impact that tourism and climate change have on each other, the recent trends and scope of future research. The objectives of this study are:

1. To explore the impact of global tourism and global climate change on each other.
2. To study the recent trends in global tourism and global climate changes.

2. Literature Review

The study on the relationship between climate and tourism has been conducted since the 1970s, but there was minimal study in the 1980s (Scott et al., 2019). An extensive examination of the subject from 1986 to 2020 revealed 1,290 pertinent articles, with the bulk being published between 2010 and 2020 (Scott & Gössling, 2022). The majority of research contributions have been produced by writers located in North America, Oceania (Australia and New Zealand), Europe, China, and South Africa. These locations are also the main focus of the study. Significant achievements in academia include the release of the (UNWTO, 2008) scientific evaluation on the impact of tourism on climate change, as well as the latest assessment of the current state of tourism and climate change (TPCC, 2023). In the original study conducted by De Freitas (2003), it was observed that the thermal, physical, and esthetic aspects of the climate are significant factors in tourism. This led to a study on the consequences of alterations in these elements on tourist destinations (appeal), tourism operators (duration of operating season), and travelers (desire to travel, selection of destinations, satisfaction with trips (Scott et al., 2019). Research has been conducted to examine the climatic preferences of visitors and the potential effects of changes in certain factors on their demand. These factors include air temperatures in various settings such as beaches, mountains, cities, and parks. Studies also aimed to deduce the most favorable climate conditions and tolerance limits, as well as comprehend the effects of severe weather occurrences on visitor demand reactions. An associated line of inquiry examined the comprehension of destination climates, as well as the utilization and significance of weather information for making travel choices (Scott et al., 2016).

The impact of climate change on skiing and winter tourism has been extensively studied. Researchers have expressed worries about the potential negative effects, such as less snow cover, shorter seasons, and increased expenses for snowmaking (Scott et al., 2012). Mountain tourism, particularly ski tourism, is now the most extensively studied area in climate risk research because of its ongoing development and the effects of climate change on it (Steiger et al., 2019). The loss of significant assets for tourism encompassed considerations of sea level rise and its effect on beach erosion and coastal infrastructure, the implications of global water security and water stress for island destinations and tourism activities, and the decline in biodiversity (Scott et al., 2012). Another area of study that examines "last chance tourism" is concerned with disappearing locations and attractions (Lemelin et al., 2010).

The impact of tourism on climate change was not addressed until the mid-1990s when more focus was placed on the emissions produced by airplanes at high altitudes and their impact on the chemistry and physics of the atmosphere. The initial evaluation of the influence of global tourism on climate change was introduced in the early 2000s by Gössling (2002). The reciprocal connection between tourism and climate change was subsequently recognized in the "Djerba Declaration" of 2003 by the United Nations World Tourism Organization (UNWTO). Successive nationwide studies were conducted to evaluate the level of tourist emissions on a country-wide scale. The selection of system boundaries (resident-based accounting versus destination-based accounting), the inclusion of specific subsectors (transport, accommodation, activities, shopping, and food), the consideration of lifecycles and supply chains, and the coverage of both short and long-lived greenhouse gases posed challenges. According to Lenzen et al. (2018), the worldwide impact of tourism on CO_2-equivalent emissions in 2013 was estimated to be around 8%, and there are indications of ongoing significant expansion. Subsequently, the COVID-19 pandemic has resulted in a significant decline in tourism and the corresponding emissions. However, there is presently a resurgence in tourism that is expected to surpass the levels seen in 2019, making 2024 the first year to exceed pre-pandemic tourist activity.

The declaration of the UNWTO in 2008 acknowledged the increasing clarity regarding the impact of tourism on climate change and identified climate change as the most significant obstacle to achieving sustainable tourism. Scott et al. (2019) conducted a comprehensive worldwide vulnerability assessment on tourism, examining 27 factors that encompass assets, operational costs, demand, host nation deterrents, adaptive capacity, and host country adaptive capacity. The data indicates that economically reliant impoverished nations, who anticipate significant increases in tourism, are also the most susceptible to the impacts of climate change.

3. Current Trends

Climate change will harm the tourist industry through two primary channels: climate hazards, which encompass both physical and socio-economic consequences, and carbon risks, which will result in higher travel expenses (Scott & Gössling, 2022). Both factors will have an impact on the availability and desire for tourism, and their relationship is intricate. Physical climate hazards include phenomena such as severe weather events, rising sea levels, the loss of biodiversity, and threats to human health. Socio-economic repercussions encompass the consequences of an event or situation that result in disturbances to social structures and alterations to economic prospects and employment. Carbon hazards are linked to the objectives of reducing emissions and the cost of carbon, which are relevant for tourism goods that have a high carbon footprint. Transport providers will face risks while transitioning to new technologies and fuels such as electrification, e-fuels, and biomass-based fuels. The aviation industry, in particular, has obstacles due to the inherent difficulty and high expense involved in achieving decarbonization. These topics may be categorized into many areas of study that are currently being examined.

3.1. Extreme Weather Events, Loss of Assets, and the Accumulation of Climate-Related Risks

Significant scientific endeavors have been focused on comprehending extreme weather phenomena, such as intense heat and fire exposure, reduction in snow cover, the consequences of rising sea levels, and the loss of (TPCC, 2023). These methods have been employed to create more intricate vulnerability indices for worldwide destinations (Scott et al., 2019) or to examine the effects on specific components of the tourism system, such as the snowmaking needs in destinations reliant on skiing, the consequences of an increase in the number of hot days for urban tourism, or the risks of rising sea levels for World Heritage Sites. Small island developing states (SIDS) encounter a range of challenges including more intense and slower-moving storms, periodic droughts, coral bleaching, rising sea levels, excessive heat, and the expenses associated with reducing carbon emissions. These changes will result in alterations in the competitiveness and demand of different destinations.

Mitigation is closely linked to the issue of emission sources and expansion. This has resulted in the recognition of distinct levels of analysis, such as global, national, regional, and individual viewpoints. In 2019, tourist emissions amounted to 5.4 gigatonnes of CO_2-equivalent, accounting for 8–11% of worldwide greenhouse gas emissions. This percentage is significantly greater in wealthy nations, where tourism contributes significantly to consumption-related emissions. The United States, China, India, and Germany are the primary countries responsible for significant emissions. It is important to acknowledge that the wealthiest citizens in these countries have the most significant impact on national emissions. Tourism demand growth is the primary factor behind the ongoing increase in emissions.

3.2. Technological Advancements

An increasing number of scholarly articles are focused on examining the technological, economic, social, and environmental consequences of changes in transportation systems, as the whole tourist industry relies on transportation. Regarding surface-bound transport, the emphasis is on implementing electrification and utilizing renewable sources of power. In the aviation sector, the goal is to advance sustainable biofuels, synthetic e-fuels, hydrogen, and battery technology. As for shipping, the focus is on employing batteries for smaller vessels, as well as utilizing methanol or synthetic LNG. Every technology is interconnected with space since it necessitates the development of new production areas (such as wind power and industries), the use of suitable biomass (like biofuels), or the need for extra transportation infrastructure (such as highways, ports, and airports). The demand for land will increase when low-carbon energies replace fossil fuels, particularly due to the high power consumption needed to produce smaller amounts of synthetic fuels in some energy sources known as e-fuels. Land use is a significant concern associated with the growth of the tourism industry. This includes constructing resorts, transportation infrastructure, and producing food. However, this expansion of tourism is happening at the expense of diminishing natural resources such as snow, beaches, and coral reefs, which are essential for tourism.

3.3. The Impact of Climate Change on Tourism in Certain Geographic Locations

Tourism Geographies focuses on examining the many effects of climate change on travelers, tourism stakeholders, and destinations. It highlights the pressing requirement for implementing adaptive and mitigative measures to address the growing environmental threats. The texts discuss various spatial perspectives on the vulnerability of nature-based tourism in the Nepalese Himalayas, the perceptions of Christmas tourists regarding climate change in Finland, the impact of climate change on Tyrol's ski tourism, and the perceptions and responses of Scotland's ski industry. These studies highlight the need for management, concerns over the authenticity of winter experiences, challenges for sustainability, and the varying readiness to adapt in the face of climate change. Within the realm of winter, snow, and mountain research, there has been a significant focus on disappearing glaciers. Various articles have explored topics such as adaptive strategies (Stewart et al., 2016), last-chance phenomena (Salim & Ravanel, 2023), and the emotional and ethical aspects of studying melting glaciers (Dannevig & Rusdal, 2023). Additional subjects that have been explored include the consequences of climate change on the development of second homes in coastal regions (Steffansen et al., 2023) as well as strategies for reducing climate change risks in tourist destinations (Loehr & Becken, 2023). These studies jointly emphasize the weaknesses and necessary adjustments caused by climate change.

3.4. Administration

As our understanding of the causes of greenhouse gas emissions increases, the study of governance is becoming increasingly relevant. However, the tourist sector is not well represented in global climate policy. The tourism policy in the majority of countries is incongruous with climate policy. Only a small number of nations have included tourism in their Nationally Determined Contributions (NDCs), and an even smaller number have specified steps. However, there is no evidence to suggest that any of these initiatives are substantial enough to effectively cut absolute emissions under the goals of the Paris Climate Agreement (TPCC, 2023). These findings reveal specific spatial characteristics. For instance, the most vulnerable destinations, such as small islands, are also the ones that offer tourism products with the highest carbon emissions. Although some of these destinations have committed to reducing emissions domestically, the issue of energy-intensive transportation and long-haul markets is still unresolved and, in most cases, not addressed in national tourism policies. The conclusion is that unless expansion, especially in air transport, is restricted and global visitor flows are restructured, achieving decarbonization in tourism is becoming increasingly improbable.

4. Impact of Climate Change on the Tourism Sector

The tourism sector is seeing growth in both its scale and economic significance. Several formerly inaccessible locations have recently gained popularity as vacation spots. Nevertheless, the ecosystems of several resorts are exceptionally susceptible to the impacts of climate change. Global and regional temperatures are increasing. The year 1998 was the most excessively hot year of the millennium, while the 1990s were the most heated decade. The years 1998, 1997, 1995, and

1990 are ranked as the four hottest globally, listed in descending order of significance. Climate models indicate that there will be a gradual increase in temperature of 0.2–0.3°C per decade, and it is projected that sea levels will rise at a pace of 4 to 10 cm every decade.

The effects of climate change on tourism are expected to appear in various ways depending on the specific circumstances of each location. These consequences will primarily arise indirectly as a result of heightened pressures exerted on environmental systems. The tiny island states will experience the most severe consequences due to the influence of rising sea levels. The Maldives, a burgeoning tourism hotspot, faces significant susceptibility to the rising sea levels.

Climate change is anticipated to heighten the vulnerability to disease in many regions around the globe, hence discouraging tourism. The Eastern Mediterranean resorts will have more frequent spells of excessive heat, leading to increased discomfort as the frequency of days with temperatures above 40°C rises. The reduction in cloud cover in Australia will lead to heightened exposure to the sun's detrimental radiation, while there is a high probability of malaria resurfacing in Spain, which is the preferred vacation spot for UK travelers.

Winter tourism might be impacted when the Alps and other skiing locations witness reduced snowfall and shorter skiing seasons. The lower-lying ski destinations, such as Garmish-Partenkirchen in Germany, and the already marginally profitable resorts in the Scottish Highlands will have significant repercussions. In another instance, the glaciers of New Zealand's Aoraki/Mt Cook National Park, which border Westland Tai Poutini National Park, provide visitors with opportunities for glacier tourism. The Tasman Glacier is the largest of the four glaciers in the park; the others are Hooker, Mueller, and Murchison. Significant effects of environmental changes have already been felt in the growth of proglacial lake tourism activities on the Tasman Glacier, as well as on tourist access and scenic amenities.

The golf industry is becoming a more significant part of the global tourism economy and one of the largest recreation sectors globally. Climate change is expected to have a significant impact on golf because it will change the length of the golf season, affect demand for golf both seasonally and overall, and necessitate more turf management (i.e., irrigation, turf grass selection, turf disease, and pest management), and endanger some of the world's historic coastal courses due to erosion and sea level rise. Future travel patterns may alter as a result of climatic conditions. The natural elements that often draw tourists to particular locations are dwindling, such as the bleaching of coral reefs in locations like the Great Barrier Reef. Emission of greenhouse gases intensifies climate factors like rainfall, temperature, and meteorological events that influence the epidemiology of disease thus impacting travels around.

Since consumer decisions are influenced by perceptions of environmental changes or climate conditions as much as actual conditions, it is anticipated that perceptions of the future effects of climate change will play a major role in the decision-making process for both tourists and tourism investors. Nevertheless, the image is not wholly bad. One of the main draws of several vacation destinations is the promise of reliable sunlight and high temperatures. An overall warmer environment would be advantageous for tourist places like the UK, where summer weather conditions are now very unpredictable. As the weather becomes warmer, more UK people may choose to take vacations within the country. As the UK's climate becomes more Mediterranean, we may expect to see growth not just in the local market but also in the incoming foreign market.

5. Limitations of the Research

According to a recent analysis, all travel destinations and tourist markets must adjust to the direct and indirect effects of rapidly advancing climate change. This adaptation is necessary to either reduce the adverse effects or maximize the potential benefits (Scott & Gössling, 2022). Currently, studies are scarce on the potential, which are predominantly taken advantage of by tiny innovative firms, as well as a few prominent destinations and destination marketing organizations. Occasionally, governments have played a significant role in facilitating transitions, as seen by Turkey's compulsory environmental certification program for all hotels nationwide. Valuable knowledge may be acquired by thoroughly examining enterprises and locations that have made efforts to either adjust to climate change or contribute to mitigation through significant transformations. The preceding sections have outlined the tourism issues associated with climate change, namely the dangers connected to climate and carbon. However, further study is needed to address these challenges. None of the destinations, regardless of their size, have conducted a thorough evaluation of their overall climate and carbon risk, which hampers their ability to take meaningful action against climate change. The present understanding of climate change adaptation and the effectiveness of existing adaptations in dealing with future climate change is still lacking, as is the ability to scale up and maintain the sustainability of current adaptation efforts and any additional advantages they may bring to destinations (Scott & Gössling, 2022). While climate change remains the most significant concern for mankind, it is crucial to recognize that its repercussions are influenced and intensified by other concurrent events. These include three main factors: (i) humanity's significant impact on the limits of the planet; (ii) the emergence of a new geopolitical landscape due to the Ukraine war; and (iii) the presence of inequality patterns that have led to populism and division in many nations. Tourism is influenced by and contributes to alterations in several aspects of the environment, including biosphere integrity, land system change, freshwater consumption, ocean acidification, biogeochemical fluxes, stratospheric ozone depletion, atmospheric aerosol loading, and chemical pollution. However, the relationship between tourism and global environmental change is not fully comprehended, as previous research has only examined the aspects that directly impact the tourism industry, such as biodiversity, land use, and freshwater. Furthermore, even for these aspects, the global situation is still uncertain.

Climate change mitigation continues to be a significant area of research, with connections to underexplored areas of sub-sector-specific carbon budgets, management, technology innovation, fuel transition, society, governance, and finances. The economic elements are quite significant, given the substantial cost burden associated with the shift to net zero. Carbon taxes and allowances have been the subject of discussion for many years, although they have only been implemented in a few countries and on a small scale (World Bank, 2023).

To inform and accelerate the decarbonization of the tourist sector at a pace compatible with science-based sectoral targets (−50% by 2030, net-zero by 2050), new research contributions are desperately needed at all scales (traveler, organization, destination, sectoral). Leading tourism organizations, such as the International Civil Aviation Organization, the United Nations World Tourism Organization, national tourism organizations, and destination management organizations, need to critically analyze how their pro-growth and pro-climate discourses and policies are incompatible (Becken, 2019; Loehr & Becken, 2021). Subsequent studies need to concentrate on assessing the efficiency of current policy structures and governance systems in advancing environmentally conscious travel. Sustainable tourism practices also require the creation of comprehensive, context-specific rules that strike a balance between economic interests and environmental conservation. In addition, sustainable tourism practices also require the creation of comprehensive, context-specific rules that strike a balance between economic interests and environmental conservation (Leung, 2021).

Lastly, it is crucial to recognize that climate change research is venturing into novel areas within the geography literature that link geography, climate change, and social dynamics. This research examines several factors such as gender, discourse, and identity. Barnett (2020) provides an analysis of the political economies that contribute to vulnerability to climate change. The author emphasizes the importance of comprehending power dynamics and the underlying processes that create vulnerability. This perspective is also echoed by Mahony and Hulme (2018). Undoubtedly, a significant obstacle to altering the global tourism system is the unwavering insistence of influential individuals or groups on maintaining the existing state of affairs, which perpetuates the system and magnifies its effects. Mahony and Hulme (2018) has highlighted the implications of adaptation attempts in the context of São Tomé and Príncipe. To get beyond these dominant systems of power, studies on the spatial aspects of personal identity have examined how government control and social activism influence public participation and decision-making (Dowling, 2010). This line of research has become particularly relevant in the context of contemporary populism and increasing divisions within society. In their study, Jeffrey and Dyson (2021) focus on prefigurative politics, which involves embodying the desired change. This concept is very applicable to reshaping power dynamics in various contexts, including academic cultures (Higham & Font, 2020).

6. Conclusion

The hazards posed by climate change to the tourist industry, as well as the industry's role in contributing to these risks, are now generally comprehended, despite the existence of some research gaps and the constant emergence of new ones. Physical hazards are expected to become more significant shortly, posing a threat to the sustainability of tourism or specialized tourist activities at an expanding number of sites both locally and across time. The impact of tourism on climate change is increasing, as indicated by continuous growth patterns in emissions that are in direct opposition to net-zero objectives. It is important to analyze these risks about other concerns, such as the polycrisis. This implies that the stability of the global tourism system will rely on addressing and taking action on all dimensions of risk. As discussed, climate change brings in both physical and socio-economic consequences. In physical the area and the time period available for the tourism leisure activities get affected and the tourism industry and the population engaged get affected by the loss of revenue. Changes in climate accelerate the wearing of infrastructure, affecting railway, air and road travels. Domestic and international tourism gets disrupted by the extreme weather conditions intruding on the travel plans. When climate change intensifies, the industry's long-term survival depends on its ability to adapt and minimize its environmental impact. The need for stakeholders in the tourist system, including corporations, destination marketing organizations, national tourism organizations, national governments, and organizations like the IC AO, to take action on mitigation has been emphasized for many years. However, authorities and organizations have displayed limited motivation to actively participate in mitigation endeavors.

Although experts in the tourism industry are increasingly in opinion that concentrating on domestic travel will help cut down on carbon emissions, foreign travel is nevertheless flourishing. Tourist are changing their habits in response to climate change, such as remaining indoors during periods of high heat and going on excursions during the evenings when it is cooler. That being said, there are signs that travelers are growing more vigilant of extreme conditions of weather. Consequently, the global tourist industry will probably continue to grow until a significant reduction in global carbon emissions occurs, which will ultimately reshape the geographical landscape of tourism worldwide.

References

Barnett, J. (2020). Global environmental change II: Political economies of vulnerability to climate change. *Progress in Human Geography*, *44*(6), 1172–1184. https://doi.org/10.1177/0309132519898254

Becken, S. (2019). Decarbonising tourism: Mission impossible? *Tourism Recreation Research*, *44*(4), 419–433. https://doi.org/10.1080/02508281.2019.1598042

Dannevig, H., & Rusdal, T. (2023). Caring for melting glaciers. *Tourism Geographies*, *25*(6), 1679–1695. https://doi.org/10.1080/14616688.2023.2278762

De Freitas, C. R. (2003). Tourism climatology: Evaluating environmental information for decision making and business planning in the recreation and tourism sector. *International Journal of Biometeorology*, *48*(1), 45–54. https://doi.org/10.1007/s00484-003-0177-z

Dowling, R. (2010). Geographies of identity: Climate change, governmentality and activism. *Progress in Human Geography*, *34*(4), 488–495. https://doi.org/10.1177/0309132509348427

Gössling, S. (2002). Global environmental consequences of tourism. *Global Environmental Change*, *12*(4), 283–302. https://doi.org/10.1016/S0959-3780(02)00044-4

Gössling, S., Scott, D., Hall, C. M., Ceron, J. P., & Dubois, G. (2012). Consumer behaviour and demand response of tourists to climate change. *Annals of Tourism Research, 39*(1), 36–58. https://doi.org/10.1016/j.annals.2011.11.002

Higham, J., & Font, X. (2020). Decarbonising academia: Confronting our climate hypocrisy. *Journal of Sustainable Tourism, 28*(1), 1–9. https://doi.org/10.1080/09669582.2019.1695132

Jeffrey, C., & Dyson, J. (2021). Geographies of the future: Prefigurative politics. *Progress in Human Geography, 45*(4), 641–658. https://doi.org/10.1177/0309132520926569

Lemelin, H., Dawson, J., Stewart, E. J., Maher, P., & Lueck, M. (2010). Last-chance tourism: The boom, doom, and gloom of visiting vanishing destinations. *Current Issues in Tourism, 13*(5), 477–493. https://doi.org/10.1080/13683500903406367

Lenzen, M., Sun, Y. Y., Faturay, F., Ting, Y. P., Geschke, A., & Malik, A. (2018). The carbon footprint of global tourism. *Nature Climate Change, 8*(6), 522–528. https://doi.org/10.1038/s41558-018-0141-x

Leung, A. (2021). Sustainable tourism practices: Balancing economic growth and environmental conservation. *International Journal of Business, Management and Visuals, 4*(2), 1240–1253.

Loehr, J., & Becken, S. (2021). The tourism climate change knowledge system. *Annals of Tourism Research, 86*, 103073. https://doi.org/10.1016/j.annals.2020.103073

Loehr, J., & Becken, S. (2023). Leverage points to address climate change risk in destinations. *Tourism Geographies, 25*(2–3), 820–842. https://doi.org/10.1080/14616688.2021.2009017

Mahony, M., & Hulme, M. (2018). Epistemic geographies of climate change: Science, space and politics. *Progress in Human Geography, 42*(3), 395–424. https://doi.org/10.1177/0309132516681485

Pereira, M. G., Parente, J., Amraoui, M., Oliveira, A., & Fernandes, P. M. (2019). The role of weather and climate conditions on extreme wildfires. *Extreme Wildfire Events and Disasters: Root Causes and New Management Strategies*, 55–72. https://doi.org/10.1016/B978-0-12-815721-3.00003-5

Salim, E., & Ravanel, L. (2023). Last chance to see the ice: Visitor motivation at Montenvers-Mer-de-Glace, French Alps. *Tourism Geographies, 25*(1), 72–94. https://doi.org/10.1080/14616688.2020.1833971

Scott, D., & Gössling, S. (2022). A review of research into tourism and climate change: Launching the annals of tourism research curated collection on tourism and climate change. *Annals of Tourism Research, 95*, 1–13. https://doi.org/10.1016/j.annals.2022.103409

Scott, D., Hall, C. M., & Gössling, S. (2016). A report on the Paris climate change agreement and its implications for tourism: Why we will always have Paris. *Journal of Sustainable Tourism, 24*(7), 933–948. https://doi.org/10.1080/09669582.2016.1187623

Scott, D., Hall, C. M., & Gössling, S. (2019). Global tourism vulnerability to climate change. *Annals of Tourism Research, 77*(May), 49–61. https://doi.org/10.1016/j.annals.2019.05.007

Scott, D., Hall, C. M., & Stefan, G. (2012). *Tourism and climate change: Impacts, adaptation and mitigation*. Routledge.

Steffansen, R. N., Staunstrup, J. K., Sørensen, M. T., & Hjalager, A. M. (2023). Sinking into the ocean? Climate change risks and second home planning. *Tourism Geographies, 25*(6), 1638–1656. https://doi.org/10.1080/14616688.2023.2269555

Steiger, R., Scott, D., Abegg, B., Pons, M., & Aall, C. (2019). A critical review of climate change risk for ski tourism. *Current Issues in Tourism, 22*(11), 1343–1379. https://doi.org/10.1080/13683500.2017.1410110

Stewart, E. J., Wilson, J., Espiner, S., Purdie, H., Lemieux, C., & Dawson, J. (2016). Implications of climate change for glacier tourism. *Tourism Geographies*, *18*(4), 377–398. https://doi.org/10.1080/14616688.2016.1198416

TPCC (Tourism Panel on Climate Change). (2023). In S. Becken & D. Scott (Eds.), *Tourism and climate change stocktake*. https://tpcc.info/

UNWTO. (2008). *Climate change and tourism: Responding to global challenges*. https://www.unwto.org/archive/global/news/2011-08-16/climate-change-and-tourism-responding-global-challenges

World Bank. (2023). *State and trends of carbon pricing 2023*. http://hdl.handle.net/10986/39796

Printed and bound by CPI Group (UK) Ltd, Croydon, CR0 4YY

13/09/2026

14961063-0007